THE EDITOR

ALLEN W. STOKES is Professor of Animal Behavior at Utah State University, where he has taught since 1952. He earned his B.S. in Chemistry from Haverford College and his M.A. in that field from Harvard University. In 1952, he received his Ph.D. in Wildlife-Zoology from the University of Wisconsin.

Professor Stokes spent the 1958–1959 academic year at Cambridge University, under the auspices of a National Science Foundation Fellowship, working in the Subdepartment of Animal Behavior with R. A. Hinde and W. H. Thorpe, He is a past president and fellow of the Animal Behavior Society and the editor of *Animal Behavior in Laboratory and Field* (W. H. Freeman, 1968).

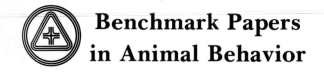

Benchmark Papers
in Animal Behavior

Series Editor: Martin W. Schein
West Virginia University

Published Volumes and Volumes in Preparation

HORMONES AND SEXUAL BEHAVIOR
  Carol Sue Carter
TERRITORY
  Allen W. Stokes
SOCIAL HIERARCHY AND DOMINANCE
  Martin W. Schein
CRITICAL PERIODS
  J. P. Scott
MIMICRY
  Joseph A. Marshall
IMPRINTING
  E. H. Hess
VERTEBRATE SOCIAL ORGANIZATION
  Edwin M. Banks
PLAY
  Dietland Müller-Schwarze
HUMAN SEXUALITY
  Milton Diamond
ABORTION
  Milton Diamond

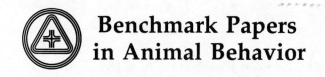

Benchmark Papers
in Animal Behavior

—— A *BENCHMARK* ® Books Series ——

TERRITORY

Edited by
ALLEN W. STOKES
Utah State University

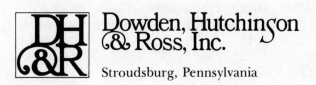

Dowden, Hutchinson
& Ross, Inc.

Stroudsburg, Pennsylvania

Copyright © 1974 by **Dowden, Hutchinson & Ross, Inc.**
Benchmark Papers in Animal Behavior, Volume 2
Library of Congress Catalog Card Number: 73–18327
ISBN: 0–87933–113–5

Manufactured in the United States of America.

Exclusive distributor outside the United States and
Canada: John Wiley & Sons, Inc.

74 75 76   5 4 3 2 1

Library of Congress Cataloging in Publication Data

Stokes, Allen W   comp.
   Territory.

   (Benchmark papers in animal behavior, v. 2)
   Includes bibliographical references.
   1. Territoriality (Zoology)--Addresses, essays,
lectures. I. Title.
QL756.2.S76   591.5'24   73-18327
ISBN 0-87933-113-5

Acknowledgments
and Permissions

ACKNOWLEDGMENTS

American Ornithologists' Union—*Auk*
  "Removal and Repopulation of Breeding Birds in a Spruce–Fir Forest Community"
  "Territorial Behavior: The Main Controlling Factor of a Local Song Sparrow Population"

American Society of Mammalogists—*Journal of Mammalogy*
  "Reproductive Behavior of the Alaska Fur Seal, *Callorhinus ursinus*"

Linnaean Society of New York
  Proceedings of the Linnaean Society of New York
    "Bernard Altum and the Territory Theory"
  Transactions of the Linnaean Society of New York
    "Studies in the Life History of the Song Sparrow I"

The Royal Irish Academy—*Irish Naturalist*
  "The Spring Rivalry of Birds"

Society for the Study of Evolution—*Evolution*
  "Evolution of Pair Cooperation in a Tropical Hummingbird"
  "Evolution of Polygamy in the Long-Billed Marsh Wren"

13th International Ornithological Congress—*Proceedings of the 13th International Ornithological Congress*
  "Ecological Significance of Territory in the Australian Magpie, *Gymnorhina tibicen*"

The Wilson Ornithological Society—*The Wilson Bulletin*
  "The Evolution of Diversity in Avian Territorial Systems"

PERMISSIONS

The following papers have been reprinted with the permission of the authors and copyright holders.

American Association for the Advancement of Science—*Science*
  "Territorial Behavior in Uganda Kob"

American Association on Mental Deficiency—*American Journal of Mental Deficiency*
  "Territorial Behavior as an Indicator of Changes in Clinical Behavioral Condition of Severely Retarded Boys"

Bailliere, Tindall & Cassell Ltd.—*Animal Behaviour*
  "Booming Territory Size and Mating Success of the Greater Prairie Chicken (*Tympanuchus cupido pinnatus*)"

E. J. Brill, Leiden, Netherlands
  Acta Biotheoretica
    "On Territorial Behavior and Other Factors Influencing Habitat Distribution in Birds: I. Theoretical Development"
  Archives Neerlandaises de Zoologie
    "Territory and the Regulation of Density in Titmice"

Behaviour
  "Variations in Territorial Behavior of Uganda Kob *Adenota kob thomasi*"
  "Aggressiveness, Territoriality, and Sexual Behavior in Field Crickets (Orthoptera: Gryllidae)"
Behaviour (Supplement)
  "Territory in the Three-Spined Stickleback *Gasterosteus aculeatus* L."
Proceedings of the 15th International Ornithological Congress
  "A Current Model of Population Dynamics in Red Grouse"

The British Ecological Society—*Journal of Animal Ecology*
  "A Long-Term Study of the Great Tit (*Parus major*)"

British Ornithologists' Union—*Ibis*
  "The Biological Significance of the Territories of Birds"

Wm. Collins & Co. Ltd.—*Territory in Birds*

Duke University Press (for the Ecological Society of America)—*Ecological Monographs*
  "The Ecology of Blackbird (*Agelaius*) Social Systems"

New York Academy of Sciences—*Transactions of the New York Academy of Sciences*
  "Density, Social Structure, and Nonsocial Environment in House-Mouse Populations and the Implications for Regulation of Numbers"

A. D. Peters and Company (agents for the work of Julian Huxley)
  Territory in Bird Life
    "Foreword"
  British Birds
    "A Natural Experiment on the Territorial Instinct"

The Zoological Society of London—*Symposium of the Zoological Society of London*
  "The Adaptive Significance of Avian Social Organizations"

Series Editor's Preface

It was not too many years ago that virtually all research publications dealing with animal behavior could be housed within the covers of a very few hard-bound volumes that were easily accessible to the few workers in the field. Times have changed! The present-day student of animal behavior has all he can do to keep abreast of developments within his own area of special interest, let alone in the field as a whole. And of course we have long since given up attempts to maintain more than a superficial awareness of what is happening "in biology," "in psychology," "in sociology," or in any of the broad fields touching upon or encompassing the behavioral sciences.

It was even fewer years ago that those who taught animal behavior courses could easily choose a suitable textbook from among the very few that were available; all "covered" the field, according to the bias of the author. Students working on a special project used *the* text and *the* journal as reference sources and, for the most part, successfully covered their assigned topics. Times have changed! The present-day teacher of animal behavior is confronted with a bewildering array of books from which to choose, some purporting to be all-encompassing, others confessing to strictly delimited coverage, and still others being simply collections of recent and profound writings.

In response to the problem of the steadily increasing and overwhelming volume of information in the area, the Benchmark Papers in Animal Behavior was launched as a series of single-topic volumes designed to be some things to some people. Each volume contains a collection of what an expert considers to be the significant research papers in a given area. Each volume, then, serves several purposes. For the teacher, a Benchmark volume serves as a supplement to other written materials assigned to students; it permits in-depth consideration of a particular topic while confronting the student (often for the first time) with original research papers of outstanding quality. For the researcher, a Benchmark volume saves countless hours of digging through various journals to find *the* basic articles in his area of interest, journals that are often not easily available. For the student, a Benchmark volume provides a readily accessible set of original papers on the topic, a set that forms the core of the more extensive bibliography that he is likely to compile; it also permits him to see at first hand what an "expert" thinks is important in the area and to react accordingly. Finally, for the librarian, a Benchmark volume represents a collection of important papers from many diverse sources, making readily available materials that it might otherwise be economically impossible to obtain or physically impossible to keep in stock.

The ordering of topics to be covered in the series was no easy matter. Eventually, of course, as many topics as possible will be included in what must be an open-ended series. However, it was decided at the outset to aim primarily at the more fundamental topic areas, topics that erupted into print some years ago and have now subsided into a semidormant, but still rumbling, state. Territory is just such a topic: the great work in this area was done several decades ago, and, while there are still many pieces to be picked up, territorialism is an accepted and classical concept explaining the behaviors of individuals of many animal species. It is fitting, then, that one of the first books in the Benchmark series should be devoted to the topic of territory.

Since the merit of the volume rests so heavily upon the judgment of the volume editor, it was important to ensure that volume editors are truly experts in their areas, as well as persons of sound judgment and scholarship. I think this volume by Professor Stokes is evidence of our success in these endeavors.

Martin W. Schein

Preface

When invited to prepare this volume on territory for the Benchmark Papers in Animal Behavior, I welcomed the opportunity. My first serious field study was on the breeding behavior of the American goldfinch. There I was to observe, first hand, territorial behavior in action. At that time, I first read Margaret Nice's *Studies in the Life History of the Song Sparrow* and was fascinated at the seeming order among song sparrows in their territorial arrangements. Subsequently I have studied the social organization of numerous birds and mammals and have moved from the mere description of territory and hierarchy to inquiry into the function and adaptiveness of the organization to the individuals.

The purpose of the Benchmark Papers is to publish original papers that have made their mark in developing concepts in a particular field. The papers should have provoked controversy or shown new insight into the subject. My first thought was to confine myself to those early classics to which one is often referred but which are rarely found in today's libraries; to share in the excitement of these early field biologists as they observed territorial behavior and wrestled with the problems of classification and determination of the functions of territory. Some of these early writings were books themselves, so in the end it was necessary to select only those portions of the works of Bernard Altum, Eliot Howard, and Margaret Nice that present the highlights of their study and the flavor of their writing.

As in so many other areas of scientific endeavor, once an exciting new field has been discovered, there develops a tendency for "bandwagon research." Scores of young investigators began life history studies of animals, especially birds, attempting to see how the territorial behavior of their species matched that of Howard's warblers or Nice's song sparrows. These studies filled the nooks and crannies in our knowledge but did little to develop new concepts. By the 1950s the study of territory had become "old hat" and was in danger of being relegated to textbook treatment. Fortunately a new burst of interest appeared, particularly oriented to an understanding of the evolution and adaptiveness of social organizations, especially as it relates to man's own social organization. For this reason I have included a number of recent papers that show the exciting new directions that the study of territorial behavior is taking. These papers, in turn, contain numerous references. Hopefully, the reader will be stimulated to delve more deeply into these new frontiers either through reading or research.

The originators of this series on animal behavior probably selected "territory" because it seemed such a tight, homogeneous package. So did I. Yet upon delving into this box I find it like that of Pandora, containing all sorts of unruly aspects that spill over into genetics, bioenergetics, evolution of mating systems, and even soil science. Territory is also inseparable from social hierarchy, for there are dominance–subordinance relationships in all territorial behavior. The reader will certainly want to read the companion volume, *Social Hierarchy and Dominance*, by M. W. Schein.

In selecting the papers to use for this volume I have tried to use those that first developed a concept. At times this was not possible because the concepts were inextricably buried in much longer papers. In such instances I have had to use other, shorter papers that developed the same concept even though at a later date. In such an event, I have tried to mention the earlier papers in my introductory comments to each part of the book.

As I have taught animal behavior over the years, I have found the subject of social organization and territory of paramount interest to students and laymen alike. They see the social behavior of animals reflected so similarly in their own lives. Man's increasing concern as to how he can control his own aggression on both a personal and a global scale has made him more interested in seeing how other animals handle this problem. Man has become more fully aware that his behavior is not liberated from its biological roots and that an understanding of territorial behavior in other animals may shed greater understanding on our own behavior.

Allen W. Stokes

Contents

V. INTRASPECIFIC VARIATION IN SOCIAL SYSTEMS

VI. FOOD AND ENERGY RELATIONSHIPS IN TERRITORIAL SYSTEMS

VII. THE EVOLUTION OF TERRITORIAL AND OTHER SOCIAL SYSTEMS

VIII. EXPERIMENTAL APPROACHES TO TERRITORY

Contents by Author

Introduction

Territory, like ecology, is a household word today. Worldwide concern over the dangers of a global conflagration has aroused keen interest in man's tendency to protect his resources. This interest has been abetted by the spate of popular books and articles on aggression and territorial behavior. Yet man has come to this general recognition of the phenomenon of territory only after centuries, if not millennia, of study. Man has known from time immemorial that animals fight and dominate others to gain access to limited resources, whether food, shelter, or the opportunity to breed. From Aristotle on, numerous persons have noted how animals may defend an area against intruders. Some have identified this as the defense of a territory in which an animal spends much of its life and prevents others of its kind from intruding. This is not surprising; one need not spend much time observing birds in springtime without becoming aware that they fight and that they tend to concentrate this fighting in the vicinity of the nest. But it was not until 1868, when Bernard Altum published the first edition of *Der Vogel und sein Leben*, that a general theory of territory became known.

Universal awareness of territoriality was to be a long time developing. Caxton's invention of the printing press has been no guarantee of rapid dissemination of knowledge; barriers of space and language still prove strong. Moffat, completely unaware of Altum's work, published his elegant observations, "The spring rivalry of birds," in 1903, setting forth with great clarity the principles of territory and the methods of display used in defending it. However, even British ornithologists paid little heed to him. Therefore, Eliot Howard had to "rediscover" the concept of territory in the early 1900s through extensive observations of song and marsh birds, which culminated in the publication of his classic *Territory in Bird Life*. The Atlantic Ocean proved a formidable barrier, however, and only after Margaret Nice had published her *Studies in the Life History of the Song Sparrow* in 1937, did America fully recognize the phenomenon of territoriality in birds. It was still to be some years before

1

the widespread occurrence of territoriality in the vertebrates and arthropods was equally appreciated.

A predictable sequence of steps occurs in the development of any new behavioral concept. Of paramount importance is a careful description of behavior, followed by a search for relationships that will result in an understanding of the causes and functions of that behavior. Only much later, after a sufficient mass of knowledge has been accumulated, can there be insight into how the behavior has evolved, either genetically or ontogenetically. Our understanding of territoriality has followed these steps, and we are far from completing any of the stages. Careful descriptions of the territorial behavior of an animal is still limited to a hundred or so species. We keep finding that the original descriptions are inadequate for solving the more searching problems of function and evolution. Most studies are too short in duration to test the effects of year-to-year changes in population density and in the host of other environmental variables. Only recently has any attempt to manipulate these environmental variables been made; still less has there been any attempt to study territoriality experimentally under controlled conditions of captivity.

Students of territoriality have been, for the most part, naturalists, people with a keen interest in free-living animals. Unlike the geneticist, who selects fruit flies for the ease they offer in the study of inheritance factors, or the experimental psychologist, who picks whatever animal is convenient in order to study learning, naturalists have stumbled onto the concept of territorial behavior as they pursued the life history study of their favorite animals. It is the rare biologist, even today, who decides to study territorial phenomena and then selects a convenient animal.

A consequence of this approach is a strong provincialism. The ornithologist pursues his study of territory in birds, quite oblivious to the splendid studies on fish, mammals, and invertebrates and vice versa. One need only look at the major reviews of territoriality and the papers in this volume to see this sharp taxonomic schism.

I have divided the volume into eight parts. The first three parts are historical in approach. In Part I, Altum, Moffat, Howard, and Nice set forth their basic ideas about territory as a defended area and the functions that a territory serves in bird life. What meticulous observers they were! Without exception, they based their ideas upon long years of study of numerous species of birds. We already see in these papers controversy arising over the attempt to find a common denominator for all territories and over the function of territories. Also, there already was disagreement over the extent to which a territory guaranteed adequate food and its effectiveness in limiting breeding density.

By the 1940s studies of territoriality had mushroomed in number. Hediger (1949) was the first to recognize the widespread use of scent posts by mammals to demarcate their territories, in contrast to the song and postural displays of birds. Greenberg (1947) had studied sunfish in aquarium conditions and was among the first to manipulate the density of the fish in his tanks experimentally to observe its effects upon social organization. By now territorial systems had proven to be remarkably diverse and this contributed to disagreement over the definition of territory and its functions.

For instance, must an area be defended to be considered a territory? Is the tiny area surrounding the nest of a colonial seabird a territory? Is there a single overriding function of territory, or has territory evolved to serve different functions in different animals? Part II, a single paper by Robert Hinde, is devoted to a detailed classification of types of territories and to a critical analysis of the functions of territory.

Perhaps no aspect of territory has provoked such heated controversy as the role of territory in limiting density. David Lack (1954) had championed the role of food in limiting animal numbers. His book *The Natural Regulation of Animal Numbers* stirred up a hornet's nest. Lack's thesis, that animals increase to the limit of their food resources and that they will produce as many offspring as they can successfully rear to maturity, seemed a highly inefficient, not to say "inhumane," scheme of population regulation. Behavioral ecologists, therefore, looked for a means of self-regulation of animal numbers through territoriality and other behavioral methods. Growing concern over the consequences of unchecked human population growth has a bearing upon this controversy. Many persons now look to the methods of self-regulation of population density found in other animals for clues to its applicability to humans. Part III is devoted to this still unsettled question of the role of territory in regulating animal numbers. It includes the three classic conflicting theories on the function of territory in limiting density by Huxley, Kluyver and Tinbergen, and Lack. Papers on field studies designed to test these theories follow.

In 1956 an entire issue of *Ibis* was devoted to territoriality in birds. This issue represents a dividing line in the study of territory. Except for Hinde's review of territory (previously cited), the papers were entirely species-oriented rather than problem-oriented. They were largely descriptions of how and when territories were established and maintained. Subsequently, there has been a strong shift toward a study of the function and evolution of territory that has revitalized the field.

First, we now recognize there is no sharp distinction between territoriality and social dominance. The more dominant males in a breeding population will acquire the best territories. This gives them the advantage of better food and nesting resources and, as a result, greater likelihood of acquiring one or more mates. In arena, or colonial, breeding animals, such as grouse and Uganda kob, the central territories are held by the dominant members of the group and these males perform most of the copulations.

Second, the social system of a species proves much more variable than had previously been thought. Whether a species is territorial or hierarchical varies from place to place and depends upon density and environmental variables. Thus the social system is both environmentally and genetically determined. The mating and social system that occurs in a given area is that which is the most successful for the dominant individuals (i.e., results in their production of maximum numbers of progeny). This has led to studies of the bioenergetics involved. An individual will find it profitable to defend a territory only when the economic advantages offset the disadvantages of greater risk of predation and greater amounts of energy expended.

Finally, studies of territorial behavior, using such adaptable animals as fish and insects, have been brought into the laboratory, where social and environmental variables may be independently controlled. These new approaches are presented in Parts IV through VIII.

Of the 26 papers in this volume, 19 deal with birds, 5 with mammals, and one each with fish and insects. Shameless bias? Possibly. But ornithology has had a long lead over other disciplines in the recognition of territoriality. This gap is narrowing rapidly. Generalizations concerning mammals are now possible for primates, ungulates, carnivores, and rodents. In the long run, fish and invertebrates may be best suited for study because of the ease of maintaining them in captivity under highly controlled conditions. In any event, only through experimentation and long-term studies can we continue to extend the frontiers of knowledge of the concept of territory.

References

Greenberg, B. 1947. Some relations between territory, social hierarchy and leadership in the green sunfish (*Lepomis cyanellus*). *Physiol. Zool.*, **20**: 269–299.

Hediger, H. 1949. Säugetier-Territorien und ihre Markierung. (Mammalian territories and their demarcation.) *Bijd. Dierk.* **28**: 172–184.

Lack, D. 1954. *The Natural Regulation of Animal Numbers.* Oxford University Press, London. 343 pp.

I
Early Description and Theory
of Territory

Editor's Comments on Papers 1 Through 4

It is fitting to begin these papers with the writing of Eliot Howard. Although not the first to be aware of the concept of territory, it was he who wrote most prolifically and who stated a general theory of territory. The few pages included here reveal the flavor of Howard's delightful literary style, his meticulous observations, and the depth of his understanding of territorial behavior. [Thanks to Julian Huxley and James Fisher, Howard's *Territory in Bird Life* is now readily available (Atheneum, New York, 1964). Their Foreword is an invaluable concise summary of the previous studies of territory.] The parts I have selected show Howard's recognition that males fight to secure a territory, not to secure a mate, as had been generally believed; the manner in which males select and establish the territory; that a male recognizes the boundary of his territory and does not fight beyond these borders; that displays can substitute for actual fighting; and that song serves not only to attract a mate but as warning to rivals.

The brief biography* of Howard that follows was written by Huxley and Fisher for their 1964 edition of *Territory in Bird Life*.

> Henry Eliot Howard, who was born near Kidderminster on 13 November 1873, died in his long-loved home, Clareland at Hartlebury near Stourport-on-Severn on 26 December 1940.
>
> Educated at Eton and Mason's College, Birmingham, Eliot Howard was by profession a steelmaster, a director for many years of the great firm of Stewarts and Lloyds. By inclination he was an amateur naturalist, if by amateur we mean (correctly) an enthusiast. Indeed, in field work he was a paragon of enthusiasm, never satisfied with anything but prolonged and intensive observation, and its meticulous recording; his notes, now permanently deposited at the Edward Grey Institute of Field Ornithology at Oxford, are, as its present Director, Dr. David Lack, writes "so abbreviated that it is almost impossible for anyone else to use them." Nevertheless this retiring and philosophical naturalist, who never published more than a few early notes in scientific journals, translated these notes into five important books, of which this volume, first published in 1920, has had

*Copyright © by Sir Julian Huxley and the late James Fisher. Reprinted by permission of A. D. Peters and Company.

time to be judged by history as a classic, and regarded universally as a milestone in ornithology's path.

Howard scarcely watched a bird outside England and Ireland. His first book *The British Warblers* beautifully illustrated in colour and black-and-white was published in London (rather expensively) in ten parts between February 1907 and June 1915; it is now a formidable collector's piece. In 1920 came the present work; in 1929 *An Introduction to the Study of Bird Behavior*; in 1935 *The Nature of a Bird's World*; in the year of his death *A Waterhen's Worlds*.

In the last two decades of his life Howard's writing led him into realms of philosophical speculation in which he was often rather difficult to follow. Yet never did beauty of style, or the nice use of words desert him: one of his last writings, of the waterhen (moorhen), describes his subject as "the perceiver with power to refer; which, being interpreted, is power to perceive things in relation. But having no power to reflect he perceives no self: memories he has, but perceiving no self he has neither future nor past—and so, no time. His world is always in the present, and mostly full of joy."

History has made Eliot Howard a quiet genius who was neither in the mainstream nor a backwater of ornithology, but floated on his own secret waters. Lacking though his major works are in bibliographic citation, it is clear from internal evidence that the major influences in his thinking were psychologists such as C. Lloyd Morgan (a close friend) and physiologists like F. H. A. Marshall. Apart from one of the writers, his closest friend active in ornithology was probably the late Percy R. Lowe; although he served for a time on the Council of the British Ornithologists' Union and became a Vice-President he shunned the ordinary social and transactional exchanges of everyday bird science. Nevertheless, in *Territory in Bird Life*, he succeeded in making a generalization that has become mainstream—perhaps (to continue the metaphor) by river-capture—and not only ornithologically mainstream; not long after 1920 questions on the significance of bird territory were being set (as the other writer well remembers) in Zoology Finals at one of our senior universities.

One can make a strong case for calling Bernard Altum the true father of the concept of territory. His publications, as far back as 1868, laid out a clear theory of terrioriality, but because of a language barrier and the relative inaccessibility of his book *Der Vogel und sein Leben*, his ideas remained largely unknown outside of Germany, even to Eliot Howard. Not until Ernst Mayr translated Altum was the English-speaking world aware of the true extent of Altum's contribution.

Probably the greatest unsung hero in the history of territory is C. B. Moffat. Had he read his paper "The spring rivalry of birds" before the British Ornithologists' Union rather than the Dublin Naturalists' Field Club and published in *Ibis* rather than the little-known *Irish Naturalist*, Moffat, rather than Eliot Howard, might be acknowledged today in the English-speaking world as the father of the concept of territory. Moffat was unaware of Altum but had studied Darwin's writings critically. He believed that territorial behavior serves to reduce the birth rate, thereby contributing to population regulation. Moffat grasped the idea of density-dependent regulation, an idea that was to lie dormant for at least another generation. Moffat dared to challenge Darwin's belief that song and plumage played a vital part in acquiring and holding a territory against other males of the same species. He presents good evidence that there is always a surplus of both males and females in an area and that these surplus birds fail to breed, not because of their incompatibility, as Darwin had argued, but because the males did not have the prerequisite territory. Moffat could also lay claim to being the first modern ethologist, for he recognized that the specially colored parts of a male's plumage served as a signal to repel rivals. In effect, he was describing sign stimuli and releasers.

C. B. Moffat (1859–1945) was trained as a lawyer but spent a good share of life as a journalist for the Dublin Daily Express. Natural history was his real love and he contributed numerous papers on botany and zoology to The Irish Naturalist. George R. Humphreys, a contemporary and friend of Moffat has this to say about him: "He was a wonderful all round naturalist and botanist, always approachable and a quiet and friendly man. He never to my knowledge used fieldglasses, the more remarkable in that he was an excellent observer of the habits of even the smallest of our songbirds and especially the warblers."

Margaret M. Nice was the third of the giants of bird territory. If you ask the average American naturalist and ornithologist which of the three giants he has read, the answer will almost always be Nice's *Studies in the Life History of the Song Sparrow*. I well remember how avidly I read this at a single sitting and the whole new world it opened to me. This classic is now available in paperback from Dover Publications.

Ornithology has made some of its greatest strides through the contributions of amateurs. Margaret Nice is one of these. Married to a busy doctor and a mother, she still managed to find time to make the most detailed and long-term life history study done of any American bird. Through Ernst Mayr she became aware of the work of Eliot Howard and his Irish predecessor, C. B. Moffat.

Nice was among the first to mark birds systematically with colored bands. By this method she was able to prove how steadfastly her sparrows returned to the same location as they had the previous year to reestablish their territory. Among her other publications are *Birds of Oklahoma* (1931) and "The role of territory in bird life" (1941). She is a Fellow of the American Ornithologists' Union and a recipient of its prestigious Brewster Award.

References

Nice, M. M. 1931. *The Birds of Oklahoma*. Rev. ed. University of Oklahoma Press, Norman, Okla. 224 pp.

Nice, M. M. 1941. The role of territory in bird life. *Amer. Midland Naturalist*, **26**: 441–487.

1

Reprinted from *Territory in Bird Life*, 15–22, 33–35, 76–80, 110–112, 114–116, 128–132, Collins, London (1948)

Territory in Bird Life

ELIOT HOWARD

INTRODUCTION

IN HIS *Manual of Psychology* Dr. Stout reminds us that "Human language is especially constructed to describe the mental states of human beings, and this means that it is especially constructed so as to mislead us when we attempt to describe the working of minds that differ in a great degree from the human."

The use of the word "territory" in connection with the sexual life of birds is open to the danger which we are here asked to guard against, and I propose, therefore, before attempting to establish the theory on general grounds, to give some explanation of what the word is intended to represent and some account of the exact position that representation is supposed to occupy in the drama of bird life.

The word is capable of much expansion. There cannot be territories without boundaries of some description ; there cannot well be boundaries without disputes arising as to those boundaries ; nor, one would imagine, can there be disputes without consciousness as a factor entering into the situation; and so on, until by a simple mental process we conceive of a state in bird life analogous to that which we know to be customary amongst ourselves. Now, although the term "breeding territory," when applied to the sexual life of birds, is not altogether a happy one, it is difficult to know how otherwise to give expression to the facts observed. Let it then be clearly understood that the expression "securing a territory" is used to denote a process, or rather part of a process, which, in order to insure success to the individual in the attainment of reproduction, has been gradually evolved to meet the exigencies of diverse circumstances. Regarded thus, we avoid the risk of conceiving of the act of securing a territory as a detached event in the life of a bird, and avoid, I hope, the risk of a conception based upon the meaning of the

15

word when used to describe human as opposed to animal procedure.

Success in the attainment of reproduction is rightly considered to be the goal towards which many processes in nature are tending. But what is meant by success? Is it determined by the actual discharge of the sexual function? So many and so wonderful are the contrivances which have slowly been evolved to insure this discharge, that it is scarcely surprising to find attention focused upon this one aspect of the problem. Yet a moment's reflection will show that so limited a definition of the term "success" can only be held to apply to certain forms of life; for where the young have to be cared for, fostered, and protected from molestation for periods of varying lengths, the actual discharge of the sexual function marks but one stage in a process which can only succeed if all the contributory factors adequately meet the essential conditions of the continuance of the species.

Securing a territory is then part of a process which has for its goal the successful rearing of offspring. In this process the functioning of the primary impulse, the acquirement of a place suitable for breeding purposes, the advent of a female, the discharge of the sexual function, the construction of the nest, and the rearing of offspring follow one another in orderly sequence. But since we know so little of the organic changes which determine sexual behaviour, and have no means of ascertaining the nature of the impulse which is first aroused, we can only deal with the situation from the point at which the internal organic changes reflect themselves in the behaviour to a degree which is visible to an external observer. That point is reached when large numbers of species, forsaking the normal routine of existence to which they have been accustomed for some months, suddenly adopt a radical change in their mode of behaviour. How is this change made known to us? By vast numbers of individuals hurrying from one part of the globe to another, from one country to another, and even from mid-ocean to the coasts; by detachments travelling from one district to another; by isolated individuals deserting this place for that; by all those movements, in fact, which the term migration, widely applied, is held to denote. Now the impulse which prompts these travelling hosts must be similar in kind whether the journey be long or short; and it were better, one would think, to regard such movements as a whole than to fix the attention on some one particular journey which fills us with amazement on account of the magnitude of the distance traversed or the nature of the difficulties overcome. For, after all, what does each individual seek? There may be some immature birds which, though they have not reached the necessary stage of development, happen to fall in with others in whom the impulse is strong and are led by them —they know not where. But the majority seek neither continent nor country, neither district nor locality is their aim, but a place wherein the rearing of offspring can be safely accomplished; and the search for this place is the earliest visible manifestation in many species of the reawakening of the sexual instinct.

The movements of each individual are then directed towards a similar goal, namely, the occupation of a definite station; and this involves for many species a distinct change in the routine of behaviour to which previously they had been accustomed. Observe, for example, one of the numerous flocks of Finches that roam about the fields throughout the winter. Though it may be composed of large numbers of individuals of different kinds, yet the various units form an amicable society actuated by one motive—the procuring of food. And since it is to the advantage of all that the individual should be subordinated to the welfare of the community as a whole there is no dissension, apart from an occasional quarrel here and there.

In response, however, to some internal organic change, which occurs early in the season, individuality emerges as a factor in the developing situation, and one by one the males betake themselves to secluded positions, where each one, occupying a limited area, isolates itself from companions. Thereafter we no longer find that certain fields are tenanted by flocks of greater or less dimensions, while acres of land are uninhabited, but we observe that the hedgerows and thickets are divided up into so many territories, each one of which contains its owner. This procedure, with of course varying detail, is typical of that of many species

T.B.L. B

that breed in Western Europe. And since such a radical departure from the normal routine of behaviour could scarcely appear generation after generation in so many widely divergent forms, and still be so uniform in occurrence each returning season, if it were not founded upon some congenital basis, it is probable that the journey, whether it be the extensive one of the Warbler or the short one of the Reed-Bunting, is undertaken in response to some inherited disposition, and probable also that the disposition bears some relation to the few acres in which the bird ultimately finds a resting place. Whilst for the purpose of the theory I shall give expression to this behaviour in terms of that theory, and speak of it as a disposition to secure a territory, using the word disposition, which has been rendered current in recent discussion, for that part of the inherited nature which has been organised to subserve a specific biological purpose—strict compliance with the rules of psychological analysis requires a simpler definition; let us therefore say "disposition to remain in a particular place in a particular environment."

But even granting that this disposition forms part of the hereditary equipment of the bird, how is the process of reproduction furthered? The mere fact of remaining in or about a particular spot cannot render the attainment of reproduction any less arduous, and may indeed add to the difficulties, for any number of individuals might congregate together and mutually affect one another's interests. A second disposition comes, however, into functional activity at much the same stage of sexual development, and manifests itself in the male's intolerance of other individuals. And the two combined open up an avenue through which the individual can approach the goal of reproduction. In terms of the theory I shall refer to this second disposition as the one which is concerned with the defence of the territory.

Broadly speaking, these two dispositions may be regarded as the basis upon which the breeding territory is founded. Yet inasmuch as the survival value of the dispositions themselves must have depended upon the success of the process as a whole, it is manifest that peculiar significance must not be attached to just the area occupied, which happens to be so susceptible of observation; other contributory factors must also receive attention, for the process is but an order of relationships in which the various units have each had their share in determining the nature and course of subsequent process, so that, as Dr. Stout says, when they were modified, it was modified.

Now the male inherits a disposition which leads it to remain in a restricted area, but the disposition cannot determine the extent of that area. How then are the boundaries fixed? That they are sometimes adhered to with remarkable precision, that they can only be encroached upon at the risk of a conflict—all of this can be observed with little difficulty. But if we regard them as so many lines definitely delimiting an area of which the bird is cognisant, we place the whole behaviour on a different level of mental development, and incidentally alter the complexion of the whole process. It would be a mistake, I think, to do this. Though conscious intention as a factor may enter the situation, there is no necessity for it to do so; there is no necessity, that is to say, for the bird to form a mental image of the area to be occupied and shape its course accordingly. The same result can be obtained without our having recourse to so complex a principle of explanation, and that by the law of habit formation. In common with other animals, birds are subject to this law in a marked degree. An acquired mode of activity becomes by repetition ingrained in the life of the individual, so that an action performed to-day is liable to be repeated to-morrow so long as it does not prejudice the existence or annul the fertility of the individual.

Let us see how this may have operated in determining the limits of the area required, and for this purpose let us suppose that we are observing a male Reed-Bunting recently established in some secluded piece of marsh land. Scattered about this particular marsh are a number of small willows and young alder trees, each one of which is capable of providing plenty of branches suitable for the bird to perch upon, and all are in a like favourable position so far as the outlook therefrom is concerned. Well, we should expect to find that each respective tree would be made use of according to the position in which the bird happened to find itself. But what actually do we find—one tree singled out and resorted to with ever-increasing certainty until it becomes an important point in relation to the occupied area, a headquarters from which the bird advertises its presence by song, keeps watch upon the movements of its neighbours, and sets out for the purpose of securing food. We then take note of its wanderings in the immediate vicinity of the headquarters, especially as regards the direction, frequency, and extent of the journeys; and we discover not only that these journeys proceed from and terminate in the special tree, but that there is a sameness about the actual path that is followed. The bird takes a short flight, searches a bush here and some rushes there, returns, and after a while repeats the performance; we on our part mark the extreme limits reached in each direction, and by continued observation discover that these limits are seldom exceeded, that definition grows more and more pronounced, and that by degrees the movements of the bird are confined within a restricted area. In outline, this is what happens in a host of cases. By repetition certain performances become stereotyped, certain paths fixed, and a routine is thus established which becomes increasingly definite as the season advances.

But while it would be quite untrue to say that this routine is never departed from, and equally profitless to attempt to find a point beyond which the bird will under no circumstances wander, yet there is enough definition and more than enough to answer the purpose for which the territory has, I believe, been evolved, that is to say the biological end of reproduction. Again, however, the process of adjustment is a complex one. Habit plays its part in determining the boundaries in a rough and ready manner, but the congenital basis, which is to be found in the behaviour adapted to a particular environment, is an important factor in the situation. For example, if instead of resting content with just a bare position sufficient for the purpose of reproduction, the Guillemot were to hustle its neighbours from adjoining ledges, the Guillemot as a species would probably disappear; or if instead of securing an area capable of supplying sufficient food both for itself and its young, the Chiffchaff were to confine itself to a single tree, and, after the manner of the Guillemot, trust to spasmodic excursions into neutral ground for the purpose of obtaining food, the Chiffchaff as a species would probably not endure. All such adjustments have, however, been brought about by relationships which have gradually become interwoven in the tissue of the race.

The intolerance that the male displays towards other individuals, usually of the same sex, leads to a vast amount of strife. Nowhere in the animal world are conflicts more frequent, more prolonged, and more determined than in the sexual life of birds; and though they are acknowledged to be an important factor in the life of the individual, yet there is much difference of opinion as to the exact position they occupy in the drama of bird life. Partly because they frequently happen to be in evidence, partly because they are numerically inferior, and partly, I suppose, because the competition thus created would be a means of maintaining efficiency, the females, by common consent, are supposed to supply the condition under which the pugnacious nature of the male is rendered susceptible to appropriate stimulation. And so long as the evidence seemed to show that battles were confined to the male sex, so long were there grounds for hoping that their origin might be traced to such competition. But female fights with female, pair with pair, and, which is still more remarkable, a pair will attack a single male or a single female; moreover, males that reach their destination in advance of their prospective mates engage in serious warfare. How then is it possible to look upon the individuals of one sex as directly responsible for the strife amongst those of the other, or how can the female supply the necessary condition? As long as an attempt is made to explain it in terms of the female, the fighting will appear to be of a confused order; regard it, however, as part of a larger process which demands, amongst other essential conditions of the breeding situation, the occupation of a definite territory, and order will reign in place of confusion.

But even supposing that the male inherits a disposition to acquire a suitable area, even supposing that it inherits a dis-

15

position which results indirectly in the defence of that area, how does it obtain a mate? If the female behaved in a like manner, if she, too, were to isolate herself and remain in one place definitely, that would only add to the difficulties of mutual discovery. We find, however, in the migrants, that the males are earlier than the females in reaching the breeding grounds, and, in resident species, that they desert the females and retire alone to their prospective territories, so that there is a difference in the behaviour of the sexes at the very commencement of the sexual process. What is the immediate consequence? Since the male isolates itself, it follows, if the union of the sexes is to be effected, that the discovery of a mate must rest largely with the female. This of course reverses the accepted course of procedure. But after all, what reason is there to suppose that the male seeks the female, or that a mutual search takes place; what reason to think that this part of the process is subject to no control except such as may be supplied by the laws of chance?

* * * * * * *

I have mentioned the Reed-Bunting; let us take it as our first example and try to follow its movements when the influence exerted by the internal secretions begins to be reflected on the course of its behaviour. First, it will be necessary to discover the exact localities in any given district to which the species habitually returns for the purpose of procreation; otherwise the earlier symptoms of any disposition to secure a territory may quite possibly be overlooked in the search for its breeding haunts.

In open weather Reed-Buntings pass the winter either singly, in twos or threes, or in small flocks, on bare arable ground, upon seed fields, or in the vicinity of water-courses; but in the breeding season they resort to marshy ground where the *Juncus communis* grows in abundance, to the dense masses of the common reed (*Arundo phragmites*), and such-like places. During the winter, the male's routine of existence is of a somewhat monotonous order, limited to the necessary search for food during the few short hours of daylight and enforced inactivity during the longer hours of darkness. But towards the middle of February a distinct change manifests itself in the bird's behaviour. Observe what then happens. When they leave the reed-bed in the morning, instead of flying with their companions to the accustomed feeding grounds, the males isolate themselves and scatter in different directions. The purpose of their behaviour is not, however, to find fresh feeding grounds, nor even to search for food as they have been wont to do, but rather to discover stations suitable for the purpose of breeding; and, having done so, each male behaves in a like manner—it selects some willow, alder, or prominent reed, and, perching thereon, leads a quiet life, singing or preening its feathers. Now if the movements of one particular male are kept in view, it will be noticed that only part of its time is spent in its territory. At intervals it disappears. I do not mean that one merely loses sight of it, but that it actually deserts its territory. As if seized with a sudden impulse it rises into the air and flies away, often for a considerable distance and often in the same direction, and is absent for a period which may vary in length from a few minutes to an hour or even more. But these periodical desertions become progressively less and less frequent in occurrence until the whole of its life is spent in the few acres in which it has established itself.

The behaviour of the Yellow Bunting is similar. In any roadside hedge two or more males can generally be found within a short distance of one another, and in such a place their movements can be closely and conveniently followed. Under normal conditions the ordinary winter routine continues until early in February; but the male then deserts the flock, seeks a position of its own, and becomes isolated from its companions. Now the position which it selects does not, as a rule, embrace a very large area—a few acres perhaps at the most. But there is always some one point which is singled out and resorted to with marked frequency—a tree, a bush, a gate-post, a railing, anything in fact which can form a convenient perch, and eventually it becomes a central part of the bird's environment. Here it spends the greater part of its time, here it utters its song persistently, and here it keeps watch upon intruders. The process of establishment is nevertheless a gradual one. The male does not appear in its few acres suddenly and remain there permanently as does the migrant; at first it may not even roost in the prospective territory. The course of procedure is somewhat as follows: at dawn it arrives and for a while utters its song, preens its feathers, or searches for food; then it vanishes, rising into the air and flying in one fixed direction as far as the eye can follow, until it becomes a speck upon the horizon and is ultimately lost to view. During these excursions it rejoins the small composite flocks which still frequent the fields and farm buildings. For a time the hedgerow is deserted and the bird remains with its companions. But one does not have to wait long for the return; it reappears as suddenly as it vanished, flying straight back to the few acres which constitute its territory, back even to the same gate-post or railing, where it again sings. This simple routine may be repeated quite a number of times during the first two hours or so of daylight, with, of course, a certain amount of variation; on one occasion the bird may be away for a few minutes only, on another for perhaps half an hour, whilst sometimes it will fly for a few hundred yards, hesitate, and then return—all of which shows clearly enough that these few acres possess some peculiar significance and are capable of exercising a powerful influence upon the course of its behaviour. And so the disposition in relation to the territory becomes dominant in the life of the bird.

* * * * * * *

What then is the meaning of all this warfare? The process of reproduction is a complex one, built up of a number of different parts forming one inter-related whole; it is not merely a question of "battle," or of "territory," or of "song," or of "emotional manifestation," but of all these together. The fighting is thus one link in a chain of events whose end is the attainment of reproduction; it is a relationship in an inter-related process, and to speak of it as being even directly related to the territory is scarcely sufficient, for it is intimately associated with the disposition which is manifested in the isolation of the male from its companions, and forms therewith an *imperium in imperio* from which our concept of breeding territory is taken. But let me say

18

at once that it is no easy matter to prove this, for since so many modes of behaviour, which can be interpreted as lending support to this view, are likewise interpretable on the view that the presence of a female is a necessary condition of the fighting, it is difficult to find just the sort of evidence that is required. Nevertheless, after hearing the whole of the evidence and at the same time keeping in mind the conclusion which we have already reached, I venture to think that the close relationship between the warfare on the one hand and the territory on the other will be fully admitted.

Formerly I deemed the spring rivalry to be the result of accidental encounters, and I believed that an issue to a struggle was only reached when one of the combatants succumbed or disappeared from the locality, a view which neither recognised method nor admitted control. Recent experience has shown, however, that I was wrong, and that there is a very definite control over and above that which is supplied by the physical capabilities of the birds.

Let us take some common species, the Willow-Warbler being our first example; and, having found three adjoining territories occupied by unpaired males, let us study the conflicts at each stage in the sexual life of the three individuals, observing them before females have arrived upon the scene, again when one or two of the three males have secured mates, and yet again when all three have paired. Now we shall find that the conditions which lead up to and which terminate the conflicts are remarkably alike at each of these periods. A male intrudes, and the intrusion evokes an immediate display of irritation on the part of the owner of the territory, who, rapidly uttering its song and jerking its wings, begins hostilities. Flying towards the intruder, it attacks viciously, and there follows much fluttering of wings and snapping or clicking of bills. At one moment the birds are in the tree-tops, at another in the air, and sometimes even on the ground, and fighting thus they gradually approach and pass beyond the limits of the territory. Whereupon a change comes over the scene; the male whose territory was intruded upon and who all along had displayed such animosity, betrays no further interest in the conflict—it ceases to attack, searches around for food, or sings, and slowly makes its way back towards the centre of the territory.

Scenes of this kind are of almost daily occurrence wherever a species is so common, or the environment to which it is adapted so limited in extent, that males are obliged to occupy adjacent ground. The Moor-Hen abounds on all suitable sheets of water, and it is a bird that can be conveniently studied because, as a rule, there is nothing, except the rushes that fringe the pool, to hinder us from obtaining a panoramic view of the whole proceedings, and moreover the area occupied by each individual is comparatively small. Towards the middle of February, symptoms of sexual organic change make themselves apparent, and the pool is then no longer the resort of a peaceable community; quarrels become frequent, and as different portions of the surface of the water are gradually appropriated, so the fighting becomes more incessant and more severe. Each individual has its own particular territory, embracing a piece of open water as well as a part of the rush-covered fringe, within which it moves and lives. But in the early part of the season, when the territories are still in process of being established, and definiteness has still to be acquired, trespassing is of frequent occurrence, and the conflicts are often conspicuous for their severity.

Now these conflicts are not confined to unpaired individuals, nor to one sex, nor to one member of a pair—every individual that has settled upon the pool for the purpose of breeding will at one time or another be involved in a struggle with its neighbour. If then we single out certain pairs and day by day observe their actions and their attitude towards intruders, we shall notice that, instead of their routine of existence consisting, as a casual acquaintance with the pool and its inmates might lead us to believe, of an endless series of meaningless disputes, the behaviour of each individual is directed towards a similar goal—the increasing of the security of its possession; and further, if we pay particular attention to the circumstances which lead up to the quarrels and the circumstances under which such quarrels come to an end, we shall find, when we have accumulated a sufficient body of observations, that the disputes always originate in trespass, and that hostilities always cease when the trespasser returns again to its own territory. By careful observation it is possible to make oneself acquainted with the boundaries—I know not what other term to use—which separate this territory from that; and it is the conduct of the birds on or near these boundaries to which attention must be drawn. A bird may be feeding quietly in one corner of its territory when an intruder enters. Becoming aware of what is happening it ceases to search for food, and approaching the intruder, at first swimming slowly but gradually increasing its pace, it finally rises and attacks with wings and beak, and drives its rival back again beyond the boundary. Thereupon its attitude undergoes a remarkable change; ceasing to attack, but remaining standing for a few moments as if still keeping guard, it betrays no further interest in the bird with which a few seconds previously it was fighting furiously. On one occasion I watched a trespasser settle upon a conspicuous clump of rushes situated near the boundary. The owner, who was at the moment some distance away, approached in the usual manner, and, having driven off the trespasser, returned immediately to the clump, where it remained erect and motionless.

* * * * * * *

Thus it seems clear that the proximate end to which the fighting is directed is not necessarily the defeat of the intruder, but its removal from a certain position. And inasmuch as this result will be obtained whether the retreat is brought about by fear of an opponent or by physical exhaustion, it is manifest that too much significance need not be attached to the amount of injury inflicted. It is necessary to bear this in mind, because it is held by some, who have carefully observed the actions of various species, that overmuch importance is attached to the conflicts, that in a large number of instances they are mere "bickerings" and lead to nothing, and that they are now only "formal," which means, I suppose, that they are vestigial—fragments of warfare that determined the survival of the species in bygone ages. But if the conclusion at which we have just arrived be correct, if we can recognise a single aim passing through the whole of the warfare—and that one the removal of an intruder from a certain position, then we need no longer concern ourselves as to the degree of severity of the battles—we see it all in true perspective. Neither exhaustion nor physical inability are the sole factors which determine the nature and extent of the fighting; there is a more important factor still—position. According, that is to say, to the position which a bird occupies whilst fighting is in progress, so its pugnacious nature gains or loses susceptibility, and it is this gain or loss of susceptibility which I refer to when I speak of the fighting as being controlled.

* * * * * * *

What, first of all, are the conditions in the life behaviour during the season of reproduction that make the intervention of the voice a consideration of such importance? The general result of our investigation might be summed up thus: we found that the male inherits a disposition to secure a territory, that at the proper season this disposition comes into functional activity and leads to its establishment in a definite place, and that it cannot search for a mate because its freedom of action in this respect is forbidden by law; that the female inherits no such disposition, that she is free to move from place to place, free to satisfy her predominant inclination, and to seek a mate where she wills; and, since the appropriate organic condition which leads to pairing must coincide with appropriate conditions in the environment, that the union of the sexes must be accomplished without undue delay. Furthermore, we found that a territory is essential if the offspring are to be successfully reared; that, since the available breeding ground is limited, competition for it is severe, and that the male is precluded from leaving the ground which he has selected, and is obliged, in order to secure a mate, to make himself conspicuous. That was our general result. Now there are two ways by which the male can make himself conspicuous —by occupying such a position that he can be readily seen, or by producing some special sound which will be audible to the female and direct her to the spot. The former, by itself, is insufficient; in the dim light of the early dawn, when life is at its highest, and mating proceeds apace, what aid would it be for a male to perch on the topmost branch of a tree, how slender a guide in the depth of the forest? But whether in the twilight or in the dark, in the thicket or the jungle, on the mountain or on the moor, the voice can always be heard—and the voice is the principal medium through which the sexes are brought into contact.

Well now, we come back to the question, why, if all species have a serviceable recognition call, that call should not be sufficient for the purpose, just as, without a doubt, it is adequate for all purposes at other seasons? The answer is, I think, clear. The recognition call is not confined to one sex, nor only to breeding birds; it is the common property of all the individuals of the species, and if the female were to rely upon it as a guide she might at one moment pursue another female, at another a non-breeding male; she might even be guided to a paired female or to a paired male, and time would be wasted and much confusion arise. So that no matter how much a male might advertise himself by cries and calls which were common alike to all the individuals of the species, it would not assist the biological end which we have in view. Something else is therefore required to meet the peculiar circumstances, some special sound bearing a definite meaning by which the female can recognise, amongst the host of individuals of no consequence to her, just those particular males in a position to breed and ready to receive mates. Hence the vocal powers, the power of producing sounds instrumentally, and the power of flight, have been organised to subserve the biological end of "recognition."

* * * * * * *

The proximate end of the male's behaviour is isolation—how is it to be obtained? If, after having occupied a territory, the bird were to remain silent, it would run the risk of being approached by rivals; if, on the other hand, it were merely to utter the recognition call of the species, it would but attract them. In neither case would the end in view be furthered, and isolation would solely depend upon alertness and the capacity to eject intruders. Supposing, however, that the song, just as it serves to attract the females, serves to repel other males, a new element is introduced deserving of recognition; for those males that had established themselves would not only be spared the necessity of many a conflict, but they would be spared also the necessity of

23

constant watchfulness, and so, being free to pursue their normal routine—to seek food, to rest, and, if migrants, to recover from the fatigue of the journey—they would be better fitted to withstand the strain of reproduction; and those that were still seeking isolation in an appropriate environment, instead of settling first here and then there only to find themselves forestalled, would avoid and pass by positions that were occupied, establishing themselves without loss of time in those that were vacant. Without the aid of something beyond mere physical encounter to regulate dispersal, it is difficult to imagine how in the short time at disposal anything approaching uniformity of distribution could be obtained. Hence, both in the direction of limiting combat, of insuring accommodation for the maximum number of pairs in the minimum area, and of conserving energy, the song, by conveying a warning, plays an important part in the whole scheme.

And if this be so, if the song repels instead of attracting, it follows that the more distinct the sounds, the less likelihood will there be of confusion; for supposing that different species were to develop similar songs, whole areas might be left without their complement of pairs just because this male mistook the voice of that, and avoided it when there was no necessity for doing so. So that just as from the point of view of "recognition" each female must be able to distinguish the voice of its own kind, so likewise the warning can only be adequate providing that the sounds are specifically distinct. A point, however, arises here in regard to closely related forms. Some species require similar food and live under similar conditions of existence; they meet in competition and fight with one another; and, if they did not do so, the food supply of a given area would be inadequate to support the offspring of all the pairs inhabiting that area. Generally speaking, the more closely related the forms happen to be, the more severe the competition tends to become; and it may be argued that in such cases a similar song would contribute to more effective distribution and in some measure provide against the necessity of physical encounter; that, in fact, it would stand in like relation to the success of all the individuals concerned, as does the song to the individuals of the same species.

But we must bear in mind that the primary purpose of song is to direct the females to those males that are in a position to breed; and to risk the possibility of prompt recognition in order that the males of closely related species should fight the less, would be to sacrifice that which is indispensable for a more remote and less important advantage.

* * * * * * *

CHAPTER FIVE

THE RELATION OF THE TERRITORY TO THE SYSTEM OF REPRODUCTION

In the first two chapters I tried to show that the inherited nature of the male leads it to remain in a definite place at a definite season and to become intolerant of the approach of members of its own sex, and that a result is thus attained which the word "territory" in some measure describes. But the use of this word is nevertheless open to criticism, for it denotes a human end upon which the highest faculties have been brought to bear, and consequently we have to be on our guard lest our conception of the "territory" should tend to soar upwards into regions which require a level of mental development not attained by the bird. It is necessary to bear this in mind now we have come to consider the meaning of the territory, or rather the position that it occupies in the whole scheme of reproduction.

Relationship to a territory within the inter-related whole of a bird's life serves more than one purpose, and not always the same purpose in the case of every species. We have only to glance at the life-histories of divergent forms to see that the territory has been gradually adjusted to suit their respective needs—limited in size here, expanded there, to meet new conditions as they arose. Now some may think that the theory would be more likely to be true if the territory had but one purpose to fulfil, and that one the same for every species; and they may see nothing but weakness in the multiplication of ways in which I shall suggest it may be serviceable. But such an objection, if it were raised, would arise from a mistaken conception, a conception which, instead of starting with a relationship and working up to the "territory," sees in the "territory" something of the bird's own selection and thence works back to its origin. Holding the view that it is nothing but a term in a complex relationship

128

which has gradually become interwoven in the history of the individual, I see no reason why the fact of its serving a double or a treble purpose should not be a stronger argument for its survival. I now propose to examine the various ways in which the territory may have been of use in furthering the life of the individual, and the circumstances in the inorganic world which have helped to determine its survival.

The purpose that it serves depends largely upon the conditions in the external environment—the climate, the supply of food, the supply of breeding stations, and the presence of enemies. Hence its purpose varies with varying conditions of existence. But before we proceed to examine the particular ways in which it has been modified to suit the needs of particular classes of species, and the reason for such modifications, we must inquire whether there is not some way in which it has been serviceable alike to every species, or at least to a large majority of them.

Success in the attainment of reproduction depends upon the successful discharge of the sexual function; and the discharge of the sexual function depends primarily upon an individual of one sex coming into contact with one of the opposite sex at the appropriate season and when its appropriate organic condition arises. Now the power of locomotion is so highly developed in birds that it may seem unreasonable to suppose that males and females would have any difficulty in meeting when their inherited nature required that they should do so, still less reasonable to suggest that this power might even act as a hindrance to success-ful mating. Nevertheless, if we try to picture to ourselves the conditions which would obtain if the movements of both sexes were in no wise controlled, and mating were solely dependent upon fortuitous gatherings, we shall come, I fancy, to no other conclusion than that much loss of valuable time and needless waste of energy would often be incurred in the search, and that many an individual would fail to breed just because its wanderings took it into districts in which, at the time, there happened to be too many of this sex or too few of that. And as the power of locomotion increased and the distribution of the sexes became more and more irregular, so the opportunity would be afforded

T.B.L.   I for the development of any variation which would have tended to facilitate the process of pairing, and by so doing have conferred upon the individuals possessing it some slight advantage over their fellows.

What would have been the most likely direction for variation to have taken? Any restriction upon the freedom of movement of both sexes would only have added to the difficulties of mating; but if restriction had been imposed upon one sex, whilst the other had been left free to wander, some order would have been introduced into the process. That the territory serves to restrict the movements of the males and to distribute them uniformly throughout all suitable localities, there can be no question; and since the instinctive behaviour in relation to it is timed to appear at a very early stage in the seasonal sexual process, the males are in a position to receive mates before the impulse to mate begins to assert itself in the female.

* * * * * * *

The advantage of this territorial system is therefore apparent. Instead of this district being overcrowded and that one deserted; instead of there being too many of one sex here and too few of the other sex there; instead of a high percentage of individuals failing to procreate their kind, just because circumstances over which they have no control prevent their discovering one another at the appropriate time—each sex has its allotted part to play, each district has its allotted number of inhabitants, and the waste of energy and the loss of time incurred in the process of mating is reduced to a minimum.

Let us return again to the question of fortuitous mating, and consider the position of a male and female that have discovered one another by accident and have paired; what will be the subsequent course of their behaviour? We are assuming, of course, that a territory forms no part of their life-history. If the discharge of the sexual function takes place immediately and the ovaries of the female are in an advanced state of seasonal development, the construction of the nest will proceed without delay—and the nest will answer the same purpose as the territory in so far as it serves to restrict the movements of the birds and tends to make them remain in, or return to, its vicinity; but if not, there will be an interval during which both sexes will continue to wander as before, guided only by the scarcity or abundance of food. In the first case, there will be the attraction of the nest to prevent any untimely separation; in the second, there will be nothing in the external environment to induce them to remain in any particular spot. Now if we turn to any common species and observe the sequence of events in the life of different pairs, we shall find that pairing is seldom followed by an immediate attempt to build; that an interval of inactivity is the rule rather than the exception, and that this interval varies in different species, in different individuals, and in different seasons. Our imaginary male and female will therefore be faced with considerable difficulty; for with nothing in the external environment to attract them and with no restriction imposed upon the direction or extent of their flight, their union will continue to be, as it began by being, fortuitous. Next, let us consider their position were a disposition to establish a territory to form part of the inherited nature of the male. Each one will then be free to seek food when and where it wills and to associate with other individuals without the risk of permanent separation from its mate; and no matter how long an interval may elapse between mating and nest-building, each one will be in a position to find the other when the appropriate moment for doing so arrives. Hence, while preserving freedom of movement for each individual, the territory will render their future, as a pair, secure.

2

Reprinted from *Proc. Linnaean Soc. N.Y.*, Nos. 45–46: 24–30 (1935)

Bernard Altum and the Territory Theory
By Ernst Mayr

In 1920 Eliot Howard published a book, "Territory in Bird-life," which caused more discussions among ornithologists than any other recent publication. It has also been responsible in a high degree for the rapid spreading of a new method of bird-study: the life history of individual birds. For ten years or more it was generally believed among the English speaking ornithologists that Howard's theory was something entirely novel, but recently it has been shown that he had several forerunners (Mousley[1], Moffat[2], and others) who had made similar observations, but had not developed a detailed theory. It was Meise (1930, p. 52) who called attention to the fact that in Germany the essentials of Howard's theory were common knowledge of the field ornithologists. Bernard Altum had developed, as early as 1868, in great detail a territory concept which in some ways was even superior to that of Howard by largely avoiding generalizations concerning the occurrence of territory among birds. These ideas were published in a book called: "Der Vogel und sein Leben" (The Bird and his Life), which had a remarkable success in Germany, where seven editions were printed between 1868 and 1903.

Since this book is now out of print, it was suggested to me by several British ornithologists at the International Ornithological Congress at Oxford that I publish a translation. The idea was much encouraged by some of my American friends, and I finally undertook the work. Mrs. Margaret M. Nice and Mr. William Vogt were so kind as to check my translation, and I owe to them, as well as to Dr. Austin Rand, many helpful suggestions which I want to acknowledge with my sincere appreciation.

As an introduction, it may be opportune to say a few words about the author.

Johann Bernard Theodor Altum was born in Münster (Westfalia) on December 31, 1824. His father, a small artisan, instilled in his son a love for nature and gave him his first instructions in the collecting and study of birds. Altum entered a Catholic college in 1845 and four years later he became a priest in one of the Münster churches.

[1] Auk, 1919, p. 339; and Auk, 1921, p. 321.
[2] 1903, Irish Naturalist, XII, p. 152.

His great love for science, however, influenced him to continue his studies at the university, and in 1853 he went to Berlin to complete his work in zoology and philology. In 1855 he received his doctor's degree, and after having held an assistantship under Lichtenstein at the Zoological Museum, he returned to Münster where he obtained a teaching position.

In 1859 he became associate professor at the Münster university after giving a lecture on climatical varieties of animals. The subsequent ten years comprised his most active period of field ornithology, at the end of which he published his well-known book: "Der Vogel und sein Leben" (1868). The following year Altum was appointed professor of zoology at the forestry college at Eberswalde. In this position he became the father of economic ornithology. The results of his research on the interrelation between birds and insects are embodied in his three volume work "Forstzoologie,"[1] the "classic" of this field.

In the course of these investigations he came to the conclusion that it was futile to classify birds into beneficial and harmful species. He was, perhaps, the first conservationist to proclaim the protection of birds primarily on the basis of their aesthetic value (1881, etc.), a point of view which he defended vigorously at legislative meetings and scientific congresses against opponents who proclaimed the protection of birds merely on the basis of their economic importance.

Since this is not the place to list the many-sided activities of this gifted man, I will only mention that he was president of the German Ornithological Society for many years and his writings were popular all over Germany. More detailed information can be found in his obituaries[2]. Altum died February 1, 1900, at the age of 76 years.

The first edition of "Birds and their Life" (Der Vogel und sein Leben) was published in 1868, originally more or less as an answer to A. Brehm's book with the same name (Das Leben der Vögel, 1861). Brehm had given a very sentimental and anthropomorphic picture of the bird's life, surmising that birds felt and thought very much like humans. Altum's principle, however, was, "animal non agit, agitur[3],"

[1]Forstzoologie, 3 vol., 2nd ed., Berlin 1880.
[2]Ornith. Monatsber. 1900, p. 49-54; Schalow, Beitr. Vogelfauna Brandenburg, 1919, p. 555-569.
[3]Translated: An animal does not act, but is being acted upon, or more freely: An animal does not act by its own volition, but reacts to stimuli (drives).

a remarkably modern point of view. To support this theory he gathered one piece of evidence after another, and he also developed his territory theory in this connection.

The following translation is based on the sixth edition (1898) of Altum's work of which more copies are available than of any of the earlier editions. It is, however, a practically unaltered reprint of the first edition (1868) in those chapters in which we are particularly interested. The only real difference is the insertion on page 101 of a statement relating to the birds of prey. I have tried to follow the original as closely as possible which will account for some of the foreign sounding phrases.

Altum treats territory in his chapter on "Song," explaining that song not only coincides seasonally with the reproductive activities, but that it is a necessary and integral part of them. It serves to bring together the pairs and to fix the territory borders:

"/97[1]/ The Fixing of the Territory Borders.

"It is impossible among a great many species of birds, for numerous pairs to nest close together, but individual pairs must settle at precisely fixed distances from each other. The reason for this necessity is the amount and kind of food they have to gather for themselves and their young, together with the methods by which they secure it. All the species of birds which have specialized diets and which, in searching for food— mostly animal matter—for themselves and their young, limit their wanderings to small areas, can not and ought not to settle close to other pairs because of the danger of starvation. They need a territory (*Brutrevier*) of a definite size, which varies according to the productivity of any given locality.

"In order to determine the daily food consumption of my caged Blue-throat (*Cyanosylvia*), I once counted the 'ant-eggs' I fed it, and discovered that it devoured during an average day about 1,200 of them in addition to eight meal-worms. Now let us suppose that even a smaller quantity, perhaps 1,000 insects of the same size, would have been sufficient, /p. 98/ and let us assume that the daily food of five young in the nest equals that of the two adults; such a pair of Blue-throats, with their young, would need under natural conditions about 4,000 insects of 'ant-egg' size each day. Other investigators have ar-

---

[1]Figures refer to the pagination of the sixth edition.

rived at similar figures. All Warblers, Redstarts, the Nightingale, the Robin, the Hedge-sparrow, the Titmice, in short, all such birds, require a more or less similar quantity; and even most of the Finches and Buntings, though they are primarily seed-eaters, feed their young mostly with insects. They search for this almost incredible quantity of food, not at a distance, but in the immediate vicinity of the nest. Many insects, such as large and hard beetles, hairy caterpillars, and others, are, however, not fit for food, and it is only because numbers of insects are emerging every day that it is possible for the adult birds to satisfy themselves and their offspring. It is self-evident that a half dozen pairs of such similar species of birds cannot settle in the same immediate vicinity, for each pair must have its own territory if it is to avoid starvation or, at best, a very miserable existence.

"If a locality produces a great deal of food, the result of favorable soil, vegetation, and climatic conditions, the size of territories may be reduced to some extent. We call such localities excellent Warbler, Nightingale, etc., terrain, but even here territorial boundaries cannot be absent. It is not at all remarkable that for each species of bird the size of these necessary territories is adjusted to its exact ecological requirements and its specific food. While, for example, the Sea-eagle has a territory an hour's walk in diameter, a small wood lot is sufficient for the Woodpecker, and a single acre of brush for the Warbler. All this is well-balanced and well-contrived. Anyone who spends a bright morning during the breeding season in a country rich with birds can easily learn the size of the territories from the distances between the singing males; and it gives no little pleasure to be able to determine, in this manner, the number and approximate position of the nests of so many species and pairs of birds. One acquires, in this way, a clear insight into the economy and purposeful distribution of the whole.

"/p. 99/ Birds of different species, however, can establish their nests close together without the danger of a considerable scarcity of food, because they rarely compete with each other. Even though all of them, with their young, live on insects, the manner of securing them usually differs as well as the kind of insects on which they feed. While one species of birds catches flying imagos, another searches for animals crawling along the ground, a third gathers them from twigs and buds, a fourth and fifth get them from the cracks in the bark or even hack open the wood, others combine several methods of capture or live chiefly on the insects peculiar to certain plants. Such species of birds either do not compete with one another or do so only to a slight degree. I observed several years ago a most remarkable instance of four different pairs of birds—a Redstart, a Wagtail, a Wren and a Blue-tit, all having their nests and eggs at the same time in the same beehive. Pairs of different species frequently nest close together. Several pairs of the same species, however, which live on the same kind of food and employ the same manner of securing it, cannot nest together; they must necessarily be separated by established boundaries.

"What separates them? It is, of course, natural that the most suitable localities will be the most sought by the species preferring them. Large numbers will gather in such places, overcrowding them, while other available territories would be empty if pairs were not kept apart by force. This force is used by a male as soon as another gets too close during the breeding season. The interloper is immediately attacked in the most violent manner, and driven to a distance that is determined by the size of the required territory. I shall discuss below, in more detail, this fighting between the males and therefore will not describe its purpose now.

"But many of my readers will ask what is the connection of song, the main subject of this chapter, with the question of territory? A bird must be able to perceive another bird in order to know of his presence. The majority of the above-mentioned birds, however, for example, all of our Sylviae, Phylloscopi, Reed-warblers, Larks, Kinglets, Titmice, etc., live in dense scrubs or thick luxurious vegetation /p. 100/. Since birds cannot detect each other by their olfactory sense, as mammals do, they must make themselves conspicuous in some other manner, i.e., by their mating call, their song. If birds were more or less mute, a too close approach would rarely be noticed by them, and even if it were noticed and the intruder driven the proper distance away, he could soon come back silently and unnoticed. As it is, however, since all birds sing jealously, this is not possible, for every time a bird approaches too closely, it is at once attacked again. They sing day after day, in the morning and the evening continuously, and by this song the boundaries of the territory are fixed. In fact, the real fight, the mutual attack of the males, is frequently begun while they sing, and the song continues during the battle." [Discussion of this pp. 100,101, concluding in this sentence:] "The song, or mating call, is thus the necessary means for the required separation of the territories.

"Some species of birds, however, have no definite territories. On a single tower a hundred pairs of Jackdaws may nest, one Martin nest may be built on the side of another, and under the roof of a large house we can find dozens of nests of House Sparrows and Swifts. The arctic sea birds, Auks, Murres, Puffins, Gulls, and Terns, and also the Cormorants, Herons, Rooks, Starlings, and others nest more or less colonially, some of these in groups of many thousands.

"Birds of prey[1] of the same species usually nest far apart. But our charming Kestrel may have 10 or even 20 nests close to each other in the woods which adjoin his hunting ground, the wide open fields.[2] On the Werbellin Lake (near Berlin) 12 pairs of Ospreys had built their nests on a few old oaks, one of the trees holding even two occupied nests. The hunting territory included the above mentioned lake and a few smaller waters. A mutual restriction of the hunting territory and of the prey of the individual pairs was in this case impossible.

"/p. 102/ The above mentioned species differ in their way of living from those which maintain a strict territory by not searching for their food in the immediate vicinity of the nest, but rather in the wide sea or on a wide, open field. Or these species are omnivorous, as the House Sparrow, and will never be lacking in food for their young. These species, even those which belong anatomically to the Oscines, do not have a regular song, . . . but all of them have a specific mating call, so that the males are made aware of each other and may fight for the nesting place [the location of the nest], but definite larger distances are not necessary for them."

"/p. 128/ The Fight of the Males.

"We have learned in the important previous chapter [pp. 97-102] that the song and mating call of the male birds has among other things not only the purpose of indicating from the distance to the females, (which react only to the voice of the males of their own species), their often rather hidden station, but it also serves in many species

---

[1] The 1st edition does not contain the following 8 lines.—E.M.
[2] See also Jourdain, 1927, Brit. Birds XXI, p. 71, 100-103.—E.M.

35

as a mutual signal, to fix the distance of the nests and thus the required size of the territories /p. 129/, since males which happen to get too close to each other fight until one of the combatants has retreated to the required distance.

"This statement, however, does not hold good for all species of birds. As mentioned, those are not required to maintain a strict territory that fly far for their food (as Swallow, Swifts, Jackdaws, etc.,) or are omnivorous (as the House Sparrow), but nest sociably; yet among the individual males there is a great deal of fighting at the beginning of the breeding season. We also find these frequently furious fights in those species which move around with their young immediately after hatching, not being tied down to a restricted nesting place, to a territory which would exclusively support them (as, for example, many gallinaceous, marsh, and water birds). It is therefore clear that there must be a further necessity for these fights, in addition to the reason already given [competition for food supply]."

This ends the statements that refer directly to territory. Altum then adds, on pp. 129-132, some remarks on sexual selection, summarizing his opinion in the sentence:

/p. 132/ "In short even the expression 'fighting of the males over females' is false: The males fight to fix the size of the territory, little as they may realize the vital necessity of this, and also to select the healthiest individuals for reproduction, but for nothing else."

Reprinted from *Irish Naturalist*, **12**, 152–166 (1903)

THE SPRING RIVALRY OF BIRDS.

SOME VIEWS ON THE LIMIT TO MULTIPLICATION.

BY C. B. MOFFAT.

(Read before the Dublin Naturalists' Field Club, 10th March, 1903.)

IN the present paper I propose to put forward an opinion which I have held for a considerable time, as to what is the real reason why birds—and, perhaps, the higher vertebrate animals generally—do not increase in number from year to year. This question has, I think, always puzzled field observers, the answers commonly given being of that class which look well on paper, but which somehow don't carry conviction so readily when we close our books and look all round us for that visible and tangible evidence which ought, one would think, to be forthcoming. That great tragedy, the "struggle for existence," as pictured for us by Darwin, requires such a death-rate among the young in their first year, or before they are of age to mate, as could not be less than 90 per cent. in the case of most of our finches and other common small birds For my part, I cannot believe that the theory of Natural Selection—for which I have a great respect, and which I must carefully guard myself against appearing for a moment to call in question—requires this sacrifice, or anything like it. Such a mortality—in fact, a far greater mortality—may very well exist among the young of cannibal fishes, or of reckless multipliers like the insects. But as regards birds I am altogether unable to find grounds for believing in so great a death-rate, at any rate in our own land. It seems to me that, despite all perils, a large proportion, amounting in some species to a majority of those that leave their nest, live. For a number of years I kept count of a small isolated colony of House-martins, from the year in which a single pair bred for the first time until they had become too numerous to be counted. So long as those martins could be counted, my census continued to show that the number which returned in spring was approximately the number which had departed in autumn. This is entirely contrary to what has generally been supposed the common rule ; but it satisfies me that the "perils of migration," in the case of the swallow tribe, are altogether spasmodic.   A terrible storm, during the short time of their passage, *may* destroy them in multitudes, just as a terrible winter may wreak enormous havoc among our resident birds at home.   But these catastrophes do not occur with such frequency as is needed to account for birds not increasing in "geometrical ratio," and when we look to more ordinary checks on multiplication, while I admit to the full our very meagre knowledge of what those checks are, I say we ought not to accept an unverified assumption that they always work by *killing*.

I think that Darwin, with all his clear-sightedness, did assume this.   He kept harping on the doctrine, which seemed to him self-evident, that "more individuals are produced than can possibly survive."   In one passage of the *Origin of Species* we are told that "every being which during its natural life produces several eggs or seeds, must suffer destruction during some period of its life, and during some season or occasional year; otherwise, on the principle of geometrical increase, its numbers would quickly become so inordinately great that no country could support the product." "A struggle for existence inevitably follows from the high rate at which all organic beings tend to increase."   It is true that in another passage we read that the term "struggle for existence" is used in a "large and metaphorical sense," the real object being "not only the life of the individual, but success in leaving progeny."   That distinction is important, and it is one that Darwin never lost sight of as regards plants, which he well knew might live for a long time unfertilised. But it seems to me that it never occurred to him to give it its due consideration as regards animals.   He does not consider the possibility of any appreciable number of animals living the lives of old bachelors and old maids.   In fact, he explicitly states that there can, in the animal world, be "no prudential restraint from marriage."   This, I venture to claim, was too summary a mode of dismissing the question.   The main object of my present paper is to show—or at least to make it seem probable—that there are checks of a prudential kind on the marriage of birds, and that these checks may be a very important factor in keeping the number of birds absolutely permanent.

Birds-nesters are, I think, aware that we seldom find in close proximity to each other two nests belonging to the same species of bird. Of course I exclude from this statement birds whose custom is to nest in communities, like the Rook and Swallow. We are aware, too, that cock birds in early spring spend a great deal of their time in fighting one another. No doubt, we are in the habit, when we see one of these fights in progress, of taking for granted that each of the antagonists is violently in love with the same lady. It is not, however, generally supposed that the battles condemn the loser to lasting bachelorhood. Failing to win the bride he has been fighting for would, of course, be something of a disappointment; but the number of hen-birds, so far as we can gather, is fully equal to the number of cock-birds, so that when all the fighting is over, there is nothing to prevent all the birds from marrying and settling down to "live happily ever after." We may suppose, in our sentimental way of looking at things, that even then the poor beaten cock-bird suffers from a certain amount of depression when he thinks of the greater charms of her to whom he first paid court. But we have the assurance of experts that no such thing happens; that one hen-bird is quite as good as another, and that every cock-bird is perfectly content with the first mate he can get. That is sometimes laid down as the reason why hen-birds, as a rule, have not developed bright nuptial colours or melodious voices. The difficulty is to explain why, on such a view, the cock-birds need fight at all. If one mate is as good as another, and there are mates enough for all, the advantage of winning a battle seems hardly worth the discomfort. It has been alleged that the victor gains an advantage in point of time, and so he does, *if it is* an advantage to begin nesting a little earlier than his rivals. But that, it seems to me, is not always any advantage; as far as the progeny are concerned, it may, in a backward season, prove the reverse. Consequently, we are still without adequate explanation why the birds fight one another so hard, unless they have some other object in view than merely to win a mate.

The case would, however, be entirely different if the result of the battle were frequently to prevent the defeated bird from rearing a family in the neighbourhood at all. This would mean to the families of the successful birds, in the early days of their life, an economic advantage of the greatest value. And this is what seems to me to happen. Birds may, or may not, realise the importance of protecting their future families against the ills of congestion; but they certainly seem to have an instinctive feeling that the patch of ground on which a pair is nesting belongs to that pair, and that no other pair of the same species of bird has any right to attempt to nest upon it. And, as land is a limited commodity, the cock birds in spring have to fight one another to settle the question, which shall possess a particular plot. After each of these battles the beaten bird is driven away; and, unless he succeeds in dislodging another cock from another homestead, all his hopes of matrimony for the remainder of the season seem to be blighted. It is not that he can't find a mate, but that he has no home to offer her; and all his other attractions are, under the circumstances, completely thrown away. The most tuneful of our birds of song—the Lark, Blackbird, Thrush, and Willow-wren, amongst others—always seem to take the result of a stand-up fight as final. The female Willow-wren, for example, may sometimes be seen sitting by, watching the combat between two males of her species, at the close of which the victor drives the vanquished away, and the lady then throws in her lot with the conqueror as a matter of course. When we see courtship on this mechanical method always triumphant, even in the case of so beautiful a songster, I think we must infer that there is very little free choice or æsthetic selection, and that the hen bird is mainly guided by prudential motives in accepting the owner of the soil.

Before going further on this subject, I would like to quote a few instances which prove what a violent objection cock birds have in spring to the mere presence of other cock birds of their own species in certain spots. One morning in March, 1898, at my home in county Wexford, I was told that a Blackbird had lately been behaving in a very extraordinary way. It used to come every morning to the kitchen window, and continue for five or six hours at a stretch dashing itself against the glass. At first people thought it would break the window, but as this didn't happen they got quite used to it; and the "thudding" went on day after day, all through the forenoon and early afternoon, as monotonously as clockwork. I went to see this wonderful Blackbird, and found that its action was exactly that which cock birds adopt in fighting. In fact, it was obviously doing battle with its own reflection in the glass. For this purpose it repaired to the same window every morning during the whole of March, and the greater part of April. It never, so far as we could make out, noticed itself, or looked for itself in any other window—but used all its energies against this particular one. I need scarcely say it was a fine cock Blackbird, and we found that there was a Blackbird's nest about ten yards away from the spot where these daily battles went on.

Towards the end of April the Blackbird made peace with his similitude in the glass; but before the end of February, 1899, we found that hostilities had begun again, and in March of that year another feud broke out of the very same kind. One morning in March, 1899, I was surprised to hear what sounded suspiciously like the Blackbird battering himself at a window at the other side of the house, and after several unsuccessful attempts to stalk the performer at this window, I found it was a cock Chaffinch, who had an antipathy to his reflection showing itself in the window of a storeroom. So all through the spring of 1899 we had two daily battles going on. And in the third spring, the spring of 1900, it was exactly the same, the "crazy Blackbird"—as he was called—fighting himself at one side of the house, and the equally infatuated Chaffinch doing the same at the other. After that year, I regret to say, we saw the Blackbird no more; but the cock Chaffinch resumed his campaign at the storeroom window with unabated energy in the spring of 1901.

Now, in these two cases it is evident that the imaginary enemies, on whom so much fury was expended, were guilty of no crime beyond that of being in the spot where they were. They had paid no presumptuous address, had uttered no song, challenge, or love-note, and had never taken the slightest notice of the fair partner of either of their angry antagonists. The whole sum and front of their offending was being where they were. The imaginary Blackbird was in the demesne attached to a real Blackbird's nest; the imaginary Chaffinch was in the demesne attached to a real Chaffinch's nest. The real Chaffinch and the real Blackbird were resolutely determined to expel the trespassers; and the trespassers, although they never took the initiative in attack, were stubborn in defence, and would not go.

These two cases—within one very small area—show plainly, I think, what would occur if two pairs of Blackbirds or two pairs of Finches attempted to build very near to one another. The cock birds would almost certainly fight, and the weaker would either be killed or retreat. This practically is the view which was taken a hundred and thirty years ago by Gilbert White, when he wrote that in spring " such a jealousy prevails between the male birds that they can hardly bear to be together in the same hedge or field " ; and "it is to this spirit of jealousy," added White, " that I chiefly attribute the equal dispersion of birds in the spring over the face of the country." This was written before much attention had been drawn to the all-important question, why does not the multiplication of living creatures proceed at such a pace as to overstock the earth. But I venture to say that so far as birds are concerned it supplies in itself an adequate answer. For, in course of time, the country—or the parts of it suitable for nidification—would come to be completely parcelled out between the birds, each parcel of land belonging to a particular pair :—I mean, as against any other pair of the same *kind*. And, once that happy state was arrived at, the number of nesting pairs each year would be exactly the same, the number of nests and the average number of young birds reared would be exactly the same ; and whether there was a large mortality in winter, or a small mortality in winter, the total number of birds in the country would remain exactly the same. As long as the annual birth rate, or rather *number* of births, is constant, and has been so for a given number of years, it must be balanced by the annual death rate, and further increase of the species becomes impossible.

Suppose, for example, that ten pairs of Chaffinches have nested every year in one orchard, and that every suitable nesting tree in the orchard stands on ground belonging to one or other of the ten. If this has gone on for a period equal to the average life of a Chaffinch, it follows that the number of Chaffinches of the original stock that die every year of old age would exactly equal the number hatched every year, supposing that no mortality at all had taken place among young birds. Thus, though we may suppose that about forty young would be reared every spring by the ten pairs, there would be no increase in the total number, because each year forty old birds would have reached the end of their span of life, and the loss would, therefore, exactly counterbalance the gain.

I now approach what seems to me the crucial point involved in this question. If I am right in thinking the country is "parcelled out" in the way I have described, it appears to follow that we must have a very large number of non-breeding birds of both sexes, prevented from breeding simply by the fact that they have no suitable ground. Have we any evidence that this large reserve exists? And if we have, can it be explained on any other hypothesis than the one I have suggested?

Nearly every standard work on ornithology contains some curious cases of the great facility with which a bird that has been deprived of its mate in the nesting season gets another. Gilbert White tells Pennant that at Selborne he found it useless trying to check the usurpations of the Sparrows on his House-martins' nests by shooting the offending birds; for the one which was left, he says, "be it cock or hen, presently procured a mate, and so for several times following." Our great Irish naturalist, Thompson, relates of the Peregrine Falcon in this country, that "if either an old male or female be killed in the breeding season (not, he adds, an uncommon circumstance), another mate is found within a very few days, so that the eyries are sure to turn out their complement of young." Dr. Jenner records how one of a pair of Magpies was shot from a particular nest no less than seven times on consecutive days, but all to no purpose, and the last pair reared their young. A similar story has, I think, been more recently recorded of the Carrion Crow. Darwin was told that Sir John Lubbock's game-keeper had repeatedly shot one of a pair of Jays, and had never failed shortly afterwards to find the survivor re-mated. "I could add," continued Darwin, "analogous cases relating to the Chaffinch, Nightingale, and Redstart." The illustrious author of the *Origin of Species* (in whose work on *The Descent of Man* most of the foregoing examples are circumstantially noticed), then proceeds to quote what I must call the most remarkable instance of the whole remarkable series. He had been informed by his correspondent, Mr. Engleheart, that

that gentleman "used during several years to shoot one of a pair of Starlings which built in a hole in a house at Blackheath, but found that the loss was always immediately repaired. During one year Mr. Engleheart kept an account, and found that he had shot thirty-five birds from the same nest; these consisted of both males and females, but in what proportion he could not say. Nevertheless, after all this destruction a brood was reared."

I think it appears to be established by such records as these that there is a reserve of non-breeding birds, and, moreover, of birds perfectly willing to breed—and that they are of both sexes. The puzzling question is, why don't they breed until vacancies occur in the partnerships already existing?

This question occurred to Darwin when he was working at the subject of "Sexual Selection." "How is it that there are birds enough ready," he asks, "to replace immediately a lost mate of either sex?" He considers several possible reasons, such as that some birds are mateless through having had their nests destroyed, or their partners killed; and that some have mates for whom they do not particularly care. But these explanations, as Darwin at once saw, were of little value; for if, as he goes on to express it, "so many males and females" are "always ready to repair the loss of a mated bird," it is impossible to refrain from asking, "Why do not such spare birds immediately pair together?" And to this question Darwin suggests the answer, which he gives as his final clue to the difficulty, that these unmated birds—though all of them willing for matrimony in the abstract—are not individually pleasing to one another, and, therefore, will not have one another for husband or wife.

Now this is a rather romantic explanation, which, I fear, will not stand the cold light of arithmetic; because it is as certain as any arithmetical fact can be, that if you shoot *one* of a pair of nesting birds thirty-five times, or even half-a-dozen times, during the season, and kill birds of both sexes, you will have killed both the original members of the pair, and therefore the two who are living together, a happy husband and wife, at the close of the season, must be two of the very lot who "*ex hypothesi*" (as the Mathematicians say) were "not pleasing to one another." So the romance of the situation seems to vanish. Not only, we find, are these non-breeding birds willing for matrimony in the abstract, but they are equally willing, provided certain conditions occur, to marry one another.

To take an illustration : A pair of Starlings, whom we will call A and B, nest in a particular crevice. Five other Starlings, C, D, E, F, and G, live in the vicinity unmated, because, though of different sexes, they all fail to please one another. One morning a cruel man shoots A ; and B, the same evening, has found a new mate, who, according to our supposition, must be either C, D, E, F, or G. Suppose it to be C. There is nothing remarkable so far, because, though C didn't please D or F, he may well enough please B. But the next morning the cruel man shoots B ; and C, before sunset, has a mate in B's place. Now this must be either D, E, F, or G ; but only yesterday morning C was living unmated, because of his inability to please D, E, F, or G. How is he able to please them now, when he wasn't able a day or two ago ? I contend that we must give the very unromantic answer—he is able to please them now, because he has a bit of land.

I admit, however, that the case of the Starling, if it stood alone, would be a bad instance, because Starlings are, to a certain extent, sociable in the breeding season, and therefore the competition which evidently occurs between different individuals or different pairs may rather be for access to a particular nesting-hole, than for proprietary rights in the surrounding area. Where birds learn to breed in communities the form of their rivalry, of course, becomes modified ; but that it is still territorial seems to me the only natural explanation of many observed facts. In old established bird-communities the accommodation is often obviously limited. The individuals belonging to the community cannot all nest in the space occupied. For example, we have a small rookery, confined to two trees, on the lawn at Ballyhyland. In a spring in which the number of nests in this rookery did not exceed thirty-five, and before any young birds of the year were fledged, I have several times put out of these two trees, by clapping my hands underneath, flocks of more than two hundred rooks. Within the rookery itself, then, there must be non-breeding birds, and there must be competition for space. In the instance mentioned the non-breeders must have been twice as numerous as the breeders. But again, outside the rookery, there would be rivalry if another rookery was started within a certain radius. That has frequently happened, and I remember a case myself in which some Rooks attempted for several years to form a new rookery which was always pulled to pieces by the inhabitants of a more anciently established one. In another case, however, not far from the same spot, a few pairs of rooks succeeded, one spring, in building without molestation in a new site. Of this experiment, two facts deserve notice. Firstly, the nests were not built till about the end of April, when the birds of the older colony were busy feeding their full-fledged young; and secondly, the success was very short-lived, for at the commencement of the following spring the rooks of the old rookery came in force and carried away the sticks from the new one to rebuild their own nests, and thus the infant colony came to an end. Whatever was the motive here, it shows that to found a new rookery near an old one is no easy matter.

I will now mention another case of a bird that failed in its matrimonial hopes, although the failure was not exactly illustrative of the sort of competition I consider to be the common rule. At Ballyhyland we have no Sparrows. Consequently, the arrival of a pair of these birds in the farm-yard in the spring of 1898 excited some interest; more especially as on the very first day of their visit they attempted to gain possession of a House-martin's nest with the obvious intention of making it their own. The House-martin is not so powerful a bird as the Sparrow, but there were four pairs of House-martins nesting in the yard, and the eight Martins at once combined and beat the Sparrows away. On the following day the two Sparrows were still ranging about the place, but no sooner did the cock-bird show himself in the neighbourhood of the row of Martins' nests than he was again attacked and mobbed so severely that he retired to the other end of the yard. He then set covetous eyes on another nesting site, but that happened to be occupied by a pair of Blue Titmice. I regret to say that I was not present as a spectator of the engagement which followed, but I am told by one who was—and I can well believe the assurance—that the battle between the Titmice and the cock Sparrow was very fierce. The upshot, however, was that the aggressive cock Sparrow once more suffered a bad defeat. And from that day the *hen* Sparrow was seen in the yard no more. She had not participated in the fighting, but she saw no use, it would seem, in remaining with a husband who couldn't win a nesting site for her. The unfortunate cock Sparrow remained in the yard all through the spring and summer, perched usually on a high roof-top where he evidently regarded himself as possessing a small domain, and where we used to hear him chirping a plaintive challenge all through the day. Why, it may be asked, did he not follow his mate? I can only suggest that he saw more chance of securing a nesting site where he was than by going elsewhere to ground already in the occupation of other Sparrows. Probably he was right, if he had only used his opportunities, when he had them, a little more judiciously; but the incident shows that competition for territory sometimes occurs between birds of different species, and in that form undoubtedly prevents certain individuals from breeding at all.

It would not be hard to collect a good many cases more or less parallel to the one I have just mentioned. For example, many cases are on record of rivalry between Herons and Rooks for possession of trees in which to breed. At Tintern, in county Wexford, a war which lasted, I am told, for a considerable number of years was waged between these two species for a clump of old trees which they both wanted to build in. Hundreds of Rooks, I am assured, were killed by the Herons before the question was settled, but in the end it was settled by the victory of the persevering Rooks. Again, the Starling has to defend its territory against the Swift, and I remember coming suddenly on a pair of Starlings which had struck a full grown Swift down on a grass plot and appeared to be on the point of killing it when my arrival interrupted them. But a more remarkable case of enmity is that between the Missel-thrush and the Blackbird. On two separate occasions I have seen a Missel-thrush during the nesting season flying along carrying a murdered or half murdered Blackbird in his talons, just as a Hawk would do; in each case the bird was carried several hundred yards, and then dropped in the middle of a field—in one case dead, in the other mangled beyond hope of recovery. The Blackbird's only imaginable crime was that he had intruded, or perhaps attempted to sing, on a part of the Missel-thrush's property. When such is the animosity shown by birds in spring against individuals who are not of their own species at all, it is not hard to understand how the rivalry of those which are of the same species gradually results in a certain parcelling out of the country, and substitutes arithmetical for geometrical progression as the normal avian birth-rate.

There remains a collateral point on which I think I should touch. The advocates of what is called Sexual Selection will probably say, in opposition to what I have tried to advance, that it offers no explanation of the beautifully ornamental plumage in which so many male birds are arrayed, or of the sweet flow of song with which, in a still larger number of cases they delight our ears in spring. If the great difficulty of a cock-bird in spring is to get a plot of ground, and his prospects of matrimony are dependent on that, why, it will be demanded, should he need to have a fine voice or fine feathers? If his difficulty is to make himself pleasing to somebody else, the explanation is simple. The song and bright plumage of a cock-bird are, on this theory, the charms by which he establishes his place in the heart of her to whom he pays court. But how do they help him to win a plot of ground?

Well, as regards song, it has long been a subject of controversy whether or not it is addressed to the female at all. So far as outward indications go, male birds appear to address their songs primarily to one another. We hear them answering one another from field to field, sometimes from hill to hill, and the song partakes so strongly of the nature of a challenge that a great many birds actually sing while they are fighting. The Robin almost habitually does so ; and the Wood-pigeon's peaceful " coo "—as it sounds to our ears—is often uttered in the midst of a deadly combat. I have also heard Thrushes singing their loudest in the thick of a fray ; and Mr. Charles Witchell, the author of the *Evolution of Bird-Song* has noticed the same habit in the Tree-pipit, Chiffchaff, Willow-warbler, and Golden-crested Wren. But it is still more usual for song to provoke to combat, and here I think the general rule is accurately stated by the ornithologist Couch, who observes that "in a wild state birds of the same species will not sing near each other, and if the approach be too close, and the courage equal, a battle follows." I need scarcely add that bird-catchers take advantage of this fact to catch male birds in spring; and it is by playing upon the same propensity that boys in the country decoy the male Corncrake from his cover, by imitating his "crake" with a stick drawn across a comb. That it is the male, not the female, Corncrake who is thus decoyed, is shown by the fact of its craking in reply to the challenge, up to the very moment of leaving cover.

The chief and primary use of song, then, as I conceive it, is to advertise the presence in a certain area of an unvanquished cock-bird, who claims that area as his, and will allow no other cock-bird to enter it without a battle.

The question may be asked, is not bird-song too elaborate to be thus accounted for? Would not simpler notes answer the purpose equally well? I think it may be shown that the more elaborate singers do, however, obtain an advantage in this manner. Since only the unvanquished bird has an area to sing in, the vanquished birds must, after their defeat, observe silence, while the conqueror, secure in his holding, sings triumphantly on and on, his voice and his powers of expression naturally improving by practice, and thus proclaiming to all who come within the charmed circle of his audience what a lot of practice he has had. Thus, all over the country, the finest singers are known from their superior vocal power to be those with the longest record of success in life, and the poorer singers are naturally afraid to start competition with them. This, of course, renders their tenure of power more secure than ever.

And, in the matter of plumage, although it has been shown by careful accumulation of evidence that birds of polygamous character—and even some monogamous species— behave very much as if the female had a taste for the beautiful; and it seems unreasonable to doubt that the bright tints acquired by the more richly ornamented males are agreeable to her eye, it does not follow that that is their primary value. Darwin, in summarizing his evidence on this subject, makes the interesting statement on Mr. Jenner Weir's authority that "all male birds with rich or strongly characterised plumage are more quarrelsome than the dull coloured species belonging to the same groups." Have we not here some ground afforded us for suspecting that the bright plumage may have been originally evolved as " war paint ? " In other words, as a sort of " warning colouration" to rival males, rather than attractive colouration to dazzle the females ? I cannot, of course, go into elaborate argument on this question, but I wish to observe that I never, to my recollection, saw a conflict between two brightly-plumaged birds, in which the bright feathers were not brought into prominence in some striking manner during the fray.   The Robin so faces his opponent as to make the fullest display of his red front ; the cock Golden-crested Wren lowers its head like a bull, and flashes its crest right in the enemy's face.   The distinguished points in the Water-hen's nuptial dress are not so much in the plumage of this bird as in its bright orange bill, and the bright scarlet-orange band round its leg, which is often called its garter ; and it is significant that when two Water-hens fight in the breeding season they sit back on their hind quarters and strike at one another with their feet.   The male of the Night-jar is distinguished by the white patches on his wings and tail, which only show when these are extended ; and two birds of this species whom I once watched fighting used to lie on the ground, menacing one another with deep frog-like noises, and then to rise with vehement beating of wings, which showed to perfection—even in the dim light—the white ornamentation of their plumage.   I need hardly mention the notorious case of the Ruff, whose decoration is really useful as a shield in his very celebrated battles.   These are just a few illustrations of what seems to me to constitute a general rule, and they serve to show that bright colours have another object than to please the hen bird's eye.   When we pass in spring over a gorsy Irish moor, and see those splendid little birds, the cock Stonechats, perched conspicuously on the tops of the furze bushes, does not each of them remind us of a bright little flag, put up—as it were—to mark that such and such an area is under such and such a dominion ? If the cock bird shared the dull plumage of the hen, the signal would be less useful in two respects ; it would not be seen so far, to begin with, nor would it show—when it was seen— that the bird belonged to the fighting sex, and was of full age to maintain his right.   Without, then, wishing to push argument on this subject too far, I say these bright colours — apart from what is called Sexual Selection—are means to a definite end; they are means by which cock birds impress certain lessons on one another, and if they do not help a bird to *win* his plot of ground, they, at any rate, render his subsequent possession of it less liable to disturbance. In other words, it appears to me a general conclusion from the facts on which I have sought to lay stress, that those effects which are commonly ascribed to Sexual Selection are capable of being explained by a form of Natural Selection, and that Natural Selection—on the other hand—does not, so far as birds are concerned, require, as Darwin took for granted, a wholesale annihilation of the weaker young, but can, and probably does, largely work by condemning to unproductiveness the less powerful adults.

Ballyhyland, Wexford.

Reprinted from *Trans. Linnaean Soc. N. Y.*, **4**, 57–83 (1937)

Studies in the Life History of the Song Sparrow I

MARGARET M. NICE

Chapter VI: Territory Establishment

Melospiza melodia, in my experience, is a typically territorial bird, behaving very much as does Howard's classic example—the Reed Bunting (*Emberiza schoeniclus*). Territory is of fundamental importance to the Song Sparrow on Interpont—the basis of its individual and social life for more than half of the year. Special ceremonies are concerned in the establishment of territory; the matter of song is closely bound up with territory, while males show a strong and lasting attachment to their individual territories.

A. THE ESTABLISHMENT OF TERRITORY

That territorial behavior is deeply ingrained in my birds is evidenced from two things: the elaborate ceremonies that are involved in its maintenance, and the part it plays in the change from juvenal to adult singing.

When a new male Song Sparrow arrives in spring, the neighboring males at once try to drive him off. If he is a transient, he flies, but if a candidate for a territory, he stands his ground—and then the "territory establishment" begins.

The complete procedure consists of five parts: assuming the role; staking out the claim; the chase; the fight; and finally the proclamation of ownership of each bird on his own bit of land.

In the first part the two birds show diametrically opposed behavior. The invader—puffed out into the shape of a ball, and often holding one wing straight up in the air and fluttering it—sings constantly but rather softly, the songs being given in rapid succession and often being incomplete. The defender, silent and with shoulders hunched in menacing attitude, closely follows every move of the other bird.

The newcomer continues to sing flying in this peculiar puffed out shape from bush to bush that he wants to claim. Soon the owner begins to chase the intruder, but the latter, if determined, always returns to the spot he wants to claim. The chasing continues and at last finishes with a fight on the ground. After this the new bird is either routed or both males retire to their respective territories, and sing loud and long, answering each other.

In less serious encounters the chasing and fight are omitted, the first and last parts only being indulged in. When, however, affairs are in deadly earnest, as in the spring when a summer resident returns and finds a resident has adopted his old territory, there is little wing fluttering and puffing, merely the singing, chasing and fight.

With a thickly-settled Song Sparrow population, territory establishment ceremonies of all degrees of seriousness may be seen throughout the year except during the molt; in fall and winter they are not common and occur only on mild days. At these seasons a bird will go some distance to start a "territory establishment" with another male with whom there is no question of real conflict over boundaries. In such cases the roles of despot and underling are freely interchanged. Excluding the very mildest territory establishment manifestations that are indulged in only by a young bird in the fall on some occasions when another Song Sparrow alights on a branch above him, the less serious the encounter, the more prominent is the posturing, bluff taking the place of action. As Howard, *86*, p. *37*, says "violent wing-action and violent contortions of the body are associated with postponed reaction."

When a summer resident returns to find his old territory preempted by another bird, at first the new arrival takes the role of the invader and is pursued by the bird in possession, but it does not take long for an old bird to reverse matters; after a fight or two he becomes the defender and drives his rival. Burkitt, *29*, tells of an old Robin Redbreast (*Erithacus rubecula*) being driven from his territory by a young bird; but this has not happened to my knowledge with the Song Sparrows; with them the old bird usually drives off the interloper, although sometimes he will take a neighboring territory. But as this sometimes happens under no pressure from other males, we cannot be sure that the old male was really intimidated by the young one.

Territory establishment ceremonies have not been worked out in such detail with any other species so far as I know. Howard writes of "butterfly-like" and "moth-like" flights, and of rapidly vibrated wings, and Pickwell, *147*, describes the boundary quarrels of the Prairie Horned Lark (*Otocoris alpestris praticola*) which show much resemblance to those of my Song Sparrows, except that the fight takes place in the air. But neither of these authors clearly differentiates between the behavior of the two participants, perhaps because they worked with unbanded birds. The Micheners describe what they think may be "a ceremony marking territorial lines" with Mockingbirds (*Mimus polyglottos leucopterus*), where one of the owners of the territory "came to the fence and approached the unbanded bird facing it and bowing and bobbing. One would step forward and the other back and then they would reverse," *123*, p. 126. Closely similar behavior is reported by Laskey, *100a*, with *Mimus p. polyglottos*.

It is reasonable to expect strongly territorial species to have special instinctive reactions by which territory questions can be settled. In order to observe and understand these, however, one must have individuals plainly differentiated; one must study the birds from the first taking up of territory; one must study two or three pairs intensively at first and finally there must be a sizeable population, so that territory establishment behavior can be shown. In 1935, for instance, when there were very few Song Sparrows, I saw almost no activity of this nature, although I was especially on the look out for it.

B. Territory and the Development of Song

Volumes could be written on the matter of song and territory, but I will confine myself to a brief treatment of two features.

With *Melospiza melodia* song is the chief means of proclaiming territory; the taking up of territory in late winter and the beginning of zealous singing coincide; while the main season of Song Sparrow song on Interpont is in March before the arrival of the females.

Territory has a powerful influence on the development of the Song Sparrow's juvenal warble into the short separate songs of maturity. A young bird may be warbling along peacefully by himself, but the moment a territory rival appears, the singing becomes almost typically adult. In late February a young bird may warble in low situations on his territory, but when he sits high in a tree proclaiming ownership, his songs are adult in form. The young transient males that pass through in March warble freely, but I have never heard a young summer resident male warble in the spring on Interpont; upon the arrival at the nesting grounds the bird reacts as an adult. With the young residents the warble is given up in late February and never reappears, all of the late summer and fall warbling coming from young birds.

C. Summary

1. The Song Sparrow has a special ceremony consisting of posture, song and fighting for the procuring and defending of territory.

2. The new bird takes a humble, subservient role, the owner a dominating, threatening attitude.

3. The complete ceremony consists of five parts: assuming the role; staking out the claim; the chase; the fight; the subsequent proclamation of ownership.

4. Song is *Melospiza melodia's* chief means of proclaiming territory.

5. The young male has a continuous song of warbling character; but in territorial situations this is changed to the adult form of song.

CHAPTER VII

Territory Throughout the Year

The actual breeding season of the Song Sparrow lasts from 3½ to 5 months, but the territory is inhabited by the summer resident male from 6½ to 8 months and by the resident throughout the year. It is not, however, defended during the molt, nor the cold of winter, and only to a limited extent in fall.

A. Territory in the Fall

The Song Sparrows normally molt in August and September, an occasional bird not finishing till October. Because of my absences from Columbus at this season I do not have much data on the molt of the adults. Wharton, *199*, in Groton, Mass., says the molt of his local Song Sparrows begins during the second 10 days in August and lasts from 40 to 45 days, but from my scattered observations I should expect it to last longer. Magee, *118a*, states that the wing molt of Purple Finches (*Carpodacus p. purpureus*) takes 10 weeks on the average. In 1930, perhaps in some way due to the unprecedented drought, the birds started to molt the middle of July and were through molting more than two weeks before their usual time.

1. *Singing in the Fall*

With the adult males there is a recrudescence in fall, in a lessened degree, of spring behavior so far as territory is concerned. Young males that have settled unmolested during the molt of the owner, are now driven off with appropriate territory establishment procedure, although other Song Sparrows are tolerated. Singing is heard again from some of the adult residents, while others are practically silent. During normal years the singing from summer residents is of irregular occurrence, but in 1930 there was a wonderful amount from both classes of males.

With many of the birds entirely through the molt the 10th of September instead of the last of the month as usual, with fine weather in September and an extraordinarily mild early October, and with the migration not taking place until its usual time in mid-October, we enjoyed a most unusual treat of Song Sparrow music. The summer resident 1M in 1929 sang Sept. 28, 29 and Oct. 4, but in 1930 from Sept. 17 to Oct. 11. Song was recorded from another summer resident—10M —Sept. 10 to Oct. 11 in 1930; Sept. 28 and Oct. 4, 1931; Oct. 9, 1932, and Sept. 28, 1933.

4M's early morning singing has started on the following dates: Sept. 29, 1929; Sept. 10, 1930; Sept. 28, 1932 (we returned to Columbus the day before) ; Sept. 28, 1933; Sept. 30, 1934; and Sept. 29, 1935. Considerable warbling is heard from juvenals in the fall—from residents, summer residents, transients and winter residents.

2. Taking Up of Territories

Many young residents take up their territories in their first fall and keep them for the rest of their lives; others try to do the same but are driven out by the owner when he completes his molt; still others do not settle down until February. I do not know whether this difference depends on age or other factors.

Some young summer residents also choose their territories in their first fall and return to them the following spring.

185M was caught in our garden Aug. 3, 1933, in juvenal plumage and was noted warbling 50 meters to the south from Oct. 4 to 6; on Mar. 16 he returned to the very same spot. In 1931 a right-banded bird warbled constantly west of our garden on Sept. 28 and Oct. 15, but I was not able to trap him; on Feb. 27 a right-banded bird returned and took up his territory in this same spot (112M). On Oct. 1 I banded 134M and found him Oct. 6 warbling south of the third dike; on Apr. 1 he returned and took up his territory about 100 meters to the south of this place, which at this time was entirely filled by other males.

Burkitt's, *28*, young Redbreasts (*Erithacus rubecula*) took up territories in July and August; Miller, *125*, found that with California Shrikes (*Lanius ludovicianus gambeli*) fall is the main time for taking up of territories; the Micheners report that young Mockingbirds (*Mimus polyglottos leucopterus*) do so in August and September, *123*, while British Stonechats (*Saxicola torquata hibernans*) settle in pairs on their territories in October, *101a*. But all these species defend their territories throughout the year. It is interesting to find the Song Sparrow, which defends his territory only during the breeding season, settling on it so early in life.

B. Behavior in Winter

It may be largely a matter of habit that keeps the adult residents of both sexes in the vicinity of their territories throughout the winter, if sufficient food and cover are present. Similar behavior is shown by the winter residents, in a few cases for a number of years, as with W6, as told in Chapter IV.

At this season the male resident may range over an area approximately 150 by 225 meters, a district six to ten times as large as the breeding territory. In cold spells birds may come unusual distances for brief visits to my feeding station, several from 270 meters, while two traveled more than 500 meters (57M and 58M, see Maps 9 and 13).

In cold, snowy weather Song Sparrows are apt to form into small flocks, the organization of which is very loose. On Jan. 16, 1931, I watched 50M leave his regular flock in our graden and join another below the first dike, the birds here paying no special attention to him. After staying with them for five days, he returned to his former companions. These flocks on Interpont are *not made up of "family parties" nor of "neighborhood groups,"* since they are composed of both residents and winter residents, and family ties are broken with the young when the latter are a month old; while mates, even if both are resident and winter near together, apparently pay no more attention to each other in fall and winter than they do to strangers.

C. Behavior in Spring

In late January or early or mid-February, depending on the weather, the resident Song Sparrows begin to take up their territories —isolating themselves through hostility to other members of their species and making themselves conspicuous by song.

1. *Song and Temperature*

Song gradually comes to an end in November, and no matter what warm and pleasant weather may occur in December, only occasional snatches of song are heard. (There have been three warm spells in December of three days duration and one of six days during the period of this study; mean temperatures ranged from 7.2°-14.4° C. (45°-58° F.), or 7.2°-15° C. (13°-27° F.) above normal, the median temperature being 10° C. (50° F.).) But in January song usually begins again, there having been from 4 to 16 days per month on which a fair amount of song was recorded from 1930 through 1935. Table VI shows the mean temperatures at which the Song Sparrows started singing.

TABLE VI

Lowest Mean Temperatures That Started Singing

Date of Start of Singing	Mean Temperature of Day of Start and Two Previous Days						Normal Temperature of Day of Start	
	Centigrade				Fahrenheit		C.	F.
Jan. 7, 1930 - - -	3.8	9.4	12.2	39	49	54	—1.7	29
Jan. 8, 1935 - - -	8.3	10	12.2	47	50	54	—1.7	29
Jan. 13, 1930 - - -	— 2.2	5.6	8.9	28	42	48	—1.7	29
Jan. 13, 1932 - - -	0	8.9	13.3	32	48	56	—1.7	29
Jan. 19, 1933 - - -	7.2	6.6	8.9	45	44	48	—2.2	28
Jan. 21, 1934 - - -	2.2	4.4	8.3	36	40	47	—2.2	28
Jan. 24, 1931 - - -	— 2.2	1.1	6.1	28	34	43	—2.2	28
Feb. 2, 1930 - - -	— 3.3	0.6	4.4	26	33	40	—1.7	29
Feb. 2, 1932 - - -	— 6.6	—4.4	2.2	20	24	36	—1.7	29
Feb. 2, 1935 - - -	— 2.8	1.1	0	27	34	32	—1.7	29
Feb. 7, 1934 - - -	— 1.7	—3.3	—2.8	29	26	27	—1.7	29
Feb. 9, 1935 - - -	— 3.3	3.8	2.8	26	39	37	—1.1	30
Feb. 11, 1934 - - -	—16.8	—8.9	—2.2	2	16	28	—1.1	30
Feb. 14, 1936 - - -	— 6.6	1.1	0	20	34	32	—1.1	30
Feb. 24, 1936 - - -	— 5.5	2.2	8.9	22	36	48	0	32

There has been some singing on the 7th and 8th of January following two warm days, and from the 13th to 21st following one warm day. From Jan. 21 singing has started in earnest when the previous day was only 3.3° C. (6° F.) above normal; by Feb. 2 singing has been heard on the first warm day, and by the 7th may reappear after an interval of bleak weather at a temperature slightly below normal. Singing appeared Jan. 7 and 8 at temperatures 14° C. (25° F.) above normal; from the 13th to 21st at 10°-15° C. (19°-27° F.) above normal; on the 24th at 8° C. (15° F.) above; on Feb. 2 from 2°-7° C. (3°-12° F.) above and Feb. 7 and 11 at 1.2° C. (2° F.) *below*. In 1936 when there had been no previous singing, it started on Feb. 14 at 1.1° C. (2° F.) above normal, and restarted on the 24th at 9° C. (16° F.) above.

That singing appears at progressively lower temperatures is clearly shown in Chart VIII, for which Prof. Selig Hecht of Columbia University kindly drew the curve and gave me its formula.

Ts.=54.2° F. —0.7d. (12.3° C. —0.39d.).

Ts.=the temperature at which singing starts, d.=day, 0.7=the constant indicating the slope of the curve. Or in other words the threshold of singing was 54.2° F. (12.3° C.) on Jan. 7 and decreased about ¾ of a degree Fahrenheit (about 2/5 of a degree Centigrade) each day.

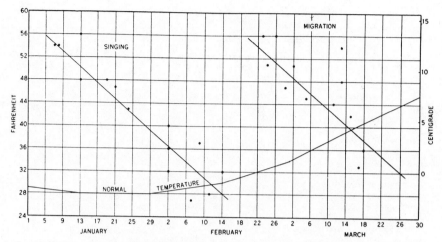

CHART VIII. *Threshold of Singing of the Residents and Migration of the Males. Dates of Start of Singing, 1930 to 1936, as shown in Table VI. Dates of Migration of Breeding Males, 1930 to 1935, as shown in Table III.*

It is of great interest that the curves for the threshold for the start of singing and for migrating should start at approximately the same temperature and have a similar slope, but the dates are a month and a half apart.

Singing and territory activity are well established the fourth week in January at a mean temperature of 6° C. (43° F.). This is also the average temperature at which the main migration of the males took place (Table III). It is of interest to note that 100 years ago De Candolle found that 6° C. or 43° F. was the threshold for growth with wheat and other plants. This "has formed the base used by Merriam (1894) in working out his life zones. This is also the base commonly used by meteorologists" (Shelford, *177*).

Temperatures at which the birds will *start* singing and those at which they *will* sing after once being well started are two very different things. If the Song Sparrows are once well started, they will sing to some extent at surprisingly low temperatures for a day or two. But a sudden drop in temperature, especially if accompanied by a bleak wind may stop singing temporarily, even as late as Mar. 6 (1932). There is also a difference between restarting and making the first start, as was shown in 1936. The birds that have been well

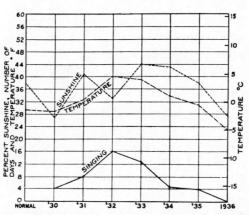

CHART IX. *Average Temperature, Percentage of Sunshine, and Number of Days on which Song Sparrows Sang in January from 1930 to 1936*

started, and then stopped by a bleak spell, begin more readily than did those in 1936 that got no chance to sing until Feb. 24, except for one day—Feb. 14. (During the last half of January, 1936, the highest mean temperature was 3.8° C. (39° F.); after that there was nothing but cold weather till Feb. 13 and 14, after which there was another cold spell lasting till the 23rd.)

Singing in January is not an automatic response to a certain temperature; it is influenced by the temperature of the previous days, and also by other weather conditions, being inhibited by strong wind, and

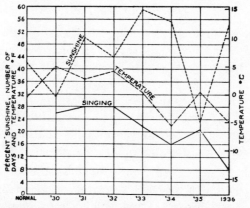

CHART X. *Average Temperature, Percentage of Sunshine, and Number of Days on which Song Sparrows Sang in February from 1930 to 1936*

sometimes apparently by cloudiness. It also depends on the individuality of the bird—some males starting to sing much earlier than others—, and upon whether or not he has already been singing.

The influence of light upon the breeding cycle has been much emphasized by Bissonnette, *22b,* Cole, *40b,* Rowan, *163,* and Witschi, *211a.* Let us see whether the percentage of sunshine appears to affect the singing of the Song Sparrows. In Chart IX the percentage of sunshine in Columbus in January from 1930 through 1936 is given, as well as the average temperature of these months and also the number of days on which singing occurred, while corresponding data for February are given in Chart X.

The amount of singing correlates very well with the average temperature of these two months throughout the seven years, but does *not* correlate with the percentage of sunshine.

An interesting case that bears on this point of the effect of temperature versus lengthening days is given by Laskey, *100a:* a certain banded Mockingbird (*Mimus polyglottos*) in Nashville, Tenn., began to sing on Feb. 26 in 1933 and on Mar. 4 in 1934, but on *Jan. 10* in 1935. "The temperature during January was unusually high and the excess for the month up to the 12th was 99 and reached 162 by the 19th." For 12 days he sang and courted his last year's mate, and then the temperature "dropped in one day from 59 to 14, followed by snow. Each bird retired to its own territory and they took no further interest in one another until Mar. 3." Excessive temperature had started courting activity at a time when the days had barely begun to lengthen.

2. *Defense of the Territory*

Hostile behavior towards territorial rivals begins at the time that singing is well established. Other Song Sparrows, that is, juvenal residents that have not yet started to sing, and winter residents, are tolerated. Perhaps this is a matter of personal acquaintance. 4M showed no hostility to two different young winter residents that stayed on his territory through February and one until Mar. 7. Warbling, being as it is, an expression of youth and of entire lack of intention to establish a territory, does not antagonize an adult male.

By March, however, all Song Sparrows are driven off, as are most other birds unless they are too large or too indifferent. House Sparrows (*Passer domesticus*) and Goldfinches (*Spinus tristis*) ignore the threats of the Song Sparrows that learn in turn to ignore these species. Field Sparrows (*Spizella pusilla*) are driven off with special vigor; nevertheless, two pairs used regularly to nest on Interpont in the midst of the Song Sparrows.

The Song Sparrow pair dominates most of the species that come into the territory; transients usually fly away, while the nesting birds merely avoid the threatened attack. The species driven off by both male and female Song Sparrows include: Juncos (*Junco hyemalis*), Tree Sparrows (*Spizella arborea*), Field Sparrows (*Spizella pusilla*), White-throated Sparrows (*Zonotrichia albicollis*), White-crowned Sparrows (*Zonotrichia leucophrys*), Fox Sparrows (*Passerella i. iliaca*), female Cardinal (*Richmondena c. cardinalis*), Red-eyed Towhee (*Pipilo e. erythrophthalmus*), Indigo Bunting (*Passerina cyanea*). Grey-cheeked Thrush (*Hylocichla minima aliciae*), Olive-backed Thrush (*Hylocichla ustulata swainsoni*), Hermit Thrush (*Hylocichla guttata faxoni*), Northern Yellow-throat (*Geothlypis trichas brachidactyla*), House Wren (*Troglodytes a. aedon*), Alder Flycatcher (*Empidonax t. trailli*), and Ruby-crowned Kinglet (*Corthylio c. calendula*). The approach of a Cowbird (*Molothrus a. ater*) is greeted with the anxiety note; if the enemy comes near the nest site it may be attacked by both of the Song Sparrows.

Territorial zeal is stated by Meise, *121,* to show a recrudescence at the beginning of each new nesting cycle, but this has not been my experience with the Song Sparrow. Territorial zeal typically diminishes as the season advances, unless a new territorial situation arises, such as the arrival of a new male, or as in the case when K2 nested outside of her mate's—1M—territory in the territory of her neighbor 4M.

D. SUMMARY

1.   Some of the male Song Sparrows sing regularly in the fall.

2.   In 1930 there was an exceptional amount of autumn singing, with all the birds through the molt two weeks or more early, and unusually warm weather in October.

3.   4M has shown a remarkable regularity in the beginning of his singing each fall from 1929 to 1935 with the exception of 1930.

4.   Some young males, both residents and summer residents, take up their territories in their first fall.

5. Song Sparrows flock to a certain extent in cold, snowy weather. These flocks are not made up of family parties nor exclusively of neighborhood groups. Range of ¼ mi.

6. The resident males start their singing and take up their territories during warm weather in late January or early February.

7. Mean temperatures of 10° C. (50° F.) on Jan. 7 and 8, and of 9° C. (48° F.) on Jan. 13 will bring some singing, while singing will be well established in late January at temperatures from 8°-6° C. (47°-43° F.), and on Feb. 2 at 2° C. (36° F.), as shown in Table VI.

8. The threshold of singing was 54.2° F. on Jan. 7 and decreased 0.7 of a degree Fahrenheit each day, as will be seen in Chart VIII. This is similar to the threshold for migration, but occurs a month and a half earlier.

9. The number of days on which singing was recorded in January and February from 1930 through 1936 correlates well with the average temperature of these months, but not with the percentage of sunshine (Charts IX and X).

10. Song Sparrows try to drive from their territories most other species except those decidedly larger.

CHAPTER VIII
The Territories from Year to Year

The question of the return of birds to their homes is one of perennial interest. How faithfully do adult birds—males and females—return to their territories? How far from their birth place do young birds settle? Over how much ground does one family scatter? Answers to these questions can be given in regard to the Song Sparrows on Interpont.

A. The Territories of the Adult Males

On Maps 2, 3, 4, 5, 6, and 7 we see the territories of the male Song Sparrows on Central Interpont during 6 seasons. The last five give the status at the beginning of the nesting season Apr. 6, but

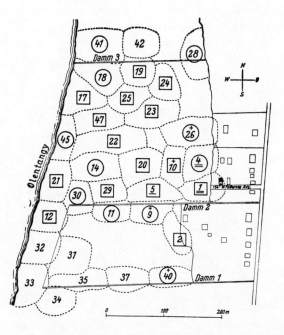

MAP 2. *Territories on Central Interpont, June, 1930. 33 males*
A circle means a resident, a square a summer resident, a cross a first-year bird.
A bird present the previous year is underlined, a line being added for each
subsequent year. (Map 2-5 by courtesy of the Journ. f. Ornithologie.)

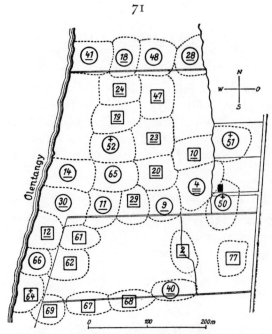

MAP 3. *Territories on Central Interpont, April 6, 1931. 31 males*

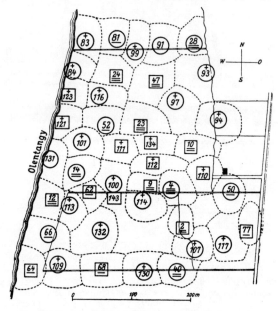

MAP 4. *Territories on Central Interpont, April 6, 1932. 44 males*

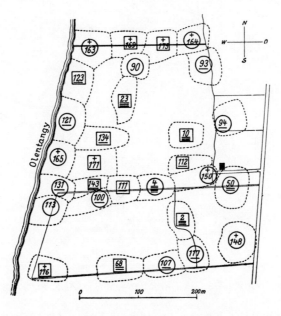

MAP 5. *Territories on Central Interpont, April 6, 1933. 29 males*

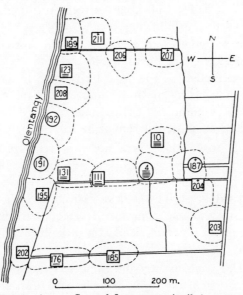

MAP 6. *Territories on Central Interpont, April 6, 1934. 19 males*

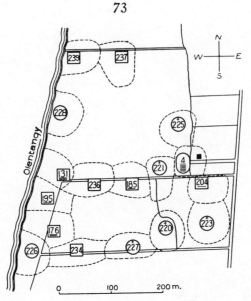

MAP 7. *Territories on Central Interpont, April 6, 1935. 17 males*

Map 2 represents conditions in June of 1930. The reason for this is that the June map shows two males that were not present in April, while the four that had disappeared between April and June of course could not be present the following years.

In the first map there are 33 males (35 in April), in the second 31, in the third 44, in the fourth 29, in the fifth 19, in the sixth 17. Unfortunately a number of the 1930 birds were never banded, hence I do not know how many of the 6 numbered from 31 to 37 are included among those numbered 61, 62, and 66-69. I believe that 42M was the same bird as 48M, but cannot be sure.

The map of 1931, when there was a scarcity of Song Sparrows and yet no destruction of cover, is of interest in showing that the birds did not spread out any more than usual. I have often noticed that a new arrival in spring—a first year bird—will try and try to establish a territory among a group of Song Sparrows, at the same time ignoring equally favorable land at a little distance, that is entirely unclaimed.

Do the males have exactly the same territories year after year? This has been true of many birds, notably 2M, 12M, 20M, 23M, 28M,

40M, 41M, 50M, 52M, 54M, 58M, 131M, and others, for periods ranging from two to four years. But some change of territory has been a
common occurrence; often the new and old partly coincide, but at
other times they do not. In some cases this may perhaps be due to
the exigencies of the situation a summer resident finds on his arrival;
occasionally an old bird apparently adopts a slightly different territory
rather than driving out the birds already established; this was true
of 24M and 47M, in 1931. But a resident or an early summer resident may shift his territory with no pressure from other birds; this
was the case with 4M, 18M, 19M, and 111M; while 185M moved from
the first to the second dike, on his second return, although only one
male was in residence along Dike I.

4M I believe to have nested in much the same territory shown in Map 3 as
early as 1928, although I did not band him until 1929. He has had a somewhat
different territory each year, and in the winter of 1931-32 he moved 30 meters
to the west, although there was no question of any Song Sparrow driving him.
In his early years he was a pugnacious bird, the tyrant of the neighborhood, and
kept 1M continually stirred up defending his boundaries. In 1932, however, 4M
spent much less energy in picking quarrels and allowed 110M—a summer resident juvenal—to settle down in 1M's former land with hardly a protest. The
next winter he moved even further west over into 9M's former territory, and
there he nested for three years, but in 1935 he came back into our garden.

As to changes of territory following destruction of cover, in
March, 1933, four banded males were driven from Upper Interpont;
two first-year birds left the region and were never seen again; but
two adults made short moves: 96M settled across the river and later
in the season returned to his old territory, while 90M moved 180 meters
to the south, settling just south of Dike 3.

Territories may range in size from 2,000 square meters to nearly
6,000 (half an acre to one and a half acres), depending partly on the
pugnacity of the owner and partly on the amount of space available.

B. The Returns of the Females

The female that has nested before tries to return to her former
home, but this is often impossible because another bird may have pre-
ëmpted her place. In that case she often settles next door, but sometimes joins a male at a distance from 200 to even 700 meters, even
though there may be unmated males near her old territory.

In 54 instances involving 41 birds I know the territories of females two years in succession; 20 of these were the same, 16 were neighboring territories, in 9 cases the birds moved about 100 meters, in 7 from 150 to 250 meters, in two 400 meters, and one 700.

On Map 8 territories are shown of 14 females: two for four years in succession, three for three years, and nine for two years. Maps 11-14 show the locations of four females two years in succession, four for three years and one for four years. There are 20 other females whose residences two years in succession are known, but these five maps present all kinds of situations from those females

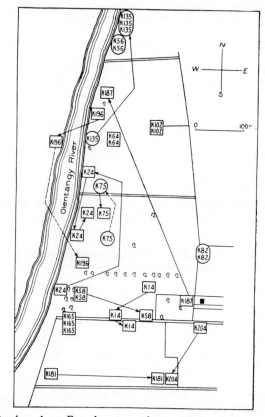

MAP 8. *Territories of 14 Females, two, three and four years in succession. A broken line indicates a change of residence during one season. K24 and K135 were present four years, K14, K58 and K165 three seasons, 9 others two seasons. K75 changed status from resident to summer resident.*

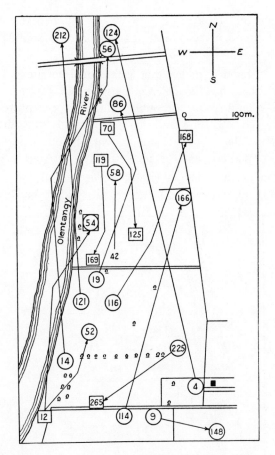

MAP 9. *Territories of 13 Males in Relation to Birth Place. 9 residents, 4 summer residents.*

that stayed on or returned to the very same territories for two or in two cases (K135 and K165) three years, to the birds that moved the longest distances, with a fair sample of short distance moves besides.

C. TERRITORIES OF THE MALES BANDED IN THE NEST

The territories of 13 young males banded in the nest are shown in Map 9 with arrows connecting these territories with their birth places. Territories of eight other young males in relation to birth place are shown on Maps 11-14. 106M was hatched just below Dike I in 1931 and found in 1933 nesting 1,400 meters south of his birth place.

Three right-banded males that established territory (two across the river and one below Lane Avenue) were not captured despite repeated efforts on my part; although all three were present in April only one survived till June and he disappeared later that summer. The parentage of these birds is not known. In one case—145M—the parentage is known, but not the bird's territory (Map 14).

The distances that the 22 young males settled from their birth places ranged from 100 to 1,400 meters, the median being 280 meters.

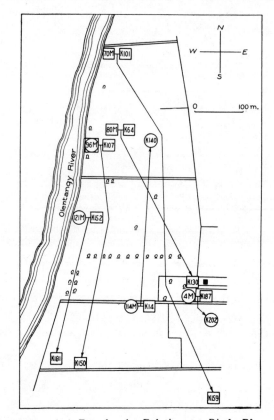

MAP 10. *Territories of 6 Females in Relation to Birth Place, 2 residents, 4 summer residents*

The 15 residents settled from 100 to 660 meters from their birth places, the median being 330 meters; the 7 summer residents settled from 155 to 1,400 meters from their birth places, the median being 270 meters.

D. Territories of the Females Banded in the Nest

Fourteen females banded as nestlings survived to start the following breeding season, but two right-banded birds disappeared before they could be captured. K66 banded in 1930 just south of Dike I was found in 1933 almost one mile south. Territories of five of these birds in relation to birth place are shown on Maps 11, 13 and 14. The territories of the other six are given on Map 10.

The distances from the birthplace of the territories on which the 12 females settled ranged from 45 to 1,300 meters, the median being 270 meters.

Of the 40 nestlings that survived to adulthood only five—four males and one female—were recaptured in our garden; all the others were located and trapped on their territories.

E. Territories of Song Sparrows Banded in Our Garden

The majority of Song Sparrows trapped in our garden, that were later found nesting, have not scattered widely. Of 20 males caught in the fall of 1931 and spring of 1932, 18 settled between 120 and 550 meters from our house, one 700 meters, and one 1,600 meters. Careful censuses over the intervening region failed to show any other banded birds. Six males trapped in the garden the following fall and winter took up territories at the following distances: one in the garden, and the others, 90, 225, 300, 450 and 900 meters, the median distance being 260 meters. Eight females captured during these same periods settled from 90 to 550 meters away. To sum up, of these 34 birds, 31, or 91 per cent, made their homes within 550 meters of our garden, while 26, or 77 per cent, did so within 360 meters. This illustrates how little these Song Sparrows wander, either in fall or spring, before settling down.

The sedentary character of this species once it has taken up its territory is shown by the fact that only three of these 34 birds were later recaptured in our garden; the others were located by repeated searches and recognized by their colored bands. Probably most of them were young when banded. *It is not possible to judge of the survival of a territorial bird like the Song Sparrow—either adult or young—by the birds retrapped at a central point.*

F. Some Family Histories

Genealogical trees of several families have been given in Chart V in Chapter IV.

Let us see where these different relatives settled on Interpont. Maps 11 to 14 show the direct descendants in each of the lines, and the mates of these descendants if any offspring are known to have survived to breed, or if anything is known of the previous or subsequent history of these mates, in which case the earlier or later territories are shown. The territories are given of eight young males and five young females, in relation to birth place; and the territories of

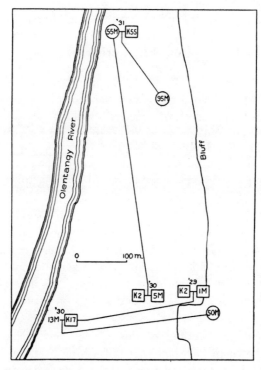

Map 11. *K2 and her Descendants. The date gives the year of nesting*

four males two years in succession, of four for three years in succession, and one for four years.

Map 11 shows the descendants of K2, a summer resident female that had two summer resident mates, a summer resident daughter, resident son, and two resident grandsons. The son (55M) died during his second summer. His son (95M) in his first winter sustained a broken leg that never healed properly; he

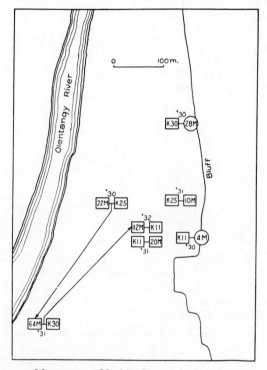

MAP 12. *22M, his Son and Grandson*

was deserted by his mate before nesting began and did not survive his second winter. The daughter (K17) raised only the first of her three broods and did not return the following year. Her son (50M) returned almost to the territory of his grandfather and here he lived to be a little over three years old. Twice I banded great-grandchildren of 1M and K2; 50M's five young in June, 1931, and three young May 17, 1933, but none of these survived, to my knowledge.

On Map 12 the territories are given of my only straight summer resident line for three generations—22M, his son and grandson, and other nesting places of the mates of each of the males. Both 64M and his son 112M nested two sea-

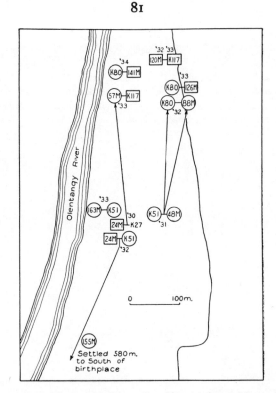

MAP 13. *K51's and 24M's Descendants. See Map 14 for 24M's and 126M's mates in 1932. K117 rejoined 120M in 1933, but after his death joined 57M*

sons, but only one brood of young was raised by either bird in the three years before we left Columbus in June.

On Map 13 there is no third generation, but a number of half brothers and one case of full brother and sister. A summer resident (24M) has had two resident sons that settled in opposite directions from home. 57M is a particularly interesting bird, because he has always been retiring, almost never singing, yet he survived to be almost six years old, obtaining mates during each of the five seasons, and raising young at least once (1932) and probably several times. (In 1934 his mate laid five eggs; two were taken by Cowbirds and the other three were sterile.) His mate in 1933 (K117) had remated with her former mate (120M), but upon his death joined 57 M. K51 was the mother of 88M and K80— the brother and sister that mated; none of the young from their first nest survived.

The descendants and different residences of K28 are given on Map 14. For two years this bird lived in a pretty, tangled spot on Central Interpont, but the third year I found her 700 meters to the south nesting on a dump below Lane

Avenue. And there were still several bachelors in the vicinity of her former home. The fourth year she returned to the dump, but disappeared soon after. Her daughter (K63) was the mother of a brood of five, three of which survived to adulthood, my only example of such a happening. 145M was caught 50 meters to the south of our grounds on Oct. 4, 1932, and never seen again; he certainly had not nested on Interpont, nor in the vicinity, unless to the east in town.

The sisters K123 and K131 settled in opposite directions from their birth place; the latter was present three years, having the same mate during the last

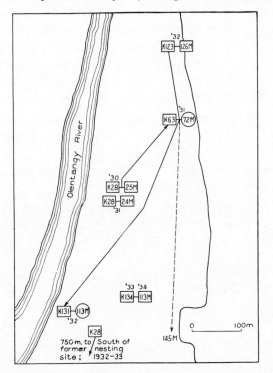

MAP 14. *K28: her Descendants and Residences during four Seasons. Nesting place of 145M unknown; trapped in place indicated, Oct. 4, 1932*

two years. In May, 1932, I banded K28's children and also two broods of her great-grandchildren, but none, unfortunately, were found in subsequent seasons.

If all the young of all the birds shown on the maps could have been banded each year, the genealogical tables undoubtedly could have been continued for some of the lines. But that was not possible and the known history of all these families has come to an end. Of all the birds shown on the four maps, not a single one is alive at the date of writing—April, 1936.

G. Summary

1. Some male Song Sparrows keep the same territory year after year, while others make slight changes, as shown in Maps 2-7 where the territories on Central Interpont are shown from 1930 to 1935.

2. Females have returned to their former nesting territories in 20 of 54 cases, have settled next door almost as often and in the other instances, have settled at distances from 100 to 700 meters (Map 8).

3. Twenty-two males banded in the nest have taken up territories from 100 to 1,400 meters from their birth places, the median distance being 280 meters (Map 9).

4. Twelve females banded in the nest have settled from 45 to 1,300 meters from their birth places, the median distance being 270 meters (Map 10).

5. Of the 40 nestlings known to have survived to adulthood, only five were captured in our garden.

6. Thirty-four Song Sparrows trapped in our garden have settled from 0 to 1,600 meters away, 77 per cent of the birds within a distance of 360 meters. Only three of these individuals have been recaptured in the garden.

7. The territories of the four families whose genealogical trees are shown in Chart V in Chapter IV are shown in Maps 11 to 14 with other territories of the mates of the birds involved.

II

The Biological Significance
of Territory

Editor's Comments on Paper 5

5   **Hinde:** The Biological Significance of the Territories of Birds
Ibis, **98**, 340–369 (1956)

By the time Margaret Nice had published her classic study of the song sparrow, zoologists were fully aware of the concept of territory in birds; far less was known, however, about territoriality in other animals. Beginning in the 1930s, there was a great surge of territorial studies. Nice (1941) gives some indication of this in her important review of territory in bird life. Of the 387 titles she lists, there were three from the seventeenth century, two from the eighteenth, and six from the nineteenth. Eleven papers came from the first decade of the twentieth century, 15 from the second, 48 from the third, and 302 from the fourth. These papers were to provide the "facts" upon which the subsequent highly controversial debates on the roles of territory were to be founded (see Part III).

Nice's review is particularly valuable for its excellent summary of early studies of territoriality in animals other than birds. Charles Elton, far better known for his work on population fluctuations in mammals, had identified territorial behavior in wood ants by 1927 (Elton, 1932). Among fish, the most detailed studies of territory have been done on the three-spined stickleback. These studies have been centered at Leiden University, beginning with the work of Pelkwijk and Tinbergen (1937) and continuing to the present with the work of Iersel (1953) and Van Den Assem (Paper 24). Cichlid fishes have also received considerable attention, beginning with the work of Baerends and Baerends-Van Roon (1950). Territoriality in fishes has been reviewed by Noble (1938) and Gerking (1953). Since fish are so difficult to observe in the wild, the role of territory in regulating density is still largely unstudied.

Territoriality had been overlooked in amphibians until recently, probably because this group of vertebrates seems so poorly adapted for acting aggressively. Emlen (1968) has now found that bullfrogs show classic territoriality. They engage in calling and postural displays to intimidate rivals, but if these fail they actually push the rival to oust him. Territoriality may well be widespread among amphibians.

Territoriality among reptiles is seen primarily in lizards (Evans, 1936; Noble, 1934).

Pioneers in the study of mammalian territory were Carpenter (1934) with his work on howler monkeys, Burt (1940) with small mammals, and Hediger (1949). Hediger emphasized the role of scent marking as a means of demarcation of mammalian territories. This explained how nocturnal animals could still defend a territory when visual signals were of no use. Detailed classifications of mammalian social organizations have been made by Eisenberg (1966) and Fisler (1969). Brown and Orians (1970) have made a brief survey of spacing patterns in animals.

I have chosen to include Robert Hinde's paper for his keen analytical statement of the subject to 1956. Besides giving a workable classification of territories, he examines the various functions that had been ascribed to territories.

Two questions concerning the function of territoriality arose soon after Howard's book appeared in 1920. First, can territoriality limit the density of a species in a particular area? (See Part III for some key papers dealing with this controversy.) Second, does the territory assure a food supply for at least some species, as Altum, Howard, and Nice proposed? Reviews by Lack and Lack (1933) and Tinbergen (1957) challenged the accepted dogma. Hinde not only summarizes the opposing arguments in these controversies but also adds his own views. Perhaps his major contribution was to recognize that it is fruitless to seek a single function of territoriality. Instead, territories, depending upon species and location, have numerous functions, which he proceeds to identify and discuss. There is no inkling in Hinde's review that territory is just part of a continuum of social organization (see Part V), nor that environment plays perhaps a greater part than genetics in determining if a species is territorial at given time and place (see Parts V and VII).

Hinde chooses to define territory as "a topographically localized defended area." He excludes from typical territory the "moving territories" that Jenkins (1944) observed in geese. As his geese moved about, they maintained an area about them within which they would not tolerate other geese. This resembles the concept of "individual distance" first described by Hediger. Individual distance is a characteristic of many animals that serves to keep others at a distance that does not interfere with their own activity. Leyhausen (1971), among others, has suggested that in sedentary animals, the intolerance toward others expressed as individual distance may increase during the breeding season to the point that it becomes a territory.

Robert Hinde is a member of the Subdepartment of Animal Behavior and a Royal Society Research Professor at Cambridge University. His monumental book *Animal Behaviour* is a synthesis of ethology and comparative psychology, drawing upon more than 2500 publications. His colleagues will always appreciate his unusual ability to provoke stimulating discussion about behavior.

References

Baerends, G. P., and J. M. Baerends-Van Roon. 1950. *An Introduction to the Ethology of Cichlid Fishes. Behaviour* (Suppl. 1). 243 pp.

Brown, J. L., and G. H. Orians. 1970. Spacing patterns in animals. *Ann. Rev. Ecol. Syst.*, **1**: 239–262.

Burt, W. H. 1940. Territorial behavior and populations of some small mammals in Southern Michigan. *Misc. Publ. Zool. Univ. Mich.*, **45**: 1–58.

Carpenter, C. R. 1934. *A Field Study of the Behavior and Social Relations of Howling Monkeys.* Comp. Psychol. Monogr. 10. 168 pp.

Eisenberg, J. F. 1966. The social organizations of mammals. *Handbuch der Zoologie.* 8 Band/39 Lieferung: 10(7), 1–92.

Elton, C. 1932. Territory among wood ants (*Formica rufa* L.) at Picket Hill. *J. Animal Ecol.*, **1**: 69–76.

Emlen, S. T. 1968. Territoriality in the bullfrog, *Rana catesbeiana. Copeia*, **1968**: 240–243.

Evans, L. T. 1936. Territorial behavior of normal and castrated females of *Anolis carolinensis. J. Genet. Psychol.*, **48**: 88–111.

Fisler, G. F. 1969. Mammalian organizational systems. *Contr. Sci. Los Angeles Co. Museum Natural History,* **167**: 1–32.

Gerking, S. D. 1953. Evidence for the concepts of home range and territory in stream fishes. *Ecology*, **34**: 347–365.

Hediger, H. 1949. Säugetier-Territorien und ihre Markierung. (Mammalian territories and their demarcation.) *Bijd. Dierk.*, **28**: 172–184.

Iersel, J. J. A. van. 1953. *An Analysis of the Parental Behaviour of the Male Three-Spined Stickleback (Gasterosteus aculeatus L.). Behaviour* (Suppl. 3). 159 pp.

Jenkins, D. W. 1944. Territory as a result of despotism and social organization in geese. *Auk*, **61**: 30–47.

Lack, D., and L. Lack. 1933. Territory reviewed. *Brit. Birds*, **27**: 179–199.

Leyhausen, P. 1971. Dominance and territoriality as complemented in mammalian social structure. *In* A. H. Esser (ed.), *Behavior and Environment: The Use of Space by Animals and Men*, 22–33. Plenum Press, New York. 411 pp.

Nice, M. M. 1941. The role of territory in bird life. *Amer. Midland Naturalist*, **26**: 441–487.

Noble, G. K. 1934. Experimenting with the courtship of lizards. *Nat. Hist.*, **34**: 3–15.

Noble, G. K. 1938. Sexual selection among fishes. *Biol. Rev.*, **13**: 133–158.

Pelkwijk, J. J. ter, and N. Tinbergen. 1937. Eine reizbiologische Analyse einiger Verhaltensweisen von *Gasterosteus aculeatus* L. *Z. Tierpsychol.*, **1**: 193–200.

Tinbergen, N. 1957. The functions of territory. *Bird Study*, **4**: 14–27.

5

Reprinted from *Ibis*, **98**, 340–369 (1956)

THE BIOLOGICAL SIGNIFICANCE OF THE TERRITORIES OF BIRDS.

By R. A. HINDE.

(*Ornithological Field Station, Madingley, Cambridge University Department of Zoology.*)

Received on 2 March 1956.

PART I. BASIC CONSIDERATIONS.

1. INTRODUCTION.

Although territorial behaviour has now been studied in many species of birds, there is still no general agreement as to its biological significance. Eliot Howard (1907–14, 1920), who first brought it to the notice of English-speaking ornithologists*, described two main functions—to facilitate pair formation and ensure an adequate supply of food for the young ; since then many others have been suggested.

The contributions to this number of the ' Ibis ' have been collected by the Editor to illustrate some of the diversity of avian territorial behaviour, and to give a selection of modern views on the problem of its function. This paper is intended to introduce (and to some extent to summarise) these species studies : it is not a comprehensive review, though most of the detailed studies of territory published since Mrs. Nice's (1941) review have been included as additional illustrations. Further references are given by Armstrong (1947). Certain topics—such as the function of winter territories (refs. in Lack 1954 b) and of flock territories (Lorenz 1938 ; Davis 1942 ; Robinson 1945 ; Wilson 1946), and the relation of polygamy to territorial behaviour (Williams 1952 ; Armstrong 1956 ; Haartmann 1956), have been omitted. My debt to previous writers on the territory problem—particularly Mrs. Nice, Dr. D. Lack and Dr. N. Tinbergen—will be obvious. I am also grateful to Dr. Tinbergen for allowing me to read an unpublished manuscript and to Drs. C. B. Goodheart, D. Lack, W. H. Thorpe, N. Tinbergen and the Editor for reading this paper in draft and much helpful discussion.

* A fairly detailed discussion had been given by Altum (1868, see Mayr (1935)) and territorial behaviour had been mentioned by a number of other writers (refs. in Nice, 1941; Lack, 1944; Armstrong, 1953).

For brevity, only the author's name and the date (1956), but not the species he studied, are given in text references to the other papers in this symposium. These are as follows :—

Andrew, Yellowhammer *Emberiza citrinella* and Corn Bunting *E. calandra* ; Armstrong, Wren *Troglodytes troglodytes* ; Blank & Ash, Partridge *Perdix perdix* ; Conder, Wheatear *Oenanthe oenanthe* ; Durango, Red-backed Shrike *Lanius collurio* ; Gibb, Tits *Parus* spp. ; Gibb, Rock Pipit *Anthus spinoletta* ; Haartman, Pied Flycatcher *Muscicapa hypoleuca* ; Hale, Redshank *Tringa totanus* ; Lanyon, Eastern Meadowlark *Sturnella magna* and Western Meadowlark *Sturnella neglecta* ; McCarten & Simmons, Great Crested Grebe *Podiceps cristatus* ; Marler, Chaffinch *Fringilla coelebs* ; Mountfort, Hawfinch *Coccothraustes coccothraustes* ; Simmons, Little Ringed Plover *Charadrius dubius* ; Snow, Blackbird *Turdus merula* ; Swanberg, Thick-billed Nutcracker *Nucifraga caryocatactes* ; Tinbergen, Black-headed Gull *Larus ridibundus*, Herring Gull *L. argentatus* and Kittiwake *Rissa tridactyla* ; Young, American Robin *Turdus migratorius*.

2. DEFINITION, CLASSIFICATION AND DIVERSITY.

Controversy about the biological significance of territory has been due partly to lack of agreement about the meaning of the term. Howard (1907–14, 1920) used it to refer to virtually any defended area, from the tiny nest territories of cliff birds to the extensive areas defended by some passerines and birds of prey. Comparing the territories of Guillemot and Ruff in a letter to Mayr (1935, quoted by Nice, 1941), he wrote "there is no difference, as far as I can see, in the congenital foundation. From this congenital foundation have grown regions of different sizes with different proximate ends ; but neither size nor end change the congenital foundation." Although this view was criticised by some ornithologists (e.g. Alexander 1921, Jourdain 1927, Nicholson 1928, D. & L. Lack 1933 : see also Durango 1956), most now agree that if territorial behaviour is to be defined descriptively, it is illogical to exclude some cases on the basis of their function. Further, evidence presented in sections 4 and 5 shows that all types of territory depend on basically similar mechanisms (see also Tinbergen 1936 ; Lorenz 1938 ; Nice 1941). Therefore, of the many definitions which have been proposed (e.g. Mayr 1935 ; Tinbergen 1936, 1939 ; Crawford 1939 ; Lack 1939 a ; Davis 1940 ; Jenkins 1944 ; Armstrong 1947), Noble's (1939) simple description of territory as " any defended area "* is suitable here. Defence may of course be by threat, song or any

---

* Armstrong (1947) criticises this definition because it does not state against whom the territory is defended—" the essential feature of territory is defence against competitors of the same species ". However, aggressive responses towards conspecific individuals, species of similar appearance and predators depend on related mechanisms. Further, interspecific territorialism occurs in a number of cases (e.g. Fuggles-Couchman 1943; Simmons 1951 (and references cited), 1956; Lanyon 1956).

2 B 2

other behaviour pattern evoking avoidance in other individuals, as well as by actual combat, and there is no implication that the area is sharply defined. This definition includes areas round roosting sites (e.g. Rankin & Rankin 1940), favourite perches etc., as well as breeding territories, and so carries no inferences as to function. Excluded are non-defended areas, such as the " winter range " of many passerines and the covey territories of the Partridge (Blank & Ash 1956).

In common usage, " territory " refers to a topographically localized defended area : regions defined by reference to a moving datum (e.g. " individual distance ", " mated female distance ", Hediger 1942 ; Jenkins 1944 ; Burckhardt 1944 ; Conder 1949) are usually excluded. However, such mobile areas are defended by behaviour similar to that used in the defence of static ones, and must frequently be taken into account in what follows.

Most of the territories considered in the specific papers of this symposium can be grouped together as breeding territories—that is, they are defended especially before or during the season of reproduction (cf. Tinbergen 1936, 1939), though in some cases defence continues throughout the year. Breeding territories are themselves diverse, and systems for classifying them have been proposed by Mayr (1935), Nice (1941), Armstrong (1947) and others. For present purposes, a simple one (similar to that of Mayr) will suffice :—

Type A. Large breeding area within which nesting, courtship and mating and most food-seeking usually occur, e.g. some warblers Sylviidae (Howard 1907–14) ; Robin *Erithacus rubecula* (Lack 1939 a).

Type B. Large breeding area, not furnishing most of the food, e.g. Nightjar *Caprimulgus europaeus* (Lack 1932) ; Crimson-crowned Bishop *Euplectes hordeacea* (Lack 1935) ; Reed Warbler *Acrocephalus scirpaceus* (Brown & Davies 1949).

Type C. Nesting territories of colonial birds—a small area round the nest (e.g. Howard 1920).

Type D. Pairing and/or mating territories. Small areas, not used for nesting, such as the pairing territories of gulls (Tinbergen 1956) and the mating territories of lek birds (references on p. 356).

These categories are intended only as an aid to discussion, for the diversity of nature can never be fitted into a system of pigeon-holes. In some species individuals occupy territories of more than one type within the same period (e.g. Greenshank *Tringa nebularia*, Nethersole-Thomson 1951 ; Tinamou *Nothoprocta ornata*, Pearson 1955), or in succession (e.g. Laridae, Tinbergen 1953, 1956) ; and in others the kind of territory varies with age, habitat or population density (e.g. Eastern Goldfinch *Spinus tristis*, Drum 1939 ; Nice 1939, 1941 ; Nickell 1951 ; Great Reed Warbler *Acrocephalus arundinaceus*,

Meylan 1938 ; Great Crested Grebe, McCarten & Simmons 1956 and references cited ; Mountfort 1956).

The types of territory defended may also differ between the two sexes of the same species : this is obvious in many lek birds, but also occurs in those passerines in which the female does not help to defend the male's territory, but does defend an area round the nest site (e.g. the Humming-bird *Archilochus colubris*, Pitelka 1942 ; Wood Warblers e.g. *Dendroica* spp., Kendeigh 1945 ; Red-winged Blackbird *Agelaius phoeniceus*, Nero & Emlen 1951 ; Durango 1956 ; Marler 1956 b). Closely related species may differ in their territorial behaviour (e.g. Blackbird and Fieldfare *Turdus pilaris*, Snow 1956 ; Red-winged and Brewer Blackbirds *Euphagus cyanocephalus*, Williams 1952 ; Black-headed Gull and Kittiwake, Tinbergen 1956 ; see also Jourdain 1921). Sometimes it is the female who is primarily responsible for territorial defence—in which case the male may be responsible for incubation, e.g. *Hydrophasianus chirurgus* (Hoffmann 1949), or may play no further part in reproduction, e.g. Mexican Violet-ear *Colibri thalassinus* (Wagner 1945). Further, territorial boundaries are often vague and their defence spasmodic (Young 1956). In a few cases, breeding territories are apparently absent, e.g. Emperor Penguin *Aptenodytes forsteri* (Stonehouse 1953) ; Redshank (Hale 1956).

In fact few statements of universal validity can be made about territory in birds and it is better perhaps to aim at some comprehension of its diversity.

3. ANALYSIS OF TERRITORIAL BEHAVIOUR.

Behaviour involved in the establishment and maintenance of a territory can be analysed into two and often three main components, as Howard (1920, 1929) fully realised :

(*a*) Restriction of some or all types of behaviour to a more or less clearly defined area.

(*b*) Defence of that area.

(*c*) Self-advertisement within the area.

Components (*a*) and (*b*) are essential to territory as here defined, but (*c*) is not, although it has an important role in the breeding territories of many song birds. (Definitions of territory given by Mayr (1935) and Lack (1939 a) make (*c*) obligatory also.) Any one of these forms of behaviour may occur without the others : thus activities may be restricted to an area which is not defended (e.g. winter range of many passerines) ; an area not topographically localized may be defended (e.g. individual distance, see p. 342) ; and, more rarely, self-advertisement may occur independently of attachment to a site (e.g. Hawfinch song, Mountfort 1956). Clearly these components may have quite different functions.

4. THE NATURE OF AGGRESSIVE BEHAVIOUR

D. & L. Lack (1933), wishing to limit use of the term " territory " to those cases in which it is accompanied by " song, plumage display and some fighting ", argued that the areas round the nest site which are defended by many passerines at the time of nesting depend on fighting of a type different from that involved in the establishment of the original breeding territory earlier in the season.   Mayr (1935) and Tinbergen (1939) supported this view.   If true, it would be an important argument against the broad definition of territory as " any defended area ".   More recent evidence on the relation between the fighting behaviour shown in different contexts must therefore be considered.

Fighting arises in many different situations—over food, perches, females, nest sites, and so on.   This led Tinbergen (1951) to propose that " there is not one instinct of combat.   There are several sub-instincts of fighting." "Instinct ", in Tinbergen's sense, refers to a mechanism, and he thus implies that the fighting seen in different contexts depends on different mechanisms. In view of the similarities and relationships between the different forms of fighting, this view is no longer tenable.   Thus :—

(i) The motor patterns used in fighting in the different contexts are fundamentally similar.

(ii) The animal which is a strong fighter in one context is usually also a strong fighter in another (Hinde 1952 ; Marler 1955).

(iii) In certain circumstances fighting (e.g. in late-summer flocks of juvenile passerines) that starts in one context (e.g. over food) is continued into fighting behaviour typical of another (e.g. over breeding territory).

(iv) The diurnal rhythms of fighting over food and fighting over perches amongst captive Chaffinches are similar, and differ from those of other activities such as feeding (Marler 1955).

(v) There is a common factor in the stimulus situation eliciting inter-specific fighting in all contexts, namely certain characters of a member of the same species.   (Fighting against other species and predators is not considered here, though there is strong evidence that it also is related to interspecific fighting.   See, e.g., Davis 1941 ;   Fitch 1950 ;   Tinbergen 1952 b ; Simmons, 1956).   Although some species patrol their territories (a type of appetitive behaviour in which the threshold for fighting is temporarily low) and although aggressive behaviour may be aroused by calls and by conditioned environmental stimuli (e.g. sites of previous encounters, Hinde 1952), normally it is evoked only by a potential rival.   Equally, whatever the circumstances of the encounter, the intensity of the aggressiveness usually diminishes, not on the performance of the fixed action pattern of striking, but on the departure of the rival. This can hardly be described as a " consummatory situation " (cf. Thorpe 1954) as it is merely the converse of the eliciting one : the concept would become valueless if used in this way.

For such reasons, the essential unity of all forms of fighting has been emphasised in an earlier paper (Hinde 1952). It was there implied that the strength of aggressiveness, while sometimes virtually independent (as in young passerines (Hinde 1953 a), and the fighting over individual distance), is usually governed by the state of some other instinct. This latter point is supported by various observations of changes in individual distance with mood (e.g. Jenkins 1944 ; Burckhardt 1944 ; Conder 1949 ; Crook, pers. comm.). Careful experiments by Marler (1956 a) and Andrew (in press) have shown, however, that a period of food deprivation does not affect the threshold of aggression in Chaffinches and Buntings at a feeding place. These experiments thus suggest that other tendencies (e.g. to feed) do not increase the strength of aggressive behaviour in a given context.

Of course this does not mean that the internal state has no influence on aggression. Quite apart from the known effects of testosterone (Shoemaker 1939), mood may be important in giving significance to an external object which is the immediate cause of the fight. While the violation of individual distance is "the basic stimulus for aggression" in some contexts such as much fighting in winter flocks (e.g. Marler 1955, 1956 a), in others it is not only the spatial relationships between the combatants, but those between the combatants and some external object, which are important. Thus a Great Tit in reproductive mood will attack a male who, though one hundred yards distant, is intruding on his territory, but tolerate him ten yards away across the boundary. Two minutes later, the same territory owner may, while feeding, ignore an intruder three feet from him inside the territory. Andrew (in press) gives two examples illustrating the same point in his studies of four-months-old Yellowhammers in captivity. Although the peck-dominance hierarchy was normally independent of position in the cage, two cases of temporary locality-linked dominance were related to the mood of the birds concerned. Of course in such cases it is impossible to be certain whether the mood in question merely affects the tendency to stay near the object in question, and thus decreases the tendency to flee, or directly affects aggressiveness.

In summary, then, there is strong evidence that aggressive behaviour which arises in different situations depends on the same mechanisms ; the essential difference between the various types of territory is thus not in the quality of the fighting itself. The strength of aggressiveness may, however, depend on the relative position of the two combatants, or on their relationship with some external object whose relevance is determined by the mood. Further the relative efficacy of stimuli (e.g. predators, conspecific males) in eliciting fighting may vary with the context : the fighting of a song-bird near the nest site is elicited by a much wider range of stimuli than the fighting in defence of the initial breeding territory (D. & L. Lack 1933).   We are thus concerned with fighting specialised to serve different functions.   Even within the category of breeding territories, the objects defended may be extremely diverse, for example :—

(*a*) Mate. Snow-bunting *Plectrophenax nivalis* (Tinbergen 1939) ; geese (Jenkins 1944) ; Grosbeak *Pheuticus melanocephalus* (Weston 1947) ; various species (Conder 1949) ; Brewer Blackbird (Williams 1952) ; California Quail *Lophortyx californica* (Genelly 1955) ; Blank & Ash 1956 ; Gibb 1956 a ; McCartan & Simmons 1956 ; Marler 1956 b ; Snow 1956.

(*b*) Song or look-out post. Red-tailed Hawk *Buteo jamaicensis* (Fitch et al. 1946) ; Swallow Tanager *Tersina viridis* (Schaefer 1953) ; Andrew 1956 ; Gibb 1956 a ; Marler 1956 b.

(*c*) Nest site or nest.   Black-headed Gull (Kirkman 1937, 1940) ; House Sparrow *Passer domesticus* (Daanje 1941) ; Prong-billed Barbet *Dictorhyncus frantzii* (Skutch 1944) ; Eastern Bluebird *Sialia sialis* (Thomas 1946) ; White Stork *Ciconia ciconia* (Haverschmidt 1949) ; Hornbill *Lophoceros suahelicus* (Ranger 1949–51) ; Eastern Goldfinch (Stokes 1950) ; Collared Flycatcher *Muscicapa albicollis* (Löhrl 1951) ; Cliff Swallow *Petrochelidon pyrrhonata* (Emlen 1952) ; Brewer Blackbird (Williams 1952) ; Pied-billed Grebe *Podilymbus podiceps* (Glover 1953) ; Zebra Finch *Poephila guttata* (Morris 1954) ; Swift *Apus apus* (D. & E. Lack 1952) ; Prothonotary Warbler *Protonotaria citrea* (Walkinshaw 1954) ; Gibb 1956 ; Haartmann 1956 ; Tinbergen 1956.

(*d*) Display-ground or bower.   Various lek species (references on p. 356) ; bower-birds Ptilonorhynchidae (Marshall 1954).

(*e*) Family.   Geese (Jenkins 1944) ; Boyd 1953 ; Blank & Ash 1956 ; Hale 1956 ; McCartan & Simmons 1956 ; Simmons 1956 ; Tinbergen 1956.

(*f*) Covey.   Howard & Emlen 1942 ; Blank & Ash 1956.

(*g*) Food source. Ruby-throated Humming-bird (Pitelka 1942) ; Swanberg 1956.

(*h*) A more or less diffuse area (e.g. Blank & Ash 1956 ; Simmons 1956 ; Tinbergen 1956).   Such cases are not basically different from the preceding ones and often some (usually central) part is defended more vigorously than the rest.

Even the defence of extensive areas shown by many passerines often depends primarily on the defence of objects or situations, and results only secondarily in the defence of a larger area.   In the Great Tit (Hinde 1952), for instance, aggression is concentrated first round the mate, then round the song-posts and later still round the nest. The boundaries, while sometimes influenced by topographical features (e.g. Conder 1956) are defined by skirmishes with neighbours and become sharper the denser the population. In this way the boundaries are learnt and the area becomes integrated as a unit. (See also Diesselhorst 1949 on the Yellowhammer ; Schaefer 1953 on the Swallow Tanager ; Armstrong 1955 on the Wren ; Marler 1956 b on the Chaffinch.) The process is probably similar, but much accelerated, in migrant species (e.g. Howard 1920)—thus Raines' (1945) observation that *Sylvia* spp. neglect the outer song-posts when the nest site is selected suggests a similar change in what is defended.

Clearly the nature of the object(s) or situations defended is of great importance in assessing the biological significance of the territory.

5. TERRITORY AND AGONISTIC BEHAVIOUR.

When a territory owner sees an intruder, he may attack it immediately. His behaviour is, however, usually complicated, for stimuli from a fellow member of the species result in a tendency to flee as well as a tendency to attack. The precise behaviour shown at any moment varies with the absolute and relative strengths of these two tendencies (Tinbergen 1952 a, 1953 ; Hinde 1952, 1953 b ; Morris 1954 ; Moynihan 1955 ; Marler, in press). Some of the factors influencing these tendencies affect both more or less equally, but others one more than the other. In fighting which has reference to an object, the relative strength of the attacking tendency varies with the distance of the object, the effect depending on other factors, including the extent of previous " ownership " and, possibly, the " drive state " at the time. In territorial behaviour, for instance, the owner usually attacks strange males inside his territory and flees from them outside, while the boundary is often first established as the line along which the two are in approximate balance (see above).

As we have seen, in many species self-advertisement plays an important part in the establishment of territory. Often the beginning of full song and the establishment of the territory are simultaneous and influenced by the same causal factors (Tinbergen 1939). Further, an intruder may flee from the song of a territory owner just as he flees from a threatening or attacking one (Lack 1939 a)—in each case his flight is equally a consequence of the territorial behaviour of the owner.

6. THE NATURE OF THE EVIDENCE CONCERNING THE BIOLOGICAL FUNCTIONS OF TERRITORY.

In general the results of territorial behaviour may be biologically dis-advantageous, neutral or beneficial to the individual's chance of ultimate reproductive success. For example, the increased susceptibility to predators involved in territory defence may be disadvantageous, but the guarantee of a suitable nest site beneficial.

In the past, two rather different meanings of " function " have been confused in discussions about territory. Often the term has been used for any advantageous consequence that results from site attachment and aggression (and not, e.g. from habitat selection). Alternatively, the meaning can be restricted to biologically significant consequences—that is, those through which selection in favour of the behaviour can act. Since selection can operate to maintain the behaviour at a given level through a certain effect only if any decrease in that effect is disadvantageous, consequences which are beneficial are not necessarily significant functions in this sense. For example, a song-bird's territory may provide a nest site, but this cannot be a significant consequence of territory ownership if such sites are superabundant. Similarly, epidemic diseases may be less dangerous to birds holding large territories, but this does not necessarily mean that the vigour of territorial behaviour is maintained at the level necessary for holding such a large area by selective forces acting through disease.

If selective forces dependent on only one effect—say that of providing a supply of some commodity—determined the vigour of territorial defence, they could never lead to the defence of an area larger than that necessary to guarantee the supply. Defence of a larger area would entail disadvantageous consequences without compensatory reward. On the other hand, even if the supply is not guaranteed for all circumstances, the behaviour may still be functional if the chances of an adequate supply are increased. In practice, the selective forces acting on territorial behaviour are often more complex than this, and the optimum size in given circumstances may be that which reduces the chances of reproductive failure for one reason $x\%$ and for another by $y\%$, both x and y being affected by a small change in territorial vigour. In this way more than one result of territorial behaviour may be biologically significant. Further, the optimum itself will change with the pressure exerted by newcomers.

In practice, reliable direct evidence about the functions of territory is scarce. In the virtual absence of quantitative data on the relations between breeding success and characteristics of the territory (but see Conder 1956) and of experimental evidence, it is necessary to rely on :—

(a) Observations on the specific characteristics of the territory—that is, on the way in which the territorial behaviour has become specialised in evolution : this ultimately involves comparison with other species. This can be a reliable source of evidence : thus the function is likely to be defence of the nest if aggression is most vigorous near it. But this still does not show whether defence of the nest forces other birds to nest at a distance and thus reduces predation, or reduces interference in nesting, or what.

(b) Circumstantial evidence that a given consequence is likely to be both significant and advantageous. Much discussion about territory falls into this category : some of it is valuable, but some must be regarded as irrelevant. For example :

(i) To argue that territory cannot have a certain function in one species because it does not have that function in another unrelated species is a *non sequitur* (see e.g. pages 359 and 362).

(ii) Observations showing that interference in copulation occurs occasionally in a given territorial species, or that it occurs in other species, do not show that it would occur in the given species to a significantly disadvantageous extent if no territory were held.

(iii) Similarly, feeding on the territory does not show that the provision of food is a significant and advantageous consequence (Lack 1954 b)—nor, if most of the food is collected outside the territory, can it necessarily be argued that the territory has no significance in relation to food (p. 360).

PART II. SUGGESTED FUNCTIONS OF TERRITORIAL BEHAVIOUR.

It is now possible to consider various views on the functions of territory. Some are concerned exclusively with site attachment, and some with both site attachment and hostility. Aggression related to an area defined by a moving datum (e.g. female, young) will not be discussed here.

1. SITE ATTACHMENT WITHOUT ASSOCIATED AGGRESSIVENESS.

Although birds, since they lay eggs and show parental care, are forced to confine some of their reproductive activities to a limited area,* the restriction is often greater than necessity appears to demand. Some activities may be limited to the territory for months before egg-laying or—more rarely—the young may remain after they are able to leave. We are thus concerned here with site attachment outside the nesting season. Two main functions have been suggested :—

(a) Familiarity with food sources, refuges from predators, etc. may enable the adults and/or young to exploit the area more efficiently. Tinbergen (1953), for instance, has recently emphasised the importance of knowledge of the terrain in escape from predators by young gulls, and Southern (1954), writing of the Tawny Owl *Strix aluco*, in finding food. Armstrong (1956), Blank & Ash (1956), Conder (1956), Durango (1956), Gibb (1956b), Snow (1956), Swanberg (1956) and Tinbergen (1956) also consider it to be important, though Andrew (1956) does not believe it assists escape in buntings. Circumstantial evidence thus suggests that familiarity is an advantageous

---

* There is a partial exception even to this—the Emperor Penguin shuffles about, taking its egg with it (Stonehouse 1953).

consequence of site attachment in many species, but direct evidence is scarce.

(b) Familiarity increases fighting potentiality. Although settled owners are occasionally evicted from their territories by newcomers (e.g. Robin (Lack 1939 a); Wren (Kluijver 1940); Blackbird (Jackson 1954, Snow 1956); Wren (Armstrong 1956)), prior ownership undoubtedly gives an advantage in aggressive encounters (cf. p. 347; Collias 1944). Early attachment to a site could thus promote any of the advantageous consequences discussed below under 2. The earlier a bird seeks out an area, the more likely he is to find a suitable one (assuming competition for the best areas), the less likely he is to be dispossessed, and perhaps the more likely he is to find a mate (see Howard 1920; Armstrong 1956; Gibb 1956b; Snow 1956; also Blank & Ash 1956)—though it may be advantageous for second-year birds to return to the site where they bred previously. Thus in the American Coot *Fulica americana* (Gullion 1953) small territories are held all winter and form a nucleus for the breeding territories, even though pair formation is independent of territory (Gullion 1954): *Fulica atra* is probably similar (Cramp 1947; Höhn 1949). Similarly, winter territories form the nucleus of breeding territories in the Robin (Lack 1939 a), some *Parus* spp. (Hinde 1952) and many other birds (e.g. Gibb 1956 b; Snow 1956; Swanberg 1956).* For many species, however, winter conditions make site attachment early in the season disadvantageous. Further, recently established first-year birds may be evicted by territory owners of the previous year (Song Sparrows *Melospiza melodia*, Nice 1941; Whinchat *Saxicola rubetra*, Schmidt & Hantge 1954; Armstrong 1956; Conder 1956), so it may be advantageous for them to establish their territories later than older birds. There will thus be an optimum time for territory establishment in each species, locality and type of individual.

2. SITE ATTACHMENT AND HOSTILITY.

(a) *General. Can territorial behaviour limit density?* Although winter flocking behaviour leads to aggregation† of individuals within the limits of their range, the maintenance of individual distance ensures over-dispersion within the limits of the flock. Similarly, in the breeding season, territorial behaviour results in over-dispersion of pairs over the nesting area (or, in lek birds, of males over the pairing area)—though in colonial species at all times, and non-colonial species when the population is low, they may be aggregated over the habitat as a whole (e.g. Whinchat, Schmidt & Hantge 1954). Of

* See Lack (1954b) for further notes on winter territories.

† Following Salt & Hollick (1946), individuals or pairs are said to be " aggregated " if they are dispersed less than they would be if distributed at random, and " over-dispersed " if more dispersed (i.e. if they are more evenly distributed than they would be if distributed at random). These terms thus refer only to the pattern of distribution, and not to the actual density.

course both aggression by established owners and avoidance by intruders contribute to this over-dispersion (see iv).

Some theories about the function of territory depend on this dispersive effect alone, others on the further consequence that over-dispersion in a homogeneous habitat results in an even distribution of resources among the breeding individuals. In either case the additional question arises—Can territory limit breeding density ? That is, can it ensure a minimum degree of isolation from other individuals or a minimum supply of the resources ? If territories were perfectly compressible, they could produce over-dispersion without limiting density.

Howard implied that density could be limited, and he has been supported by others who have witnessed the failure of new arrivals to establish themselves (e.g. Tinbergen 1939, 1956; Howard 1940; Errington 1945; Kluijver *et al.* 1940; Nice 1941; Mickey 1942; Mason 1947; Lanyon 1956), who believe that territory has a minimum size (e.g. Nice 1943; Kendeigh 1941; May 1949; Kluijver & Tinbergen 1953; Armstrong 1956; Gibb 1956 a, 1956 b; Lanyon 1956; Marler 1956 b; Swanberg 1956), or who have seen suitable unoccupied areas quickly filled (e.g. Tinbergen 1956; Lanyon 1956; Marler 1956 b). It has, however, been argued that although breeding density is related to resources, it is not by means of territorial behaviour. Thus :—

(i) The large individual variations in territory size in many species (e.g. Bishops *Euplectes*, Moreau 1938; nearly all papers in this symposium) suggest that territory size is not specific (Lack 1954 b).* This is not a valid argument because, although territory size (in common with all other biological characters) varies, different species have different mean sizes. Such size differences occur even between closely related species and there is strong, though circumstantial, evidence that differences are due to species characteristics and not merely to changes in population pressure (Hinde 1952; see also p. 358). Though the average territory size of any given species is often much the same in different areas, some variation with the habitat must be expected (e.g. Hinde 1952; Andrew 1956; Gibb 1956 b; and especially McCartan & Simmons 1956).

(ii) Referring to Huxley's view that territories, though variable in size, have only a limited compressibility, Lack (1954 b) suggests that this " is important for the present argument only if the limit in question is frequently reached in nature, and if it then corresponds to the minimum quantity of food needed . . . " We are concerned for the moment only with the first of the two points made here : the relation between territory size and food requirements will be discussed later (p. 358). Lack's reference to a " limit " which may or may not be " reached " implies something more absolute than either

---

* The methods of measurement used by many authors (e.g. Odum & Kuenzler 1955) assess not the defended area but the range.

is necessary for the regulation of density, or in fact occurs in nature. In practice it is a matter not of minimum size at which the defence of territory owners is sufficient to keep out all intruders, but of an optimum ultimately determined by the relative advantages, to both owners and newcomers, of persistent defence (or intrusion) or retreat (cf. Lack 1952). Defence of an area greater than the optimum will be uneconomical while continued compression of the area will reduce the chances of reproductive success to a point where they do not compensate for the disadvantages (e.g. physiological exhaustion, increased exposure to predators, etc.), which presumably accompany a breeding attempt.

Since the limit is not an absolute one, evidence about how closely it is approached in nature is naturally scarce; but it is certain both that birds do have to defend their territories from encroachment, and that species normally holding territories will not breed without them (Gibb 1956 a, 1956 b). Further, observations by Kluijver & Tinbergen (1953) suggest that territorial behaviour does regulate density in nature. They showed that an area of broad-leaved woodland was occupied by Great Tits each year before the poorer surrounding coniferous woodland; and that the population in the broad-leaved woodland was almost constant, while that outside fluctuated greatly. The more favourable habitat was thus occupied first, and only when a certain density had been reached there, was the pinewood occupied. Kluijver & Tinbergen believe (but cannot prove) that this critical density was set by the territorial behaviour of the first settlers. They stress, however, that their observations do not show what would happen in the broad-leaved woods if there were no vacant places in the pine woods. (Lack, e.g. 1954 b, gives examples of a similar type of distribution in other species, though he does not believe territorial behaviour is responsible.)

(iii) Lack also argues that many observers have recorded newcomers successfully establishing themselves in spite of fierce opposition from owners (Howard 1920; Reed Warbler, Brown & Davies 1949; Collared Flycatcher, Löhrl 1951; Yellowthroat *Geothlypis trichas*, Stewart 1953), and in some species territories seem almost indefinitely compressible (e.g. Young 1956). On the other hand there have also been many records of late arrivals failing to establish themselves (references on p. 351), although such events are much more difficult to observe. The important question is what conditions are necessary for a newcomer to establish himself in a given habitat. This is discussed later.

(iv) Lack (1954 a & b) points out that the colony size of the Heron *Ardea cinerea* is probably related to food supply, and that this cannot be a consequence of territorial behaviour, which affects only the structure of the colony and not the distribution of individuals between colonies. He therefore makes the important suggestion that young birds avoid joining colonies whose size is large relative to the resources. This seems a probable explanation of the observed distribution of such species.

However, Lack (1952, 1954 b) further suggests that a similar mechanism operates among song birds—i.e. that density is regulated not so much by the aggression of territory owners as by the avoidance, by newcomers, of settled areas. An aversion for densely populated areas had previously been suggested by Errington (1934), Kluijver & Tinbergen (1953) and others. The problem here is to establish the stimuli on which this aversion depends. Of course if the newcomers move on because of song or aggression by residents, this is a result of territorial behaviour (see p. 347) : the question is merely whether newcomers avoid occupied areas in the absence of aggressive and self-advertising behaviour by the residents, and on this point there is little evidence (see, e.g., Swanberg 1956). However, it seems likely that avoidance of a non-aggressive male is mediated by the same mechanisms as avoidance of an aggressive one and merely implies only a weak tendency to approach or to settle; both cases are similar in causation and function *.

Since breeding density is related to resources (Lack 1954b, 1955) there must be some mechanism which prevents settling when there are many other individuals relative to the resources. In colonial species like the Heron the stimuli indicating this high density are apparently independent of any postures or calls evolved to intimidate newcomers. A similar mechanism may operate in passerines but we have no evidence about its importance. We do know, however, that determined attempts to settle may be thwarted by the territorial behaviour of residents, and it thus seems likely that it s territorial behaviour which ultimately regulates density in a given habitat in such species.

This, of course, does not imply an absolute limit to territory size. A new-comer will settle if the habitat is sufficiently attractive to overcome his tendency to avoid the resident's song and hostility. In a favourable habitat the intruder desists only when he meets with the strong opposition consequent upon a high density of previously established birds (see, e.g., Swanberg 1956), though this opposition need not involve actual combat. In very rich areas he will persist in spite of the previous settlers (refs. on 360). On the other hand, if the sign stimuli controlling habitat selection are weak, they can be effective only when motivation is high—the more favourable sites will be settled first, the less favourable only when density in the favourable ones is

---

* Although Howard insisted that the male bird, while establishing his territory, actively " sought isolation ", his descriptions suggest that this initial isolation could come about merely through the selection of a habitat different from that occupied by the flock (see, e.g. Mountfort 1956) and avoidance of other territories, and not through avoidance of *all* other individuals. In fact there is evidence that even individuals of species with Type A territories tend to settle near each other when the population density is low—not to seek maximum over-dispersion (e.g. Schmidt & Hantge 1954, Fisher, 1954).

considerable (e.g. Kluijver & Tinbergen 1953; Conder 1956; Durango 1956; Blank & Ash 1956).*   As Lack stresses, the Dutch Great Tit data of Kluijver (1951) and Kluijver & Tinbergen (1953) suggest that it may be advantageous for newcomers to breed in sub-optimal but uncrowded habitats; this implies that natural selection will limit persistence in the face of opposition (or self-advertisement) by established owners; and that habitat selection (like nest selection in tits, Hinde 1952) in such species is subject to selection pressure promoting response to a sub-optimal stimulus when an optimal one is not available (see also Tinbergen 1956).   Further, the intensity of defence by territory owners, as well as the persistence of newcomers attempting to settle, must be regulated by natural selection, in accordance with the relative advantages of persistence and withdrawal.†

This argument implies that marginal habitats are always available for birds excluded from the better ones. What happens when all possible habitats are occupied is not known, though flocks of non-breeding individuals occur in the breeding season even amongst passerines (Howard 1920; Kendeigh 1941).   Probably there comes a time when it is better not to try to breed at all.

In summary, then, the observations that territory owners are forced to defend their territories from encroachment, that some attempts to settle are unsuccessful, and that some individuals are forced to breed in sub-optimal habitats, strongly suggest that territory regulates the density of many species in the most favoured areas.   Proof is lacking.   There is no direct evidence that territory limits the total breeding population in all habitats, but it seems reasonable to suppose that if an individual is unable, because of territorial behaviour of rivals, to acquire the resources or isolation necessary for the chances of reproductive success to outweigh the disadvantages of a breeding attempt, he will not continue to try.

We may now consider the importance of hostility linked with site attachment in the breeding season.

(b) *Facilitation of pair formation and maintenance of the pair*.   The probable importance of territory in the formation and maintenance of the pair bond, suggested by Howard, has also been stressed by later writers (e.g. D. & L. Lack 1933; Lack 1935, 1939 a & b, 1948; Mayr 1935; Tinbergen 1936, 1939; Noble 1939; Nice 1941; Armstrong 1947; etc.).

* Further, if the habitat ceases to be suitable, residents will move out (e.g. Crossbills *Loxia curvirostra*, etc., Lack 1954 b).

† Since aggression enters into the relations between male and female (refs. on p. 347), Tinbergen (1952) suggests that territorial aggressiveness is limited because very aggressive individuals would be less efficient at pairing.   This may be the case, but it is suggested here that selection may also impose a limit within the context of territorial fighting itself—both because defence of too large an area is uneconomical, and because the chances of reproductive success in an already crowded area may not justify an attempt to settle.

For the purposes of this argument additional complexities caused by the variations of clutch size with breeding density have been ignored.

In the first place, many authors suggest that adequate spacing of the males will facilitate the meeting of potential mates (e.g. Howard, 1920; Lanyon 1956; Young 1956)—though it is by no means clear why a male who isolates himself should find a mate more easily than one who stays in the flock. Second, the territory may provide the male " with a more or less prominent, isolated head-quarters where he can sing or otherwise display" (D. & L. Lack 1933). Often linked with this is the importance of the territory as a rendezvous, for rapid pair formation is unusual (but contrast Lanyon 1956; Young 1956). Usually the initial responses of each sex to the other are agonistic; and pair formation, involving a gradual increase of mutual tolerance as well as of sexual behaviour, is a slow process. A given male is thus more likely to obtain a mate if the female can find him again after a temporary separation (e.g. Tinbergen 1939, 1956; Laven 1940; May 1949; Emlen 1954; Andrew 1956; Armstrong 1956; Conder 1956; Durango 1956; Gibb 1956 b; Marler 1956 b; Simmons 1956; Tinbergen 1956). Though this will be aided by site attachment, hostility is also usually necessary to prevent the female from finding another male at the rendezvous (contrast Tinbergen 1956). Territory may even serve as a rendezvous from one season to the next (Rowan 1955; Conder 1956; Lanyon 1956; Marler 1956 b; Snow 1956).

In many species it seems also to facilitate the maintenance of the pair bond. Howard (1920), Brown & Davies (1949) and others have emphasised that in Warblers it must be large enough to contain the wanderings of the female when she first arrives (see also Andrew 1956; Gibb 1956 a; Marler 1956 b). In species where a large area is defended after pair formation this may be its main function (e.g. Graceful Warbler *Prinia gracilis*, Simmons 1954).

Tinbergen (1939) has stressed the biological importance of monogamy in many species, and shown how it is promoted by each member of the pair directing its aggressiveness primarily on individuals of the same sex (see also, e.g., Laskey 1944; Stokes 1950; Blank & Ash 1956; Gibb 1956 a; Haartman 1956; Lanyon 1956; Marler 1956 b; Simmons 1956; Snow 1956 and May 1949). Polygamy is possible in the Corn Bunting only because fighting between females is unimportant (Andrew 1956). Any behaviour of the male which helps to prevent his mate being fertilized by another male is likely to carry a great selective advantage—even in territorial species, stolen copulations are sometimes attempted (e.g. Haartman 1956; Marler 1956 b; Tinbergen 1956). Where the male assists in parental care, it may be of equal import-ance to the female to prevent his being unfaithful (e.g. Tinbergen 1939).

Specializations of territorial behaviour which suggest a role in pairing are numerous. The intra-sexual nature of the fighting has already been noted. In some species (e.g. some gulls, Noble & Wurm 1943; Tinbergen 1956; Greenshank, Nethersole-Thomson 1951) part of the territorial beahviour serves no other function, and the area (pairing territory) is abandoned after pair formation (or, alternatively, pairs may be formed only when the male has a territory—Lanyon 1956; Snow 1956). In lek birds, the territory is used only for courtship and/or copulation (references in Armstrong 1947; also Pitelka 1943; Taber 1949; Skutch 1949 b; Lindemann 1951; Andersen 1951; Bancke & Meesenburg 1952; Mildenburger 1953; Höhn 1953).

In Cowbirds *Molothrus afer* territorial defence seems entirely concerned with relations between male and female (Laskey 1950).

Often vigorous territorial defence coincides rather closely with the period when the mate must be protected from the attentions of other males—it often increases when the females arrive or the pair is formed, and decreases when the eggs are laid (e.g. Yellow Wagtail *Motacilla flava*, Smith 1950; Conder 1956; Durango 1956; Gibb 1956 a; Lanyon 1956; Simmons 1956; Snow 1956; see also Tinbergen 1939; Howard 1929) except near the nest and sometimes even there (Mountfort 1956). If the female leaves the territory to forage on a neutral area, the male may continue to protect her (Tinbergen 1939; Marler 1956 b; Snow 1956). On the other hand, those song birds which pair in the flock have only a small territory which is not established until a few days before nest-building (e.g. Pine Siskin *Spinus pinus*, Weaver & West 1943; Cape Canary *Serinus cinicollis*, Skead 1948; Cedar Waxwing *Bombycilla cedrorum*, Putnam 1949, Lea 1942; Eastern Goldfinch, Nickell 1951; Lawrence Goldfinch *Spinus lawrencei*, Linsdale 1950; Swallow Tanager, Schaefer 1953; Hawfinch, Nicholson 1929; Mountfort 1956). But the tits (*Parus* spp.), which often pair before establishing a territory, are an exception here (Gibb 1956 a).

In many species, the self-advertisement which accompanies territorial behaviour is clearly related to pair formation—the volume of song is often reduced, without a parallel fall in aggressiveness, on the arrival of a female (e.g. Howard 1920; Tinbergen 1939; Lack 1939a; Walkinshaw 1944; Durango 1956; Haartman 1956). Further, suitable song posts are often necessary for territorial establishment, and aggression is concentrated round them (p. 346).

Here, however, we must note many exceptions. Song in some *Turdus* spp. does not seem to be linked with pair formation (Lack & Light 1941; Gurr 1954; Snow 1956; Young 1956), even though it is reduced by the presence of the mate; and in the Little Ringed Plover advertisement continues until incubation (Simmons 1956). Unpaired male Wheatears " warble ", but song is not common until after pair formation (Conder 1956). Even where self-advertisement primarily repels rivals, however, it may still be important in safeguarding the mate. But self-advertisement serves diverse social functions and it is difficult to generalise. One species may have several songs (e.g. *Troglodytes* spp., Kendeigh 1941; Armstrong 1955). In the Cliff Swallow, song attracts other individuals and increases aggressiveness, but does not repel (Emlen 1954); while in the Zebra Finch song seems to be entirely sexual and not agonistic (Morris 1954). In the Hawfinch, song seems almost functionless (Mountfort 1956).

In many species, therefore, it is likely that territorial behaviour (i.e. site attachment and hostility) has an important function in the formation and maintenance of the pair bond, though direct evidence is lacking. As usual there are many exceptions, and some species form stable pairs with no territorial behaviour, except that an area round the female may be protected (e.g. Carduelines in early spring).

(c) *Reduction in interference with reproductive activities by other members of the species.* Many authors have suggested that territorial behaviour reduces interference with reproductive activities—especially with courtship, copulation and nest-building (e.g. Armstrong 1956; Andrew 1956; Lanyon 1956; Marler 1956 b; Tinbergen 1956). That such interference may occur is shown by the many examples cited by Armstrong (1947) and in this symposium (see also May 1949; Rowan 1955). In general it is likely that such interference will be reduced by territorial behaviour. Lack (1939, 1940 a) and Marler (1956) have shown that interference in copulation occurs in captivity in species in which it is rare in nature, and Tinbergen (1956) states that it occurs in the Herring Gull " clubs " but not in the larger nesting territories. Lack's (1939 b) observations on the Blackcock *Lyrurus tetrix* indicate that territory reduces interference in copulation in that species. Concentration of copulation attempts near the nest tree or the centre of the territory may be an adaptation to prevent interference (Durango 1956; Haartman 1956; McCartan & Simmons 1956; Mountfort 1956); and it is possible that waning of territorial defence during incubation may be related to the absence of a need to prevent interference in copulation. However, Rowan (1955) records that the territory of the Red-winged Starling *Onychognathus morio* is not defended effectively against flocks, though flock birds may interfere in copulation. Further, Snow (1956) rightly argues that, since we do not yet understand the significance of the interference, to suggest that territorial behaviour prevents it is begging the question.

Interference in nest-building is considered below.

(d) *Defence of nest and nest site.* For convenience, nest territories of species which do not have specialised nesting requirements, as well as those of hole-nesters, cliff-nesters etc., are discussed here.

The initial breeding territories of many song birds are unrelated to the eventual nest site. Often, however, the aggressiveness of the male later becomes organised round the nest site so that even if the female builds outside the original borders, the area round the nest is annexed and defended* (e.g. Willow Warbler *Phylloscopus trochilus*, Brock 1910; Lanyon 1956;

* Nero & Emlen (1951) found that moving the nest of a Red-winged Blackbird across the territory boundary did not result in an extension of the territory.

2 C 2

Marler 1956b; Snow 1956; Young 1956). This may be important, because interference in nest building, stealing of materials etc., certainly occur in some species (references in Armstrong 1947; see also Morris 1954). Defence of the nest site may be very vigorous and is often directed against individuals of other species as well as conspecific ones (D. & L. Lack 1933; Tinbergen 1940; Mildenberger 1943; Rowan 1955; McCartan & Simmons 1956; Marler 1956 b; Snow 1956; Swanberg 1956). In partridges, on the other hand, there is little aggression near the nest (Blank & Ash 1956).

In species with specialized nesting requirements, defence of the nest site is even more conspicuous. In many cliff-nesters, disputes over sites are frequent and birds forced into less favourable ones do not breed so successfully (Tinbergen 1956; see also Fisher & Lockley 1954). Similarly, amongst hole-nesting passerines population density may depend on the availability of sites (e.g. Pied Flycatcher, Haartman 1951; Tits, *Parus* spp; Tinbergen 1949; Kluijver 1951); and disputes over nest sites may be exceptionally vigorous and extend to other species with similar requirements. The aggressiveness is concentrated round the nest site, and falls off with distance (Gibb 1956 a; Haartman 1956)*.

In many species the nest site also plays a part in pair formation (e.g. House Sparrow, Daanje 1941; Bluebird, Thomas 1946; Collared Flycatcher, Löhrl 1951; Purple Martin *Progne subis*, Allen & Nice 1953; Starling *Sturnus vulgaris*, Wallraff 1953; Haartman 1956; Tinbergen 1956) and its defence serves a double purpose. Clearly, in species using scarce sites it is advantageous for females to pair with males who already possess one.

Defence of the nest site is thus an important function of territorial behaviour in many species—though in song birds it is often the specialized territorial behaviour appearing later in the season which is most relevant here. Of course in some species territoriality is quite unrelated to nesting (e.g. male lek birds, references on p. 356 ; humming-bird, Pitelka 1942).

(*e*) *Food*. Both Altum and Howard believed that territory was important in some species in ensuring an adequate supply of food for the young, and they have been supported in this view by many later writers (e.g. Nicholson 1929; Tinbergen 1936, 1939; Nice 1941; Durango 1956; Marler 1956 b). Authors vary, however, in their views about how much food is provided. " Howard and Nicholson claimed that food territories have value because they contain the optimum amount of food, Nice because they contain a surplus, and Tinbergen because they contain only some " (Lack 1943).

*Among *Parus* species, the large territory size of *P. atricapillus* which excavates its nest in soft wood, may be related to the greater scarcity of suitable sites compared to other species which nest in holes. It does not seem to be merely a matter of reduced population pressure, for the method of feeding is specialized in a manner which enables it to patrol a large area (Foster & Godfrey, 1950 pers. obs.). A similar difference between the wrens *Troglodytes troglodytes* and *T. aëdon* is cited by Armstrong (1956)—though here food requirements may also be important.

In a few cases, the evidence is strong.   In some humming-birds the territory is centred round a food source (e.g. *Delphinium* spp.), defended most vigorously during feeding periods and abandoned when the food supply fails. Such territories are unrelated to nests, and though they may have a minor mating function, are concerned primarily with food (Pitelka 1942, 1951). Feeding territories have also been described in the Night-heron *Nycticorax nycticorax* (Lorenz 1938), White Wagtail *Motacilla alba* (Greaves 1941), Red-tailed Hawk (Fitch et al. 1946) and shrikes (Simmons 1951).

In most species, however, it seems very improbable that territory is of food value.   Thus in cliff-nesters and other species whose feeding areas are undefended and more or less remote from their nesting territory, it is certain that territory has no relation to food supply (refs. in Lack 1954 b).   Indeed in many species it is unlikely that territorial behaviour could become specialised to protect a food supply.   Where the young are fed on prey whose daily movement is large relative to the diameter of any possible territory, or whose abundance is governed largely by the weather (e.g. aerial and aquatic plankton, many types of fish, migrant birds) or occurs aggregated in different places at different times (e.g. fish shoals and the food of many omnivorous species) a food territory would be impracticable or useless.   (There are, of course, exceptions to this—e.g. some species whose method of hunting depends on occupying favourable look-out posts etc.—but the principle seems clear).

There remains the relatively small number of species who feed mainly within their territories (i.e. those of Type A).   Although a food value has often been suggested in these cases, the evidence in its favour is circumstantial only, and there are some arguments against it :

(i) Many species nest in colonies, yet obtain adequate food.   (D. & L. Lack 1933; Lack 1948, 1954 b.)   This, however, is no argument against the food value of territory in species of Type A.

(ii) Territories are not defended against other species with apparently similar food requirements (D. & L. Lack 1933 etc.; see also Marler 1956 b; Lanyon 1956).   However, as yet we know little about the food birds give their young.   Detailed study of *Parus* spp. by Betts (1955) shows that even closely related species may take foods more different than is apparent at first sight.

(iii) Many Type A and B territorial species collect food on neutral ground or neighbour's territories (Lack 1948, 1954 b).   Further, feeding intruders are sometimes ignored although singing ones are attacked (e.g. McCown's Longspur *Rhyncophanes mccowni*, Mickey 1942; Robin, Lack 1943; Armstrong 1956; Gibb 1956 a; Marler 1956 b; but not Rock Pipit, Gibb 1956 b); and territorial defence may be reduced when the young are in the nest

(American Redstart *Setophaga ruticilla*, Sturm 1945; Armstrong 1956; Conder 1956; Haartman 1956; Marler 1956 b; Simmons 1956; Snow 1956; Young 1956), though in a few cases it continues almost unabated (Pomatorhine Skua *Stercorarius pomarinus*, Pitelka et al., 1955; Andrew 1956; Armstrong 1956; Durango 1956). The size of the territory may be reduced even when the young are in the nest (Armstrong 1953 a; Odum & Kuenzler 1955; see footnote p. 351). Such observations show that the aggressive behaviour associated with breeding territories is not specialized for the defence of food, and thus suggest that a food value of territory could not be primary (Mayr 1935)*. However, it is still possible that territorial behaviour regulates density early in the season and, since few new birds attempt to settle once nesting has started, territorial laxity later is of no consequence (e.g. Tinbergen 1939; Armstrong 1956).

(iv) Lack claims that, even in species which feed in their territories, there is no positive evidence that the territory contains approximately that quantity of food needed to raise a brood.† In many species the territory seems to contain far more food than is necessary (e.g. Kluijver 1951; Gibb 1956) while in others it contains virtually none.

As stressed on p. 348, if a small change in territory size produces no change in the availability of food, the selective forces which maintain territorial behaviour cannot be related to its food value. On the other hand, the territory may have a food value even if it does not guarantee an adequate supply, so long as it increases the chances of reproductive success. A difficulty here, of course, is to assess availability. Estimates of the amount of food present tell us little about how much a bird could find and collect in a given period; and even measures of what it does collect do not tell us how much it could. Further, in view of the indirect evidence that the amount of food collected is often limited by the time available for collecting (Lack 1954 b), data showing that a bird can only just get enough food do not show that it could get more if it had a larger territory (cf. Gibb 1956 b).

The key question is thus not whether the territory contains approximately the food needed, but whether the available food would be reduced sufficiently to affect ultimate reproductive success if territories were compressed beyond the usual (admittedly imprecise) limit in the habitat in question. The effect could be felt by partial starvation of the young, undue exhaustion of the

* Durango (1956) cites cases where an extension of breeding territory was necessary in order to ensure the food supply.

† Lack (1954 b) rightly emphasizes that a greater breeding density in areas of greater food supply is not evidence that birds can adjust the size of their territories to the food supply; indeed such a claim implies the improbable mechanism of a decrease in persistence of territorial defence with increasing suitability of the habitat. Several authors have commented on the indefensibility of rich areas (e.g. Mocking-birds *Mimus polyglottos*, Michener 1951; Gibb 1956; Young 1956).

parents in their efforts to find food, or even in over-depletion of the food stocks (Tucker 1936). Here there is absolutely no evidence (though it may be noted that evidence of this calibre is not available for any other function of territory either).

Thus a food value of Type A territories is unproven but not disproved. The significant relations between clutch size, breeding density and habitat shown for the Great Tit in the detailed studies of Kluijver (1951) and Kluijver & Tinbergen (1953) probably depend ultimately on the availability of food (Lack 1951, 1954 b; see also 1955). Indeed since habitat selection in many species of Types A and B is related to food potentialities respectively on or near the territory (cf. Conder 1956; Swanberg 1956), and since breeding density is related to the supply of food for the young (Lack, 1954 b) and is regulated by territorial behaviour (p. 351) it remains at least possible that the limit imposed by the mechanism of habitat selection in conjunction with territorial behaviour is related to that which would be imposed by the available food. This is an unsatisfactory conclusion, but the evidence for a better one is lacking.

We may thus conclude that, while territorial behaviour is primarily concerned with food in a few species (e.g. humming-birds, Pitelka 1942), in most cases the food value is not significant. Even in most species of Type A, territorial aggressiveness is not specialized for the immediate defence of food objects, and if the territory has a food value, this is secondary (Mayr 1935). However, it is still possible, though unproven, that in species of Type A territorial behaviour does help to ensure an adequate supply of food for the young.

Finally, the Thick-billed Nutcracker, which uses its territory for storing food brought from elsewhere, requires special mention. The food value of this bird's territory is quite different from the cases discussed above (Swanberg 1956). This use of territory for food storing is certainly secondary.

(*f*) *Reduction of losses to predators.* 'Tinbergen (1952 b) has suggested that over-dispersion of the nests of species in which nests, eggs or incubating female are cryptic may reduce the effects of predation. The frequency of mechanisms for over-dispersion amongst cryptic animals and the other evidence cited by Tinbergen (1956) make it probable that this is a biologically advantageous and significant consequence in many species. This view is supported by Lack (1954 b), Gibb (1956 a, 1956 b), McCartan & Simmons (1956), Marler (1956 b) Simmons (1956), Snow (1956) and Swanberg (1956) although there is little direct evidence (Lanyon 1956). Snow points out that Blackbirds do not defend their territories against Song Thrushes *Turdus ericetorum* although they have similar nests, but Simmons describes inter-specific territorial defence between Little Ringed Plovers and Kentish Plovers *Leucopolius alexandrinus*, which may reduce predation.

Armstrong, on the other hand, gives reasons for doubting its importance :

(i) Some species which have few nest predators hold territories.   This, of course, is a false use of comparison : in such species territories have other functions.

(ii) Some species which do have nest predators do not hold territories.   In at least some of the examples cited, however, the nests are dispersed (e.g. Ruff *Philo-machus pugnax*), though the mechanism by which this is achieved is unknown.

(iii) He rejects some of Tinbergen's arguments, which depend on observations of gulls, because—through human interference—European gulls nest in more vulnerable sites than they would otherwise select.   But Laridae elsewhere nest in similar situations.

(iv) Armstrong suggests that the severe victimisation of the Wren by the Cuckoo *Cuculus canorus* in parts of Germany shows that in this case wide dispersal does not avail.   But the Cuckoo searches for nests in a quite different way from the predators discussed by Tinbergen, and anyway a mechanism cannot be dismissed as in-effective because it is not perfectly effective.

Blank & Ash (1956) believe that spacing of the partridge's nests can be of little use against ground predators since the latter hunt by scent.   However, most diurnal ground predators use sight as well, and in any case there is no reason why the same principles should not apply to scent as to sight.

Thus, although no direct evidence is available, circumstantial evidence suggests that territorial behaviour will reduce predation in species with cryptic eggs, nests or females.   No doubt the significance of this effect varies between species ; and direct evidence of its importance in any one species is not available.

(g) *Reduction of time spent in aggression: reduction of despotism of other males etc.*   Nice (1941) suggests that " the protection afforded against despotism of other males . . . is one of the chief functions of all types of territory."   Arm-strong (1956) advances a similar view.   Though such protection is certainly a result of territorial behaviour, it still leaves unanswered the question why the other males are " aggressive ".

(h) *Prevention of epidemics.*   Tavistock (1931) suggested that over-dispersion may be significant in reducing epidemic diseases.   Similar sug-gestions were made by Nice (1941), Collias (1944), Haldane (1955), Arm-strong (1956), Gibb (1956 a, 1956 b) and Marler (1956 b).

There is no positive evidence in favour of this view.   As yet we know little about the extent to which pathogenic micro-organisms are specific to host species (e.g. Jennings 1954, 1955).   An outbreak of encephalomyelitis amongst hand-reared birds at the Ornithological Field Station, Madingley was lethal to both Blue Tits *P. caeruleus* and Great Tits *P. major* while not affecting other passerines; yet there is no inter-specific territorial defence between these two species except at the nest-hole.   Bacterial diseases are usually even less specific than virus diseases (e.g. *Salmonella* infections). Amongst passerines defending large territories prevention of disease is there-fore unlikely to be a significant effect of territorial behaviour.   Amongst terns, gulls and other colonial species, where juvenile mortality from epidemics is sometimes high (personal observation) the consequences of over-dispersion may be important in this way.

(*i*) *Prevention of in-breeding and cross-breeding: range extension.* Howard suggested that territorial behaviour promoted the distribution of the species and the extension of its range; Armstrong (1947) thinks that it might prevent close in-breeding. Similar suggestions were made by Collias (1944). Haartman (1956) believes that such effects may occur but doubts their importance. Since the distances from their birthplaces at which young birds settle to breed is many times greater than the diameters of their territories (references in Lack 1954 b), it seems unlikely that territorial behaviour alone is responsible for the dispersal of young birds, apart from forcing some individuals into sub-optimal habitats.

(*j*) *Conclusion.* Two points arising from the preceding sections should be emphasized. First, the available evidence on the biological significance of territory differs little in quality from that available to Howard. While detailed quantitative studies are rare, theories about function must remain unproven. It is clear, however, that the function varies enormously between species.

Second, since the territorial behaviour of any species has consequences both harmful and beneficial to the individual, the balance of selective forces is complex and its understanding requires a detailed knowledge of many other aspects of the life history. The nature and availability of nest sites, the method of pair formation, the method of hunting, the type of food and its availability both in the breeding season and in the winter, the habitat, and predation pressure are only a few of the factors which may influence the ways in which territorial fighting is specialized. The characters of behaviour, structure and physiology form a complex web; and a change anywhere may have ramifying consequences by altering the selective forces elsewhere. For example, defence of a large territory selected through its function in pair formation may also reduce disease, even though a small change in territory size will not affect the incidence of disease. Selection for resistance to disease may then be reduced—possibly to a point where territory is selected because of its effect on disease (Goodhart 1955, personal communication). On the positive side, the effects of a change in selection pressure at one point may be permissive (as hole nesting or distasteful flesh may permit a conspicuous female coloration and thus make territorial defence by females more effective), predisposing (as a woodland breeding habitat may predispose to a certain kind of self-advertisment), or necessary (as hole nesting necessitates defence of the nest site if holes are scarce). Because of this complexity, unqualified answers about the function of territory cannot be expected.

SUMMARY.

1. This paper forms an introduction to a symposium on the territorial behaviour of birds.

2. Following Noble (1939), territory is defined as " any defended area ".   A simple classification of breeding territories is given, and the diversity of territorial behaviour emphasized.

3. Territorial behaviour can be analysed into at least two components—restriction of some or all types of behaviour to a particular area, and defence of that area.   These are often accompanied by self-advertisement.

4. The aggressive behaviour which arises in different situations depends primarily on the same mechanisms, though the fighting may be specialized to serve different functions (e.g. to defend different objects).   This justifies the broad definition of territory given above.

5. The fighting involved in territorial defence is associated with tendencies both to attack and to flee from the rival.

6. The nature of the evidence concerning the biological functions of territorial behaviour is considered.   In previous discussions of this subject the term " function " has sometimes been used to refer to any advantageous consequence of the behaviour, and sometimes restricted to consequences through which selection in favour of the behaviour can act.

7. Various possible functions of territory are considered.

(a) Circumstantial evidence suggests that the familiarity with the area which results from site attachment may assist feeding, escape from predators, etc., and may also increase fighting potentiality.

(b) There is strong evidence that territorial behaviour, in addition to producing over-dispersion, can regulate density in favoured habitats.   There is no direct evidence that territory limits the total breeding population in all habitats.

(c) In many species, it facilitates the formation and maintenance of the pair-bond.

(d) It may also reduce interference in various reproductive activities by other members of the species.

(e) Defence of the nest-site is an important consequence of territorial behaviour in many species, and the aggressive behaviour is often clearly specialized to this end.

(f) In a few species territory is primarily concerned with food, but in most the food value of the territory is not significant.   Even in the species which feed on their territories, the territorial behaviour is not specialized for the defence of food objects : in these cases it is possible, but unproven, that the territorial behaviour does help to ensure an adequate supply of food for the young.

(g) In some species the over-dispersion produced by territorial behaviour may reduce predation, though direct evidence that this is the case is not available.

(h) Maintenance of a territory may reduce the despotism of other males, but this does not explain the function of territorial aggressiveness.

(i) Territorial behaviour may reduce disease, but this is unlikely to be a significant consequence except in some colonial species.

(j) It is unlikely that the prevention of inbreeding and the promotion of range extension are significant consequences of territorial behaviour.

8. The functions of territorial behaviour are extremely diverse, and the quality of the evidence available for assessing them is little different from that available to Howard. Since territorial behaviour has consequences both harmful and advantageous to the individual's chances of ultimate reproductive success, and since the inter-relations between the selective forces governing behaviour, structure and physiology are extremely complex, simple answers about the function of territory cannot be expected.

REFERENCES.

ALEXANDER, H. G. 1921. Territory in bird life. Brit. Birds 14 : 271–275.
ALLEN, R. W. & NICE, M. M. 1953. A study of the breeding biology of the Purple Martin (Progne subis). Amer. Midl. Nat. 47 : 607–665.
ALTUM, J. B. T. 1868. Der Vogel und sein Leben. Münster.
ANDERSEN, F. S. 1951. Contribution to the biology of the Ruff (Philomachus pugnax). Dansk. Orn. Foren. Tidskr. 45 : 145–173.
ANDREW, R. J. 1956. Territorial behaviour of the Yellowhammer Emberiza citrinella and Corn Bunting E. calandra. Ibis 98 : 502–505.
ARMSTRONG, E. A. 1947. Bird display and behaviour. London.
ARMSTRONG, E. A. 1953 a. The history, behaviour and breeding biology of the St. Kilda Wren. Auk 70 : 127–150.
ARMSTRONG, E. A. 1953 b. Territory and birds. Discovery, 1953 : 223–224.
ARMSTRONG, E. A. 1955. The Wren. London.
ARMSTRONG, E. A. 1956. The Wren. Ibis 98 : 430–437.
BANCKE, P. & MEESENBURG, H. 1952. A study of the display of the Ruff (Philomachus pugnax). Dansk. Orn. Foren. Tidskr. 46 : 98–109.
BETTS, M. M. 1955. The food of titmice in oak woodland. J. Anim. Ecol. 24 : 282–323.
BLANK, T. H. & ASH, J. S. 1956. The concept of territory in the Partridge Perdix p. perdix. Ibis 98 : 379–389.
BOYD. H. 1953. On encounters between wild White-fronted Geese in winter flocks. Behaviour 5 : 85–129.
BROCK, S. E. 1910. The Willow-wrens of a Lothian wood. Zoologist (4) 14 : 401–417.
BROWN, P. E. & DAVIES, M. G. 1949. Reed Warblers. East Molesey, Surrey.
BURCKHARDT, D. 1944. Möwenbeobachtungen in Basel. Orn. Beob. 41 : 49–76.
CADE, T. J. 1955. An experiment on winter territoriality of the American Kestrel (Falco sparverius). Wilson Bull. 67 : 5–17.
COLLIAS, N. E. 1944. Aggressive behaviour among vertebrate animals. Phys. Zool. 17 : 83–123.
CONDER, P. J. 1949. Individual distance. Ibis 91 : 649–655.
CONDER, P. J. 1956. The territory of the Wheatear Oenanthe oenanthe. Ibis 98 : 453–459.
CRAMP, S. 1947. Notes on territory in the Coot. Brit. Birds 40 : 194–198.
CRAWFORD, M. P. 1939. The social psychology of vertebrates. Psych. Bull. 35 : 407–466.
DAANJE, A. 1941. Über das Verhalten des Haussperlings (Passer d. domesticus). Ardea 30 : 1–42.
DAVIS, D. E. 1940. Social nesting habits of the Smooth-billed Ani. Auk 57 : 179–217.
DAVIS, D. E. 1941. The belligerency of the Kingbird. Wilson Bull. 53 : 157–168.
DAVIS, D. E. 1942. The phylogeny of social nesting habits in the Crotophaginae. Quart. Rev. Biol. 17 : 115–34.
DIESSELHORST, G. 1949. Fruhjahrsbeobachtungen an buntberingten Goldammern. Orn. Ber. 1 : 1–31.
DRUM, M. 1939. Territorial studies on the Eastern Goldfinch. Wilson Bull. 51 : 69–77.
DURANGO, S. 1956. Territory in the Red-backed Shrike. Ibis 98 : 476–484.
EMLEN, J. T. 1954. Territory, nest-building and pair-formation in the Cliff Swallow. Auk 71 : 16–35.
ERRINGTON, P. L. 1934. Vulnerability of Bob-white populations to predation. Ecology 15 : 110.
ERRINGTON, P. L. 1945. Some contributions of a fifteen-year local study of the Northern Bobwhite to a knowledge of population phenomena. Ecol. Monogs. 15 : 1–34.
FISHER, J. 1954. Evolution and Bird Sociality in Huxley, J. S. et al. Evolution as a process. London.
FISHER, J. & LOCKLEY, R. M. 1954. Sea-birds. London
FITCH, F. W. 1950. Life history and ecology of the Scissor-tailed Flycatcher (Muscivora forficata). Auk 67 : 145–168.
FITCH, H. S., SWENSON, F. & TILLOTSON, D. F. 1946. Behaviour and food habits of the Red-tailed Hawk. Condor 48 : 205–237.

FOSTER, J. & GODFREY, C. 1950.   A study of the British Willow-tit.   Brit. Birds 43 : 351–361.

FUGGLES-COUCHMAN, N. R. 1943.   A contribution to the breeding biology of two types of *Euplectes*.   Ibis 85 : 311–326.

GENELLY, R. E. 1955.   Annual cycle in a population of California Quail.   Condor 57 : 263–285.

GIBB, J. A. 1956 a.   Territory in the genus *Parus*.   Ibis 98 : 420–429.

GIBB, J. A. 1956 b.   Food, feeding habits and territory of the Rock Pipit.   Ibis 98 : 506–530.

GLOVER, F. A. 1953.   Nesting ecology of the Pied-billed Grebe in North-western Iowa.   Wilson Bull. 65 : 32–39.

GOODHART, C. B. 1955.   Natural regulation of numbers in human populations. Eugenics Review, 47 : 173–178.

GREAVES, R. H. 1941.   Behaviour of White Wagtails wintering in the Cairo district. Ibis (14) 5 : 459–462.

GULLION, G. W. 1953.   Territorial behaviour of the American Coot.   Condor 55 : 169–186.

GULLION, G. W. 1954.   The reproductive cycle of American Coots in California. Auk 71 : 366–412.

GURR, L. 1954.   A study of the Blackbird (*Turdus merula*) in New Zealand.   Ibis 96 : 225–261.

HAARTMAN, L. von 1951.   Der Trauerfliegenschnäpper.   II. Populations problem. Acta. Zool. Fenn. 67 : 1–60.

HAARTMAN, L. von 1956.   Territory in the Pied Flycatcher.   Ibis 98 : 460–475.

HALDANE, J. B. S. 1955.   [Book review.]   Ibis 97 : 375–377.

HALE, W. G. 1956.   The lack of territory in the Redshank *Tringa totanus*.   Ibis 98 : 398–400.

HAVERSCHMIDT, Fr. 1949.   The Life of the White Stork.   Leiden.

HEDIGER, H. 1942.   Wildtiere in Gefangenschaft.   Basel.

HINDE, R. A. 1952.   The behaviour of the Great Tit (*Parus major*) and some other related species.   Behaviour Suppl. No. 2.

HINDE, R. A. 1953 a.   Appetitive behaviour, consummatory act and the hierarchical organization of behaviour, with special reference to the Great Tit (*Parus major*). Behaviour 5 : 189–224.

HINDE, R. A. 1953 b.   The conflict between drives in the courtship and copulation of the Chaffinch (*Fringilla coelebs*).   Behaviour 5 : 1–31.

HOFFMANN, A. 1949.   Uber die Brutpflege des polyandrischen Wasserfasans *Hydrophasianus chirargus*.   Zool. Jahrb. Abt. Syst. 78 : 367–403.

HOFSTRETTER, F–B. 1952.   Das Verhalten einer Türkentauber-Population.   J. Orn. 93 : 295–312.

HÖHN, E. O. 1949.   Notes on sexual and territorial behaviour in the Coot.   Brit. Birds 42 : 209–210.

HÖHN, E. O. 1953.   Display and mating behaviour of the Black Grouse.   Brit. J. Anim. Behav. 1 : 48–58.

HOWARD, H. E. 1907–14.   The British Warblers.   London.

HOWARD, H. E. 1920.   Territory in Bird Life.   London.

HOWARD, H. E. 1929.   An Introduction to the Study of Bird Behaviour.   Cambridge.

HOWARD, H. E. 1940.   A Waterhen's Worlds.   Cambridge.

HOWARD, W. E. and EMLEN, J. T., 1942.   Inter-covey social relationship in the Valley Quail.   Wilson Bull. 54 : 162–170.

HUXLEY, J. S. 1934.   A natural experiment on the territorial instinct.   Brit. Birds 27 : 270–277.

JACKSON, R. D. 1954.   Territory and pair-formation in the Blackbird.   Brit. Birds 47 : 123–131.

JENKINS, D. W. 1944.   Territory as a result of despotism and social organization in Geese.   Auk 61 : 30–47.

JENNINGS, A. R. 1954.   Diseases in wild birds.   J. Comp. Path. 64 : 356–59.

JENNINGS, A. R. 1955.   Diseases in wild birds.   Bird Study 2 : 69.

JOURDAIN, F. C. R. 1921.   [Review of " Territory in Bird Life ".]   Ibis 11 (3) : 322–324.

JOURDAIN, F. C. R. 1927.   [Review of " How Birds Live ".]   Brit. Birds 21 : 71.

KENDEIGH, S. C. 1941.   Territorial and mating behaviour of the House-wren.   Ill. Biol. Monogr. 18, No. 3 : 1–120.

KENDEIGH, S. C. 1945.   Nesting behaviour of Wood Warblers.   Wilson Bull. 57 : 145–164.

KIRKMAN, F. B. 1937. Bird Behaviour. London.

KIRKMAN, F. B. 1940. The inner territory of the Black-headed Gull. Brit. Birds. 34 : 100–104.

KLUIJVER, H. N. 1951. The population ecology of the Great Tit, *Parus m. major*, L. Ardea 24 : 133–166.

KLUIJVER, H. N., LIGTVOET, J., VAN DER OUWELANT, C. & ZEGWAARD, F. 1940. De levenswijse van den Winterkonig. Limosa 13 : 1–51.

KLUIJVER, H. N. & TINBERGEN L. 1953. Territory and the regulation of density in Titmice. Arch. néerl. Zool. 10 : 265–287.

LACK, D. 1932. Some breeding habits of the European Nightjar. Ibis 13 : 266–284.

LACK, D. 1935. Territory and polygamy in a Bishop-bird *Euplectes hordeacea hordeacea* (Linn.). Ibis (13) 5 : 817–836.

LACK, D. 1939 a. The behaviour of the Robin. Pt. I. The life history with special reference to aggressive behaviour, sexual behaviour and territory. Proc. Zool. Soc. Lond. (A) 109 : 169–178.

LACK, D. 1939 b. The display of the Blackcock. Brit. Birds 32 : 290–303.

LACK, D. 1940. Observations on captive Robins. Brit. Birds 33 : 262–270.

LACK, D. 1943. The Life of the Robin. London.

LACK, D. 1944. Early references to territory in bird life. Condor 46 : 108–111.

LACK, D. 1948. Notes on the ecology of the Robin. Ibis 90 : 252–279.

LACK, D. 1952. Reproductive rate and population density in the Great Tit : Kluijver's study. Ibis 94 : 167–173.

LACK, D. 1954 a. The stability of the Heron population. Brit. Birds 47 : 111–121.

LACK, D. 1954 b. The Natural Regulation of Animal Numbers. Oxford.

LACK, D. 1955. British Tits (*Parus* spp.) in nesting boxes. Ardea 43 : 50–84.

LACK, D. & E. 1952. The breeding behaviour of the Swift. Brit. Birds 45 : 186–215.

LACK, D. & L. 1933. Territory reviewed. Brit. Birds 27 : 179–199.

LACK, D. & LIGHT, W. 1941. Notes on the spring territory of the Blackbird. Brit. Birds 35 : 47–53.

LANYON, W. E. 1956. Territory in the Meadowlarks, genus *Sturnella*. Ibis 98 : 485–489.

LASKEY, A. R. 1950. Cowbird behaviour. Wilson Bull. 62 : 157–174.

LAVEN, H. 1940. Beiträge zur Biologie des Sandregenpfeifers (*Charadrius hiaticula* L.). J. Orn. 88 : 183–288.

LEA, R. B. 1942. A study of the nesting habits of the Cedar Waxwing. Wilson Bull. 54 : 225–237.

LINDEMANN, W. 1951. Uber die Balzerscheinungen und die Fortpflanzungsbiologie beim Kampflaüfers. Z. Tierpsychol. 8 : 210–224.

LINSDALE, J. M. 1950. Observations on the Lawrence Goldfinch. Condor 52 : 255–259.

LÖHRL, H. 1951. Balz und Paarbildung beim Halsbandfliegenschnäpper. J. Orn. 93 : 41–60.

LORENZ, K. 1938. A contribution to the comparative sociology of colonial-nesting birds. Proc. 8th Int. Orn. Congr. Oxford 1934 : 207–218.

MCCARTEN, L. & SIMMONS, K. E. L. 1956. Territory in the Great Crested Grebe *Podiceps cristatus*. Ibis 98 : 370–378.

MACQUEEN, P. M. 1950. Territory and song in the Least Flycatcher. Wilson Bull. 62 : 194–205.

MARLER, P. R. 1955. Studies of fighting in Chaffinches. I. Behaviour in relation to the social hierarchy. Brit. J. Anim. Behav. 3 : 111–117.

MARLER, P. 1956 a. Studies of fighting in Chaffinches. II. Proximity as a cause of aggression. Brit. J. Anim. Behav. 4 : 23–30.

MARLER, P. 1956 b. Territory and individual distance in the Chaffinch. Ibis 98 : 496–501.

MARLER, P. (in press). The behaviour of the Chaffinch, *Fringilla coelebs*. Behaviour Supplement.

MARSHALL, A. J. 1954. Bower-birds. Oxford.

MAY, D. J. 1947. Observations on the territory and breeding behaviour of the Willow Warbler. Brit. Birds 40 : 2–11.

MAY, D. J. 1949. Studies on a community of Willow Warblers. Ibis 91 : 24–54.

MAYR, E. 1935. Bernard Altum and the territory theory. Proc. Linn. Soc. N.Y. Nos. 45, 46 : 24–38.

MASON, A. G. 1947. Territory in the Ringed Plover. Brit. Birds, 40 : 66–70.

MEYLAN, O. 1938. Premiers résultats de l'exploration ornithologique de la Dombes. Alauda 10 : 3–61. [Not consulted.]

MICHENER, J. R. 1951. Territorial behaviour and age composition in a population of Mocking-birds at a feeding station. Condor 53 : 276–283.

MICKEY, F. W. 1942. Breeding habits of the McCown's Longspur. Auk 60 : 181–209.

MILDENBERGER, H. 1943. Zur Brutbiologie des Steinschmätzers. Orn. Mschr. 51 : 6–12.

MILDENBERGER, H. 1953. Zur Fortpflanzungsbiologie des Kampfläufers. J. Orn. 94 : 128–143.

MOREAU, R. E. & W. M. 1938. Comparative breeding ecology of two species of *Euplectes* (Bishop-birds) in Usambara. J. Anim. Ecol. 7 : 314–327.

MORRIS, D. 1954. The reproductive behaviour of the Zebra Finch (*Poephila guttata*). Behaviour 6 : 271–322.

MOUNTFORT, G. 1956. The territorial behaviour of the Hawfinch (*Coccothraustes coccothraustes*). Ibis 98 : 490–495.

MOYNIHAN, M. 1955. Some aspects of the reproductive behaviour of the Black-headed Gull (*Larus ridibundus*). Behaviour Suppl. No. 4.

NERO, R. W. & EMLEN, J. T. 1951. An experimental study of territorial behaviour in breeding Red-winged Blackbirds. Condor 53 : 105–116.

NETHERSOLE-THOMSON, D. 1951. The Greenshank. London.

NICE, M. M. 1939. "Territorial song" and non-territorial behaviour of Goldfinches in Ohio. Wilson Bull. 51 : 123.

NICE, M. M. 1941. The role of territory in bird life. Amer. Midl. Nat. 26 : 441–487.

NICE, M. M. 1943. Studies in the life history of the Song Sparrow. II. Trans. Linn. Soc. N. Y. 6 : 1–328.

NICHOLSON, E. M. 1929. How Birds live. London.

NICKELL, W. P. 1951. Studies of habitats, territories and nests of the Eastern Goldfinch. Auk 68 : 447–470.

NOBLE, G. K. 1939. The role of dominance on the social life of birds. Auk 56 : 263–273.

NOBLE, G. K. & WURM, M. 1943. Social behaviour of the Laughing Gull. Ann. N.Y. Acad. Sci. 45 (5) : 179–220.

ODUM, E. P. & KUENZLER, E. J. 1955. Measurement of territory and home range size in birds. Auk 72 : 128–137.

PEARSON, A. K. & O. P. 1955. Natural history and breeding behaviour of the Tinamou (*Nothoprocta ornata*). Auk 72 : 113–127.

PITELKA, F. A. 1942. Territoriality and related problems in North American humming-birds. Condor 44 : 189–204.

PITELKA, F. A. 1943. Territoriality, display and certain ecological relationships of the American Woodcock. Wilson Bull. 55 : 88–114.

PITELKA, F. A. 1951. Ecologic overlap and interspecific strife in breeding populations of Anna and Allen Humming-birds. Ecology, 32 : 641–661.

PITELKA, F. A., TOMICH, P. Q. & TREICHEL, G. W. 1955. Breeding behaviour of jaegers and owls near Barrow, Alaska. Condor 57 : 3–18.

PUTNAM, L. S. 1949. The life history of the Cedar Waxwing. Wilson Bull. 61 : 141–182.

RAINES, R. J. 1945. Notes on the territory and breeding behaviour of the Blackcap and Garden Warbler. Brit. Birds 38 : 202–204.

RANGER, G. 1949–51. Life of the Crowned Hornbill *Lophoceros suahelicus australis*. Ostrich 20 : 54–65 ; 152–167; 21 : 2–13; 22 : 77–93.

RANKIN, M. N. & RANKIN, D. H. 1940. Additional notes on the roosting habits of the Tree-creeper. Brit. Birds 34 : 56–60.

ROBINSON, A. 1945. The application of "Territory and bird life" to some Australian birds. Emu 45 : 100–108.

ROWAN, M. K. 1955. The breeding biology and behaviour of the Red-winged Starling *Onychognathus morio*. Ibis 97 : 663–705.

SALT, G. & HOLLICK, F. S. J. 1946. Studies of wireworm populations. II. Spatial distribution. J. Exp. Biol. 23 : 1–46.

SCHAEFER, E. 1953. Contribution to the life history of the Swallow Tanager. Auk 70 : 403–460.

SCHMIDT, K. & HANTGE, E. 1954. Studien an einer farbig beringten Population des Braunkehlchens (*Saxicola rubetra*). J. Orn. 95 : 130–174.

SHOEMAKER, H. H. 1939. Effect of testosterone propionate on the behaviour of the female Canary. Proc. Soc. exp. Biol. N.Y. 41 : 299–302.

SIMMONS, K. E. L. 1951. Interspecific territorialism. Ibis 93 : 407–413.

111

SIMMONS, K. E. L. 1953.   Some aspects of the aggressive behaviour of three closely
    related Plovers.   Ibis 95 : 115–127.
SIMMONS, K. E. L. 1954.   The behaviour and general biology of the Graceful Warbler
    Prinia gracilis.   Ibis 96 : 262–292.
SIMMONS, K. E. L. 1956.   Territory in the Little Ringed Plover *Charadrius dubius*.
    Ibis 98 : 390–397.
SKEAD, C. J. 1948.   A study of the Cape Canary *Serinus canicollis*.   Ostrich 19 :
    17–44.
SKEAD, C. J. 1950.   A study of the African Hoopoe.   Ibis 92 : 434–463.
SKUTCH, A. F. 1944.   The life history of the Prong-billed Barbet.   Auk. 61 : 61–88.
SKUTCH, A. F. 1949.   Life history and ecology of the Scissor-tailed Flycatcher
    Muscivora forficata.   Auk 67 : 145–168.
SKUTCH, A. F. 1949.   Life history of the Yellow-thighed Manakin.   Auk 66 : 1–24.
SMITH, S. 1950.   The Yellow Wagtail.   London.
SNOW, D. 1956.   Territory in the Blackbird.   Ibis 98 : 438–447.
SOUTHERN, H. N. 1954.   Tawny Owls and their prey.   Ibis 96 : 384–410.
STEWART, R. E. 1953.   A life history study of the Yellow-throat.   Wilson Bull. 65 :
    99–115.
STOKES, A. W. 1950.   Breeding behaviour of the Goldfinch.   Wilson Bull. 62
    107–127.
STONEHOUSE, B. 1953.   The Emperor Penguin (*Aptenodytes forsteri*) I. Falklands
    Islands Dependency Sci. Rep. No. 6.
STURM, L. 1945.   A study of the nesting activities of the American Redstart.   Auk
    62 : 189–206.
SWANBERG, P. O. 1956.   Territory in the Thick-billed Nutcracker.   Ibis 98 : 412–419.
TABER, R. D. 1949.   Observations on the breeding behaviour of the Ring-necked
    Pheasant.   Condor 51 : 153–195.
TAVISTOCK, 1931.   The food-shortage theory.   Ibis 13 (1) : 351–354.
THOMAS, R. H. 1946.   A study of Eastern Bluebirds in Arkansas.   Wilson Bull. 58 :
    143–183.
THORPE, W. H. 1954.   Some concepts of ethology.   Nature 174 : 101.
TINBERGEN, L. 1940.   Beobachtungen über die Arbeitsteilung des Turmfalken
    während der Fortpflanzungszeit.   Ardea 29 : 63–98.
TINBERGEN, L. 1949.   Bosvogels en Insekten.   Nederlandsch Baschbouw Tijd., 4 :
    91–105.
TINBERGEN, N. 1936.   The function of sexual fighting in birds, and the problem of
    the origin of " territory ".   Bird Banding 7 : 1–8.
TINBERGEN, N. 1939.   Field observations of East Greenland Birds. II. The be-
    haviour of the Snow Bunting (*Plectrophenax nivalis subnivalis* (Brehm)) in
    spring.   Trans. Linn. Soc. N. Y. 5 : 1–94.
TINBERGEN, N. 1951.   The Study of Instinct.   Oxford.
TINBERGEN, N. 1952.   Derived activities; their causation, biological significance,
    origin and emancipation during evolution.   Quart. Rev. Biol. 27 : 1–32.
TINBERGEN, N. 1952 b.   On the significance of territory in the Herring Gull.   Ibis
    94 : 158–159.
TINBERGEN, N. 1953.   The Herring Gull's World.   London.
TINBERGEN, N. 1956.   On the functions of territory in Gulls.   Ibis 98 : 401–411.
TUCKER, B. W. 1936.   Brit. Birds 28 : 247–248.
WAGNER, H. O. 1945.   Notes on the life history of the Mexican Violet-ear.   Wilson
    Bull. 57 : 165–187.
WALKINSHAW, L. H. 1944.   The eastern Chipping Sparrow in Michigan.   Wilson
    Bull. 56 : 193–205.
WALKINSHAW, L. H.. 1953.   Life history of the Prothonotary Warbler.   Wilson Bull.
    65 : 152–168.
WALRAFF, H. G. 1953.   Beobachtungen zur Brutbiologie des Stares (*Sturnus v.
    vulgaris* L.) in Nürnberg.   J. Orn. 94 : 36–67.
WEAVER, R. L. and WEST, F. H. 1943.   Notes on the breeding of the Pine Siskin.
    Auk 60 : 492–504.
WESTON, H. G. 1947.   Breeding behaviour of the Black-headed Grosbeak.   Condor
    49 : 54–73.
WILLIAMS, L. 1952.   Breeding behaviour of the Brewer Blackbird.   Condor 54 :
    3–47.
WILSON, H. 1946.   The Life history of the western Magpie (*Gymnorhina dorsalis*).
    Emu 45 : 233–244, 271–286.
YOUNG, H. 1956.   Territorial activities of the American Robin.   Ibis 98 : 448–452.

III

Does Territory Limit Density?

Editor's Comments on Papers 6 Through 12

One of the most controversial questions concerning territory is the extent to which it limits density. The issue is by no means settled and three theories stand out in importance. Julian Huxley (Paper 6) introduced the "elastic disc" theory. Huxley proposed that an animal gives up most readily the outer borders of its territory but fights with increasing vigor against further encroachment. Presumably, a point is reached where the defender tolerates no further reduction in the size of his territory (i.e., there is some upper limit to compressibility that would eventually limit density).

Huxley's paper has an added interest because he made his observations with Eliot Howard. Huxley is one of the world's intellectual giants. He has been knighted, is a Fellow of the Royal Society, and is recipient of numerous medals, awards, and honorary degrees in recognition of his numerous publications in the fields of science and philosophy. Despite all these honors, he lists as his leisure activities bird watching, reading, and walking.

Kluyver and Tinbergen in Paper 7 present the "buffer mechanism" theory. They postulate that birds attempt to settle in the preferred habitats. Only when this habitat becomes crowded does the surplus spill over into less favorable habitats. The inability of newcomers to settle in the favored places is partly the result of their avoidance of these places and partly the repellent effect of the established territory holders. In essence, Kluyver and Tinbergen believe there is a threshold of crowding. Only when this threshold is reached do birds settle in less favorable habitat. This, they believe, accounts for the relatively stable densities of breeding birds in the most favorable habitats.

Kluyver was, until recently, Director of the Instituut voor Oecologisch Onderzoek (Dutch Institute for Ecological Research) at Arnhem, The Netherlands. His work has dealt largely with the biology and population ecology of the starling and great tit. He has been editor of both *Ardea* and *Limosa* and is an honorary member of the Dutch Ornithological Union.

Lukas Tinbergen died at the age of 39. His death was a severe loss to the study of animal ecology. He was primarily interested in parasite-prey relationships of birds and insects. He developed meticulous techniques for assessing the abundance of prey and the amounts taken by their predators. His concept of the "search image" is one of his major contributions. Tinbergen collected much of the data in this paper near his boyhood home of Hulshorst. At the time of his death he was Professor of Zoology at the University of Groningen, The Netherlands.

David Lack in Paper 8 draws upon 16 years of data on great tits in a woods near Oxford. He says evidence is lacking to prove that territoriality places a limit on the numbers a given habitat will support. Instead, all that a territorial system does is space breeding birds uniformly, not limit them. He cites as evidence a year when the numbers of great tits were 60 percent higher than ever before. But Van den Assem (Paper 24) suggests that this unusual density could be accounted for by simultaneous, rather than sequential, settling of birds upon their territories.

Lack first gained renown for his delightful book *The Life of the Robin*. This was followed by a prolific outpouring of books and papers on natural regulation of animal numbers and ecological adaptations for breeding. Until his death in 1973, he was Director of the Edward Grey Institute, Oxford, and a Fellow of the Royal Society. He was both an iconoclast par excellence of ecological dogma and a synthesizer of ecological principles.

In one of the few theoretical papers on territoriality, Fretwell and Lucas (Paper 9) examine each of these three theories and develop models for testing these in actual field observations. Fretwell and Lucas develop the concept that a bird selects a habitat in which it can maximize its fitness (i.e., its chances of producing maximum numbers of progeny). It was believed earlier that those individuals that bred in suboptimal habitats would necessarily have a lowered fitness. Fretwell and Lucas have shown that this is not necessarily true. The disadvantage of settling in a poorer habitat may be balanced by the advantages of less competition than would exist in the most favorable habitats. This raises doubts about the "buffer mechanism" hypothesis of Kluyver and Tinbergen, which implies that birds will first settle in the most favorable habitats, and only after these have become filled to capacity will individuals spill over into less favorable habitats. Orians (1971) has expanded on these ideas.

In Parts II and III of this paper (Fretwell and Calver, 1969; Fretwell, 1969), Fretwell tests his models with field studies of the dickcissel and field sparrows. Fretwell is a theoretical population ecologist at Kansas State University. He is currently continuing his studies on population regulation through both territoriality during the breeding season and aggression within flocks during the winter. Henry L. Lucas, Jr., is the Reynolds Distinguished Professor of Statistics and Director of the Biomathematics Program at North Carolina State University. His current research is in quantitative and theoretical biology, especially the area of nutrition.

John Zimmerman (1971), supported by many more field observations than Fretwell, was also able to test territorial theories. Zimmerman, also at Kansas State University, concludes, in the following quotation from his paper, that the dickcissel adheres to the elastic disc theory of Huxley:

Territoriality as a Density-Dependent Factor

I contend that territoriality in the Dickcissel has a density-dependent effect on the population size in the following way:

1. A minimum territory size that a male will defend exists. Any increase in density within a local population above that density at which this minimum territory size is reached results in utilization by some males of habitats within the area that provide less suitable vegetation (shorter and/or less dense).

2. Furthermore, as a minimum territory size means that the most suitable habitats will be close to full utilization, additional males coming into an area must settle in habitats that are less suitable (e.g., pastures).

3. As females are responsive to the height and density of the vegetation, particularly as it relates to nesting sites, males defending less suitable habitats in response to high male densities attract fewer mates.

4. This decrease in the sex ratio is reflected in a decrease in the number of active nests at high male densities and suggests that total productivity is also decreased, although direct evidence relating productivity to the density of the male population or the vegetation index of individual territories lacks statistical significance.

The remaining papers in this part contain specific examples where territory does seem to limit density. Stewart and Aldrich (Paper 10) provided the first extensive experimental study on territorial limitations to breeding. After wholesale removal of most of the original territorial males on an area, they found just as many males as at the start. This indicated to them that there had been many surplus birds in the vicinity that filled vacant territories promptly. Hence the territorial system did limit breeding density of males, and indirectly, of females. However, the study may be atypical for unusually large numbers of warblers had been attracted to the area as a result of a spruce budworm outbreak. The reader is referred to a follow-up study by Hensley and Cope (1951) in the same area.

Robert E. Stewart is a Senior Biologist with the U.S. Fish and Wildlife Service, currently at the Northern Prairie Wildlife Research Center, Jamestown, North Dakota, where his primary interest is in habitat and population investigations of waterfowl. John W. Aldrich received his Ph.D. at Western Reserve University and then worked in museums as a curator of birds for over 10 years. For many years he was Chief of the Section of Distribution of Birds and Mammals with the United States Fish and Wildlife Service.

Frank Tompa (Paper 11) studied a stable population of song sparrows that were year-round residents on a coastal island. Each year an excess of young birds left the island; this departure was timed with the onset of territorial behavior in the fall. These birds moved to other islands with less favorable habitats. A second dispersal occurred at the onset of the breeding season, when both male and female yearlings left. Since Tompa was unable to determine whether these surplus birds ever bred on adjacent islands he has not necessarily refuted Lack's theory, but he does support the elastic disc theory.

Tompa is a cosmopolitan biologist who has worked in Hungary, Canada, Scotland, Norway, and Switzerland, where he is currently Professor of Population Ecology and Dynamics at the University of Basel. His current interests are vertebrate population ecology, with particular reference to environmental adaptation and social organization in birds and mammals.

There are two major ways to look at the effects of a territorial system upon breeding density. The first approach is to see if the system does, in fact, place an upper limit upon breeding density, either in a particular habitat (Tompa) or in a larger area encompassing both good and inferior habitats. The latter is exceedingly difficult to establish. The second approach is to ask how increasing breeding density (i.e., more and smaller territories) affects breeding performance. Increasing density could result in poorer reproductive performance through smaller or fewer clutches and lower survival of young. Eventually a point would be reached where reduced reproduction and survival of the young is balanced by higher mortality and the population reaches equilibrium. Robert Carrick, known for his work on Australian and Antarctic birds and seals, showed the latter kind of regulation to be present in the Australian magpie (Paper 12). Only one-fourth of the adult population bred successfully. Other birds took up territories in less favorable areas and either failed to breed at all or bred unsuccessfully.

The magpie is of special interest because these birds occupy a group territory that all members of the group defend. Although there should be some selective advantage in larger group size for defending a territory, Carrick was unable to determine the adverse factors placing an upper limit to group size.

It now seems clear that the question "Does territoriality limit density?" should be restated. It is more critical to ask, "How does an increase in breeding density (i.e., more but smaller territories) affect the intrinsic rate of population change (r)?" If r declines as breeding density rises, there must be some density at which r becomes zero and the population reaches equilibrium. If r reaches zero, territoriality will have limited breeding density. In the Australian magpie, the best habitats produced an annual surplus (i.e., r was positive). But for the entire study area r was zero. Clearly we need more studies like those of Carrick to measure the overall effect of territoriality upon density.

An even more fruitful question is: "What conditions of food resources (both abundance and dependability) and predator pressure make it economically feasible for an individual to defend a territory?" (Brown, 1969). Answers to this question are treated in Parts VI and VII.

At this point the reader may wish to reread Hinde's review in Part II. For a somewhat different treatment and interpretation see the more recent review "Territorial behavior and population regulation in birds" by Jerram L. Brown (1969).

References

Brown, J. L. 1969. Territorial behavior and population regulation in birds. *Wilson Bull.*, **81**: 293–329.

Fretwell, S. D. 1969. On territorial behavior and other factors influencing habitat distribution in birds. III. Breeding success in a local population of field sparrows. *Acta Biotheoret.*, **19**: 45–52.

Hensley, M. M., and J. B. Cope. 1951. Further data on the removal and repopulation of the breeding birds in a spruce–fir forest community. *Auk*, **68**: 483–493.

Orians, G. H. 1971. Ecological aspects of behavior. *In* D. S. Farner and J. R. King (eds.), *Avian Biology*, Vol. 1, 513–546. Academic Press, New York. 586 pp.

Zimmerman, J. L. 1971. The territory and its density dependent effect in *Spiza americana*. *Auk*, **88**: 591–612.

Reprinted by permission of A. D. Peters and Company from *Brit. Birds*, **27**, 270–277 (1934)

(270)

A NATURAL EXPERIMENT ON THE TERRITORIAL INSTINCT.

BY

JULIAN S. HUXLEY

In view of the recent critique of the territory theory by D. and L. Lack (*antea*, p. 179) the following observations, made while I was staying with Mr. Eliot Howard, may be of interest.

Mr. Howard has been studying the behaviour of Coots (*Fulica a. atra*) on two small artificial ponds, separated by a dam, near Hartlebury, Worcestershire. On the lower pond there was, in late January, 1933, a flock of sixteen birds, which finally split up, leaving only four pairs. These divided up the pool into a series of four well-marked territorial stretches. No flock was ever seen on the upper pool, but later in the season there were two or three breeding pairs on it.

The first observation he made this season was on the day of my arrival, December 30th, 1933. There were then thirty-three Coots—twenty-six birds in a flock on the upper pond, and seven birds on the lower, arranged in three pairs occupying territorial stretches, and an unmated bird near the upper end of the water.

On December 31st we both visited the place. The number of Coots had increased to thirty-five. The flock on the upper pool now numbered twenty-eight ; on the lower pool conditions were as before. The stretch of water available to the unmated bird was much less than that of any of the three territories ; when it ventured more than about ten yards from the shore, it was chased back. The other territories were each about sixty yards long, and extended across the whole width of the pool. I will call them 1, 2 and 3 in order from the lower to the upper end.

On January 1st I visited the spot alone. There had been a sharp frost, and the whole of the upper pool and all the lower pool except a part of its upper end, considerably less than the area of a single one of the territories of the previous day, were frozen over. The whole of the Coot population of the two pools—thirty-five—was in this open space on the lower pool. What was most interesting was the fact that one pair of birds only was still behaving in what I may call the territorial manner. It occupied more than half of the open water, and the presumed male (and rarely its mate) spent much of its time chasing intruders out of this region. If they mounted the ice, even within the territorial area, its hostility ceased.

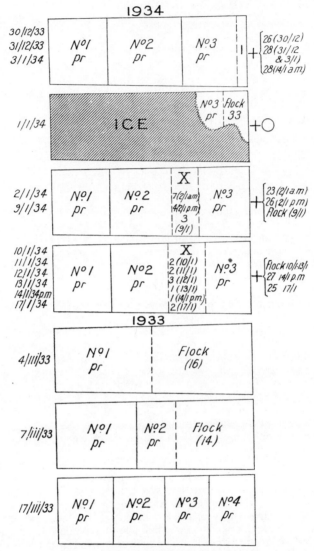

Diagrams showing approximate sizes of Coots' territories in the lower pool described in the text, on different dates in 1933 and 1934. Scale 96 yards to the inch; distances were determined by pacing. The boundaries are of course only approximate. On the left is given the date, or dates; on the right, the number of birds on the second (upper) pool, when noted.

When it was engaged in feeding it allowed a much nearer approach of other birds. Among the other birds no such deliberate hostility was observed ; there was occasional sparring, but this was always over in an instant. The typical territorial aggressive attitude, with lowered head and somewhat arched wings, was never seen among the others, but repeatedly in the male and occasionally in the female of the pugnacious pair. The same state of affairs was seen in the afternoon, save that the open water was slightly larger, and the ice no longer bore the birds' weight.

It is natural to suppose that this pair was the same as had occupied the uppermost (No. 3) of the three territories seen on the two previous days. If so, it patrolled up to the previous upper margin of its territory, but only had about one-third of its previous area of water available. In any case it appears certain that two of the three territory-occupying birds had been forced to leave their territories by reason of the ice, and that on so doing they had lost their " territorial " instinct of combativeness. This agrees with Howard's previous observations on the loss of combative instinct by Lapwings in possession of territory when on their visits to neutral ground occupied by the flock, and by Buntings and Finches in possession of territory when on their visits to neutral feeding grounds ; here, however, we have the additional point of interest that the presence on the neutral ground of birds previously in possession of territory was not voluntary, but mechanically enforced by the presence of ice. Mr. Howard informs me that, so far as he is aware, this is the first case on record of such *mechanically enforced abandonment of territory*. That low temperature was not the cause is shown by the fact that the one pair which was left by the ice in possession of part of its original open-water territory continued to show territorial activities. Territorial activity in Coots must thus be determined partly by internal state, and partly by the external fact of being actually in a staked-out territory.

Mr. Howard has kindly sent me notes on some following days, which are of great interest. By the morning of January 2nd no ice was left. The total number of Coots had increased by one to thirty-six, of which a flock of twenty-three were back on the upper pool ; among these, two brief skirmishes were noted, but no territory behaviour or prolonged pugnacity. On the lower pool three pairs of birds were again in possession of territory, and evincing territorial aggressiveness. Between the uppermost (No. 3) and the central (No. 2) territories, a flock of seven birds was feeding in a narrow zone, less than half the width of a territory. I shall call this area X. They were virtually imprisoned here, the territory-owning males on either side continually rounding them up " like sheep-dogs keeping a flock of sheep in a pen ". One bird in particular kept on trying to break away towards the upper bank, but was always prevented by the male of the uppermost territory. In the afternoon the situation on the lower pool was the same, except that only four birds were left in the " pen ".

On the morning of January 3rd the total was down to thirty-five, and the situation had returned to that of December 31st, for twenty-eight were in a flock (in which no fighting was observed) on the upper pool, and on the lower pool were three pairs in possession of territory, and showing territorial behaviour, plus an unmated bird at the extreme upper edge of the pool, which was never allowed to venture far out without being attacked. It is possible that this was the bird which had repeatedly tried to reach the bank on the previous day. The narrow zone (X), where the small flock had been feeding on January 2nd, was now shared between the upper (No. 3) and the central (No. 2) pair. The territory of No. 2 was somewhat larger than either of the other two.

The fact that three pairs were again occupying territory and showing territorial aggressiveness directly the ice disappeared strongly supports my previous conclusion, that territorial behaviour depends on two separate factors—an internal physiological state, and also an external " field of reference " in the shape of actual presence in the bird's own territory. It may, of course, perfectly well be that in other species the aggressive impulse is stronger and manifests itself, partially or fully, even outside the territory. Something like this appears to hold for the Ruff, in which the males are known to fight while on their spring migration, and not merely when on their " hills ".

The imprisonment of the remnant of the flock between two territorial pairs was presumably the result of the presence of an aggressive pair on the lower side of the upper pair, which forced that pair further up the pool ; but the situation was clearly uncomfortable for the flock, and one of unstable equilibrium, as shown by the steady passing of birds from here to the upper pool. The unmated bird appears to have had some territory behaviour developed, but, presumably owing to its being unmated, its aggressiveness was absent or very slight, and it contented itself with tenaciously clinging to a particular region.

No further observations could be made until the 9th, when three birds were seen in area X, between the territories of No. 2 and No. 3 pairs where the flock had been penned on January 2nd.   On this day it was still narrow, as on the 3rd— about 25 yards in width.   One of these was constantly being attacked by the males of the two adjoining territories.

By the 10th only two birds were in this territory (X) but it had been enlarged to a width of 40 yards, mainly at the expense of the middle territory (No. 2) which had previously been the largest.

On the 11th the situation was similar.   On the 12th there was in addition an extra bird which was at first under the top bank, but was later driven out by No. 3 into area X ; here the male in possession made a hostile gesture, but did not actually attack.   On the 13th only one bird, apparently a male, was in area X, and by the morning of the 14th the area was empty and had been reabsorbed by pairs 2 and 3. The flock on the upper pool numbered twenty-eight, making a total of thirty-four.   However, by the afternoon, it was again occupied by a single bird, and pairs 2 and 3 did not encroach upon the area.   The number on the upper pool had decreased to twenty-seven, leaving the total the same as yesterday. By the 17th a pair, but with territorial aggressiveness only poorly developed, was again in area X.

All this looks as if the territorial impulse in the male of area X was poorly developed, as shown by his leaving the territory (and presumably joining the flock) on the morning of the 13th.   It is perhaps to be presumed, though there is no proof, that he was the original odd bird previously seen under the top bank.   It is of great interest to find that the neighbour males tolerate his presence in area X in spite of his lack of aggressiveness.   This looks as if their previous area had been considerably above the normal area, which is intensely defended, and were therefore highly compressible (see below). Mr. Howard's notes for the previous season confirm this idea.

At the beginning of March, 1933, No. 1 area was territorially occupied by a pair, and this extended half-way up the pool. The rest of the pool was occupied by a flock of sixteen birds. On March 5th a second pair began showing territorial be- haviour, claiming a territory adjacent to that of No. 1 at the upper end.   From the outset·this extended further into the flock area than the original limits of No. 1.   At the lower end No. 2 pair at first only succeeded in occupying a small part of No. 1's territory, and constant fighting took place.

However, No. 1 fought rather half-heartedly, and was gradually driven back until its territory had shrunk by 25-30 yards, after which it vigorously resisted further encroachment. This state was reached on or before March 17th, when No. 1 territory was of about the same size as this season. Meanwhile two other pairs had begun to show territorial activity on March 8th, and on or before March 16th the pool was parcelled out into the four definitive territories which it supported during the breeding-season. The compressibility of the early territories as successive pairs showed the onset of territorial behaviour is well seen. Mr. Howard noted at the time : " There seems to be a minimum size of territory. If a bird owns more than the minimum he yields readily to encroachment ; if he has not the minimum he is a more persistent fighter ".

The behaviour of the Mute Swans (*Cygnus olor*) on this water was also of considerable interest. On the morning of Dec. 31st, 1933, there were eight birds on the lower pool—one family of two adults and four well-grown cygnets showing some brown in their plumage, and another adult pair. All eight were close together on our arrival. Shortly afterwards there was a commotion, and one of the pair was driven up on to the bank by the paterfamilias, and viciously pecked. Later, while out of sight on the upper pond, we again heard a commotion, and, on returning, found that the single pair had left the pool for a spot 50-100 yards away, in a meadow below the dam holding up the lower pool. They were still here in the afternoon of the same day.

On January 1st one only of this pair was in the meadow, standing just below the dam and looking towards the pool (which was out of sight over the dam). Its mate had disappeared. The family were in the open water at the upper end of the lower pool, but soon got out on to the dam between the two pools, from which my approach drove them down to the upper pool, where their weight broke the thin ice. On my returning to the lower pool, the solitary Swan had crossed the dam and was in broken ice close to the lower end. It was still there in the afternoon.

On revisiting the upper pool I found both the adult Swans, notably the male, repeatedly attacking their cygnets by biting their necks.

In the afternoon the open water had increased ; the male Swan, with arched wings, occasionally pursued his offspring, but was not able to get near enough to bite them.

On January 2nd the situation was the same, except that the unmated bird left the neighbourhood for some hours in the middle of the day. It was still the same at 9 a.m. on January 3rd, but at 9.25 the family flew back to the lower pool, and the male immediately attacked the solitary bird, driving it right on to the shore. In the afternoon the family were still on the lower pool, but there was no trace of the solitary bird.

Until the 11th inclusive the family remained on the upper pool. On the 12th the adult pair reverted to the lower pool. One of the young joined them there on the 13th, but was viciously pursued by its father; by the afternoon it had re-joined the rest of the young on the upper pool. The attitude of the male on this occasion was much more violent than on any previous day towards any of his offspring (presumably due to increasing physiological change).

In their paper the Lacks state that "there is no real evidence" that "the pugnacity of the male sets a definite limit to the number of pairs in a given area". In the case of these Swans it would certainly appear that it was doing so. The pugnacity of the male (and to a lesser extent of the female) is clearly seeing to it that one pair of Swans shall grow where two pairs grew before. The pugnacity was mainly directed towards other adults, but in some degree towards the pair's own offspring. Apparently the hostility to the young was elicited by the closer propinquity consequent upon the whole family being driven down into the very small patch of water which they broke in the ice of the upper pool.

Perhaps, however, the real point of the Lacks' statement is in the word *definite*. If so, I think everyone, including Mr. Howard himself, would agree with them. Into the determination of the density of breeding pairs in a territorial species a number of factors enter, including innate strength of territorial instinct, external conditions (temperature, etc.) affecting the strength of the instinct, availability of suitable areas, and number of competing pairs. In the instance of Reed-Buntings, quoted from Howard by the Lacks, a 3-territory area was converted into a 4-territory one by the invasion, late in the season, of a pugnacious new pair. Why not? We might easily imagine that a fifth and even a sixth pair might have succeeded in gaining entrance, but eventually a limit must have come.

The carving out by fresh pairs of territory in an already fully occupied region must continually happen early in the season, as fresh birds become subject to the internal change of state which prompts the acquisition of territory. The Coots here described provide an excellent example.

There is, indeed, a good deal of evidence that the territorial instinct is, to use a physical metaphor, compressible. If there are no neighbouring pairs close to a male in possession of territory, the instinct dies out gradually towards a certain radius from the centre. If other birds arrive the marginal zone is readily given up ; but as the edge of the territory is pushed nearer the centre, the violence of the impulse to defend it increases. As previously noted, there appears to be a minimum size of territory, any encroachment on which is bitterly resisted. Above this size, resistance to encroachment is less whole-hearted, and compressibility therefore greater. The rapid increase of territorial pugnacity as the minimum size is approached is interesting, apparently amounting almost to a discontinuity in type of behaviour.

Territories are thus partially compressible, but their compressibility is not complete. They are like elastic discs, of which there is a lower as well as a higher number which can be placed together to cover a given area. If this view is correct, territorial instinct (*i.e.*, male pugnacity while in possession of a territory) *will* be one of the more important of the factors determining the population of breeding pairs in a given area. Whether it is ever a *final limiting* factor is a theoretical question which it is impossible at the moment to answer. What seems quite clear is that it does, in conjunction with other factors, play a part in determining the actual density of breeding population in those species in which it is manifested.

Mr. Howard, I am glad to say, proposes to continue daily observation on the Coots and Swans of the two pools, so that a full account of their very interesting behaviour will be published in due course.

7

Reprinted from *Arch. Neerl. Zool.*, **10**, 265–274, 278–287 (1953)

TERRITORY AND THE
REGULATION OF DENSITY IN TITMICE

by

H. N. KLUYVER AND L. TINBERGEN

Since the publication of HOWARD's "Territory in Bird Life" the behaviour aspects of territorial practice have been studied by many authors, but we are less well informed about its ecological significance. In particular, the presumed effect of territory upon density of population is still open to discussion.

Though he does not say so in so many words HOWARD (1920, p. 286) clearly assumes that for small passerines, territories have a minimum size beyond which the birds do not allow further crowding. In HOWARD's opinion, males which arrive in an area where the population has already reached this critical level will move around until they find unoccupied ground. If they do not succeed, they will be unable to breed.

On a priori grounds some limiting effect on the increase of local populations certainly is probable. It has been shown that the frequency of territorial quarrels increases with rising density (see e.g. HUXLEY, 1934; N. TINBERGEN, 1939, p. 70). Thus an important condition for a limiting influence is fulfilled, and such effect has been assumed by several authors (e.g. NICE, 1937; KLUYVER, 1951). But this thesis has not been supported by direct evidence, as LACK (1946) stresses.

The present authors have collected some information on this problem during bird census work in Dutch woods. In this paper we will show that density of population in the more attractive habitats is buffered to a certain extent. We will examine the possible explanations of this phenomenon, among which, in our opinion, HOWARD's thesis, is the most satisfactory.

KLUYVER's observations were made under the auspices of the Phytopathological Service, Wageningen. TINBERGEN did most of his work under the "Instituut voor Toegepast-Biologisch Onderzoek in de Natuur", Oosterbeek, and continued it at the Zoological Laboratory of the University of Groningen.

We have to acknowledge valuable help from many sources. Many of the Wageningen observations were made by the late G. WOLDA. During census work in 1946–1951 TINBERGEN received much help from Miss J. C. NIJENHUIS, Miss N. CROIN, A. C. PERDECK, P. SEVENSTER, H. VELDKAMP, N. PROP, B. BENNEMA, P. GLAS, J. H. MOOK and R. SIMON THOMAS. An important series of observations was put at our disposal by L. J. KRAMER, the late A. J. SCHRAVENDIJK, and A. J. VEGTER. P. H. T. HARTLEY, R. E. MOREAU, Dr D. LACK and Dr B. GREENBERG kindly criticized our manuscript and provided linguistic help. Finally we have to express our gratitude to Prof. L. J. SMID who gave us statistical advice.

I. CENSUS DATA

Between 1941 and 1952 TINBERGEN and his collaborators determined densities of Great Tits *Parus major*, Blue Tits *Parus coeruleus* and Coal Tits *Parus ater* in the woods near Hulshorst, Guelders Prov., Netherlands. In a number of characteristic sampling areas territories were counted by mapping singing males during early morning song. The method has been described in detail by L. TINBERGEN (1946).

The Hulshorst district consists largely of almost pure woods of Scotspine *Pinus sylvestris* with a rather poor bird fauna. Among these we studied a number of lots, varying in age between 35 and 70 years. Between these pinewoods are found narrow strips of mixed wood, 100 to 200 yards wide. For the greatest part these are situated on the borders of a small brook. These mixed woods are much more attractive to many species of songbirds than the pinewoods. Nevertheless, they are also of a comparatively poor type. Undergrowth of shrubs and herbs is scarce. The tree layer mainly contains oak *Quercus robur*, beech *Fagus sylvatica*, birch *Betula* spec., Scotspine, fir *Picea excelsa*, Douglas' fir *Pseudotsuga douglasii* and smaller numbers of *Quercus* cf. *rubra*, *Alnus glutinosa* and *Larix* spec. The tallest trees measure ca 60 feet.

In both the pinewoods and the mixed woods nesting sites are present in sufficient numbers. The land owner placed nest boxes in the pinewoods which otherwise would not provide suitable nesting holes for

Fig. 1. Number of breeding territories at Hulshorst. For comparison density figures for Apeldoorn (Apd) have been added, cf. text. 1 ha = 1 hectare = ca. 2.5 acres. NOTE. 1. The area covered in the pinewood counts at Hulshorst slightly shifted from year to year. On the average 100 hectares have been investigated each year for Great Tit and Coal Tit and 80 ha for Blue Tit. Average densities per 10 ha were the following:

	Great Tit	Blue Tit	Coal Tit
Mixed Wood . . .	5.6	6.8	5.0
Pinewood	1.46	0.83	2.18

2. The Apeldoorn counts are expressed in pairs per 100 nestboxes. In some years parts of the wood have not been investigated. On the average 177 boxes have been inspected in each year. The pinewood at Apeldoorn contains a small amount of deciduous trees. Average density of Great Tits is somewhat higher than at Hulshorst.

		MIXED WOOD HULSHORST	PINEWOOD HULSHORST	APD.
		♂♂ ON 25.5 HA	♂♂ ON 100 HA	
GREAT TIT	1941	13½	18	1941 22
	1942	10	0	1942 7
	1943	13½	10	1943 24
	1946	14½	11	1946 24
	1947	14½	19	1947 28
	1948	17	11	1948 21
	1949	14½	23	1949 37
	1950	14	10	1950 22
	1951	15½	23	1951 —
	1952	17	21	1952 —
		♂♂ ON 9.7 HA	♂♂ ON 80 HA	
BLUE TIT	1941	5	3.5	1941 0.8
	1942	6½	5.2	1942 2.6
	1943	6	—	1943 7.6
	1946	9	6.1	1946 6.1
	1947	6	4.7	1947 4.5
	1948	7	10.8	1948 11.6
	1949	7½	11.7	1949 11.8
	1950	6	4.0	1950 4.9
	1951	5½	3.5	1951 —
	1952	8	9.6	1952 —
		♂♂ ON 25.5 HA	♂♂ ON 100 HA	
COAL TIT	1941	14½	34	1941 7.2
	1942	14½	36	1942 10.5
	1943	22	45	1943 16.7
	1946	10	11	1946 7.6
	1947	12½	19	1947 6.0
	1948	13½	19	1948 9.8
	1949	13	16	1949 7.3
	1950	10	13	1950 0.6
	1951	9½	13	1951 —
	1952	7½	12	1952 —

Fig. 1.

Great Tits and Blue Tits. The mixed woods contain a great number of natural tree holes and moreover some nest boxes.

Density figures for mixed wood and pinewood are summarised in Fig. 1. In the first place, density per unit of area in the mixed wood was always much higher than density in the pinewood. This applies to all three species of titmice. Apparently they have a pronounced preference for the mixed wood.

Furthermore, fluctuations from year to year are much smaller in the mixed wood than in the pinewood. This is clearest in the Great Tit. In this species density in the mixed wood fell to 10 in 1942 whereas in all other years it fluctuated only between 13 ½ and 17. The pinewood, on the other hand, had no Great Tits at all in 1942 and in the other years its population varied between 10 and 23. Apparently density in the mixed wood is buffered in some way.

This effect can be demonstrated more conveniently by plotting year for year density in the mixed wood against the accompanying value of density in the pinewood (Fig. 2A). Instead of proportionality between both figures (resulting in a straight regression line cutting the origin of the graph) one finds the points grouped on a line almost parallel with the abscissa. Thus, as density in the pinewood increases, density in the mixed wood remains almost constant. Nevertheless, it is clear that the regression line eventually must reach the point 0–0. So in its complete form it will include a part steeply rising from the origin to point 1942. This part has been extrapolated on Fig. 2A.

Density in the pinewood is an approximate index of total population in the Hulshorst district, as the mixed wood covers only a very small part of this area. Thus a gradual increase in total population hardly raises density in the mixed wood except during the very first step of the process which we only know by extrapolation.

In the Blue Tit the correlation graph for density mixed wood and density pinewood shows a slightly different picture (Fig. 2E). Unfortunately the regression line cannot be drawn very exactly as the point 1946 fits badly. It is clear, however, that it has no horizontal part like in the Great Tit. Nevertheless, density in the mixed wood increases more slowly than density in the pinewood, at least within the range of observations. So we can speak again of a buffer effect. We are inclined to explain the abnormal values for 1946 as a chance effect in sampling, but we cannot prove this.

*   *   *   *   *   *   *

At Wageningen (Guelders Prov.) KLUYVER collected a much longer series of observations on the estate "Oranje Nassau's Oord".

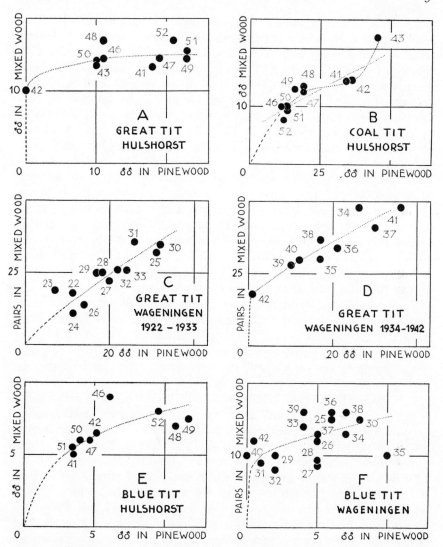

Fig. 2. Relation between numbers of titmice in mixed wood and in pinewood. Data from Fig. 1 and Table I. Regression lines drawn at sight. Dotted: years of observation.

The woods of this estate are poor in natural nest holes. Nearly all Great Tits and many Blue Tits therefore nest in boxes, which form almost optimal nesting sites for these species. Except in the years 1930–1933, boxes were present in excess throughout the estate. The number of first broods in these boxes was recorded year after year and these figures are used here as an index of population (Table I).

131

Great Tit. The Wageningen area comprises mixed wood as well as pinewood, the mixed wood being on the whole younger than that at Hulshorst, but richer in tree species. It has about the same average density of Great Tits. In earlier years, however, its vegetation was much poorer. This is clearly reflected in the census data (table I). Between 1922 and 1933 the density of Great Tits in the mixed wood was almost equal to that in the pinewood, but from 1934 onward it averaged about twice as much. Thus the Great Tits showed a clear preference for the mixed wood during these latter years but not during the earlier.

The change in 1934 was sudden. This has the following explanation. After 1930 the younger mixed plantations (which started as undergrowth in the older wood) had grown so far that they became a suitable habitat for Great Tits. There was, however, a shortage of nesting sites since there were only a few nestboxes. This situation changed in 1934, when more boxes were provided and the tits were enabled to utilise the young plantations fully. The attractiveness of the mixed part of the estate therefore increased abruptly.

The pinewood at Wageningen differs from most of the pinewoods at Hulshorst in having some shrub growth and taller trees. The average density of Great Tits is much higher than the average at Hulshorst. There, the old, well grown pinewoods have an equally dense population, but these are scarce and the average for the district depends mainly on the density of younger and slowly growing woods, which is low. Apparently the type of pinewood which prevails at Wageningen, is more attractive to Great Tits than the lower woods at Hulshorst.

In Fig. 2C and 2D density in the mixed wood at Wageningen is correlated with density in the pinewood. In the years 1922–1933, when mixed wood and pinewood were almost equally attractive to Great Tits, density in the mixed wood was directly proportional to density in the pinewood. This is shown by the regression line which passes through the origin of the graph. But from 1934 onward, when the mixed wood became more attractive than the pinewood, the relation clearly changed. The regression line for this period, when drawn as a straight line, does not cut the origin. Density in the mixed wood changed much more slowly than density in the pinewood. Thus we can speak again of a buffer effect, present since the time when the attractiveness of the mixed wood had increased. It should be noted, however, that the regression line for 1934–1942 has no horizontal part as at Hulshorst, so that the buffer effect was less pronounced at Wageningen than at Hulshorst.

*   *   *   *   *   *   *

TABLE I

Numbers of breeding pairs at Wageningen (Oranje Nassau's Oord) in 60 hectares of mixed wood and in 61 hectares of pinewood.

Year	Great Tit (*Parus major*)		Blue Tit (*Parus coeruleus*)	
	Mixed Wood	Pinewood	Mixed Wood	Pinewood
1922	18	11		
1923	19	7		
1924	11	11		
1925	32	31	15	6
1926	14	14	12	5
1927	22	20	9	5
1928	25	18	9	5
1929	25	17	10	2
1930	35	32	15	8
1931	36	26	9	1
1932	26	22	8	2
1933	26	24	14	4
Av. 1922–1933	24.1	19.4	—	—
1934	48	26	13	7
1935	30	17	10	10
1936	34	21	16	6
1937	41	30	13	5
1938	37	17	16	7
1939	28	10	16	4
1940	30	12	10	0
1941	48	36	no observations	
1942	18	1	12	1
Av. 1934–1942	34.9	18.9	—	—
Av. 1925–1942	—	—	12.2	4.6

II. THE "BUFFER MECHANISM" IN THE GREAT TIT

Both at Hulshorst and at Wageningen densities of Great and Blue Tits were high in attractive mixed woods and low in neighbouring pure stands of Scotspine. Fluctuations from year to year were small in the former habitat and considerable in the latter. As the total population in both mixed woods and pinewoods increased, density in the mixed wood rose much more slowly than density in the pinewoods. This buffer effect was very pronounced in the Great Tit at Hulshorst and in the Blue Tit at Hulshorst and Wageningen. The Great Tit records for Wageningen show it in a less marked but nevertheless convincing form. Here the effect only occurred in the period when the mixed wood was rather attractive for the tits. Finally observations on the Coal Tit in Hulshorst suggest similar relations in this species.

We must now examine the mechanism which buffers density in the mixed woods. This requires knowledge of the life history of the different species. At present, only the Great Tit has been studied in detail. We will therefore first confine our attention to this species.

* * * * * * *

Summarising we find that the distribution of Great Tits over different breeding habitats takes place in two phases. In rough outline it is already determined in autumn when many individuals settle in their final domiciles. In detail, however, it takes place in early spring when the breeding territories are established. Many birds remain faithful to the domiciles which they select in their first autumn of life. Nevertheless displacements at greater age do occur. There is, for example, a regular emigration from the mixed wood during late winter or early spring.

After this introduction, we can deal with our main point: the factors buffering density in the mixed woods.

In principle, density at a certain locality is determined by three main processes: reproduction, mortality and habitat selection (in a broad sense). We will discuss the possible influence of each of these factors in the buffer mechanism.

* * * * * * *

HABITAT SELECTION

In the preceding pages we found no reasons for assuming that either reproduction or mortality produce the density phenomena which we described. There remains only the alternative that the buffer mechanism is a component of habitat selection. We explain the census data for Hulshorst essentially in the following way: the mixed woods are "filled up" to a critical level and not further. The excess birds settle in the pinewoods. In some years, these are many, in others few.

At Wageningen, on the other hand, a fixed level of critical density cannot be distinguished but here as well the percentage of birds which settle in the pinewood increases as total population reaches higher values. A more detailed discussion of the Wageningen data will be given further below. We will first restrict our attention to the Hulshorst data.

One could give the following tentative explanation for the state of affairs at Hulshorst. Some element in the habitat, which every bird requires, is present in limited and constant supply in the mixed wood. Although many hypotheses can be made about the nature of this element, only two possibilities seem reasonable: the number of roosting holes and/or the number of nest sites is limited. Neither is the case, however. As already stated the mixed wood at Hulshorst contains a great number of tree holes (which are suitable for both roosting and nesting) as well as some nest boxes. Moreover, if the number of holes were to limit the population, this influence could be expected to act upon the numbers of Great Tits and Blue Tits taken together, because the latter species uses similar holes as the former. In other words, one would expect that only the combined population of Great *and* Blue Tits would be buffered. The percentage in which each species is represented in this sum would fluctuate in dependence on its general abundance. This clearly is not the case. The number of Great Tits in the mixed wood is almost constant despite big differences in the ratio Great Tits: Blue Tits as calculated for the district as a whole.

Other assumptions, viz. that the conditions of cover or the amount of food limit in a similar way density of the mixed wood, easily can be dismissed. Cover conditions are good throughout the mixed wood. That food supply in this habitat would attract a fixed number of Great Tits in every year seems completely improbable. In the first place the density of food varies very much from year to year. To mention one example: in the mixed wood beech mast is an important food item during the cold season. In some years it is present in enormous excess, in others it hardly can be found. In the second place, the total number of birds which seek a home is rather variable from one year to another. Even if food supply were constant and limited in the mixed wood, a great number of birds would be attracted by it in years of high population and a small number in years of low population.

So we cannot accept that the constant level of population in the mixed wood at Hulshorst is caused by the fact that the supply of things which the birds require is limited and constant in this habitat.In our opinion there is only one reasonable alternative hypothesis namely, that the density of population of the same species is a factor in habitat selection. We assume that birds which seek a home are guided by two counteracting tendencies: a preference for mixed woods as such and an aversion from densely populated localities. Birds which settle early will choose a place in the mixed wood. The more this is filled up, the less attractive it will be for other individuals. In these circumstances, the latter will prefer a habitat which per se is less attractive, but where no dense population is present. Thus as the number of individuals in the district rises, there will be an increasing disposition to settle in the pinewoods.

This hypothesis implies that density in the mixed wood will increase until the attractiveness of the habitat is counterbalanced by the repelling influence of population already present. Therefore, we can expect that density in the mixed wood will rise to a certain critical level, a "level of saturation", and not further.

That in the Great Tit many adult individuals are faithful to the domicile of the preceding year forms no serious objection against our explanation. It only means that those adults which survived already occupy part of the positions when new birds try to settle.

The next question to discuss is, in which season does the repelling influence of population density operate?

We have seen that there are two phases in habitat selection: in autumn many Great Tits settle in their domiciles, and in spring the final breeding territories are established. These are counted in our censuses. We know further that part of the birds, which have settled in autumn, are killed during winter and that others emigrate in early spring.

In principle the density effect could operate in one of the two phases of dispersal or in both. We can deduce, that it works at least in spring, but we have no observations which allow a conclusion about the autumn period.[1]

If the density effect does not operate in spring, we have to assume that it is restricted to the autumn. In winter, however, many birds die. Of course this winter mortality cannot be equal year after year. Thus the number of survivors would fluctuate from one year to another. This is in contradiction with the census data for Hulshorst, which show very small fluctuations. We therefore must conclude that the density effect works at least in spring. This implies that the birds are redistributed to a certain extent during this season.

* * * * * * *

We now must consider whether the critical level or "level of saturation" is the same in all types of habitat. In our hypothesis we supposed that the critical level represents the density value at which, according to the innate standards of the bird, the attractiveness of the habitat is counterbalanced by the repelling influence of population density. Certainly, the attractiveness of different habitats is not equal. Thus we can expect the critical level to be low in unattractive habitats and high in attractive ones. This in fact seems to be true. Parts of the woodland district at Hulshorst have a still denser population of Great Tits than the mixed woods. These are among others the gardens near houses. Apparently this habitat is still more attractive to Great Tits than the the mixed woods. As the latter are saturated in most years, the former must also be filled up to the critical level. This consequently is higher than the critical level in the mixed wood. Apparently these gardens

---

[1] KLUYVER (1951, p. 122–123) brought forward evidence for large scale emigrations of juveniles in autumn. His data strongly suggest, that these emigrations are density-dependent. It is not known, however, whether these displacements influence density in the mixed wood to a greater extent than density in the pinewoods, in other words whether they contribute to the buffer mechanism or not.

are so attractive, that the birds which inhabit them endure more members of their own species than do the birds which live in the mixed woods.

On the other hand, the pinewoods are clearly less attractive to the tits than the mixed woods. Therefore we may expect that here the critical level is lower.

Hence the important question arises, what happens when the pinewoods are also saturated to the critical level. There are no poorer habitats which then may harbour the remaining birds. There are then two alternatives: the birds may attempt to settle in any of the already saturated habitats or they may not settle at all. At present we cannot give a definite answer to this question.

If we were to assume that our explanation could also be applied to the Coal Tit (which is not proven, see below), we have one observation which might throw some light on this problem. In 1943 the population of Coal Tits at Hulshorst was very high. As compared with 1942 both the mixed woods and the pinewoods had a remarkable increase in density. This might indicate that the saturation level in pinewood was passed and that the excess birds settled in all the habitats. If this interpretation is right, population density in the mixed woods is only buffered as long as the pinewoods are undersaturated. Therefore an effect on the total size of population would be doubtful.

* * * * * * *

We have concluded that the density effect in habitat selection operates at least in spring. We must examine now by what elements of behaviour it is produced. It might be caused by aggressive action, by avoiding behaviour or by both. Now both are essential features of territorial practice, which is at a peak in spring. Therefore, we accept as the simplest explanatory hypothesis that the population effect in habitat selection is an immediate consequence of territorial behaviour. The Great Tit is a territorial bird, and in spring the mixed wood is divided completely into territories. The buffer effect implies that the size of these territories does not decrease in proportion with a rise in total population, but more slowly. It seems justifiable to regard this total population as a more or less exact expression of the number of birds which try to settle in any given year. At Hulshorst a marked increase in this number hardly lowered the average size of territories in the mixed wood. At Wageningen a definite minimum size of territory was not reached, but the decrease was relatively small.

When speaking more generally, it is of course not necessary to postulate such a relation with territorial practice. A similar effect could be reached when animals were only mutually hostile (without settling at an individual territory) or when they merely avoided each other.

The above explanation implies that we accept Howard's principle for the populations of Great Tits, but under restricted conditions. In the first place it should be added that the limit of compressibility of territories is not the same in every habitat. Further we conclude that this principle can buffer density in attractive habitats only as long as the population in less attractive environments of the same district has not yet reached the "level of saturation". We do not know what happens when more birds are present.

We should stress here that we realize we have not provided a direct proof for HOWARD's thesis. Such proof would require much more knowledge about behaviour and movements of individual birds than we have at present. We feel, however, that the facts reported here are most easily explained by Howard's hypothesis.

* * * * * * *

IV. DISCUSSION

The census data, presented in this paper, show that density of population in three species of titmice was buffered to a certain extent in densely populated mixed woods and fluctuated much in neighbouring pinewoods, where the population was low. KRÄTZIG (1939, p. 32) has found the same phenomenon. Referring to Great Tits he writes about the favourable habitats: "dass diese Typen offenbar wegen ihrer stärkeren Laubholzdurchmischung als Siedlungsgebiete bevorzugt werden. Es sind dieselben Waldteile, die auch zahlenmässig nur geringe jährliche Siedlungsveränderungen aufweisen..." and he adds that density in the neighbouring poorer habitats clearly sinks in unfavourable years.

Before KRÄTZIG, ERRINGTON (1934, 1943) and ERRINGTON and HAMERSTROM (1936) described similar relations in other species. According to these authors Bobwhite Quails (*Colinus virginianus*) and Muskrats (*Ondatra zibethica*) have a more or less constant density in the most attractive habitats whereas poorer sites are only occupied in peak years. Finally, the observations of SOUTHERN and MORLEY (1950) suggest analogous conditions in the Marsh Tit (*Parus palustris*).

When examining the nature of this buffer mechanism, we have concluded that at least in the Great Tit density of population is an important factor in habitat selection. We assume that the birds have an aversion for densely populated localities. Such a population effect in habitat selection has been suggested in several other cases.

For instance, ERRINGTON (l.c.; 1946) assumes it in his explanation of the regulation of density in Bobwhite Quail and Muskrat. SIIVONEN (1941) supposes the same factor as releaser of mass emigrations in the Waxwing (*Bombycilla garrulus*). Finally CROMBIE (1944) in his experiments with grain boring insects found a pronounced correlation between population density and the intensity of emigration among the larvae of *Rhizopertha dominica* and *Sitotroga cerealella*.

In the Great Tit movements of individuals seem to play an important rôle in the regulation of density. KLUYVER (1951) assumes that the emigrations of juveniles in autumn chiefly radiate from densely populated areas. In this paper we have postulated movements in spring which adjust density in the more attractive habitats. These findings support ERRINGTON's view that displacements and the innate standards of tolerable crowding are important factors in density regulation among higher vertebrates.

We found reasons to suppose that at least in spring the density effect in habitat selection of Great Tits is produced by territorial behaviour. Although for definite proof more observations of behaviour are required, we accept HOWARD's thesis as the most satisfactory explanation. So we assume that more birds settle in the pinewoods as territories in the mixed wood approach the limit of compressibility.

We concluded that this limit is not equal in all kinds of habitats but varies in relation with the attractiveness of the latter. During our observations the limit was never reached in the unattractive pinewoods. So we do not know, whether HOWARD's principle still works when density in the poor habitats has reached the critical level. Probably the latter condition generally is prevented by the above mentioned emigrations in autumn.

As far as concerns these autumn emigrations it is still unknown whether they have relations with territorial practice. In fact, there is a marked rise in hostility during the fall, but it is not certain whether real territories are established. A thorough study of autumn behaviour in ringed birds would be very valuable.

According to our assumption Great Tits are guided during habitat selection by two counteracting tendencies: a preference for certain habitats and an aversion for densely populated localities. The distribution over different habitats therefore is not the effect of habitat preference only, but of both dispositions.[1]

The biological significance of this principle of density regulation becomes clear, when we ask, what would happen if the birds were guided only by their preference for certain habitats and not by their aversion for crowding. The population then would be concentrated almost exclusively in the attractive habitats. When the population

[1] This point seems of importance for the interpretation of the results of bird censuses. It implies that the ratio of densities in different habitats is not a simple expression of the degree to which the birds prefer one to another. We think it probable that this effect is not restricted to the Great Tit, but is found in other species as well.

increased the densities of birds in these attractive habitats would become very great. We know that such concentrations of animals are in general unstable. There would follow an unfavourable trend in the rates of reproduction and mortality, eventually accompanied by an exhaustion of the resources of the habitat. On the other hand the surrounding unattractive habitats would be utilised to a very small extent. In the long run the species in question could only maintain a much smaller population than it does in reality. Thus the interaction of habitat preference and the aversion for concentrations prevents the development of topheavy populations in favourable habitats and ensures the utilisation of less attractive environments. The fact that individuals avoid concentrations probably will favour their survival.

This is a rather vague statement, based only on general ecological principles. But we cannot go further than this. At present we do not know at what densities the populations of the favourable habitats become "unsafe". Neither do we know whether shortages of food or some other factor (e.g. an increased risk for predation or disease) would dominate in such unfavourable development. So the hypothesis of the food value of territory (which is often linked with the thesis that territory limits the population) remains untouched in this study.

In its effect on survival the buffer mechanism has some resemblance to the emigrations of Lemmings and, probably, with similar but less conspicuous phenomena in several other species. The Lemming emigrations lower the density in favourable habitats which are heavily populated. Moreover part of the animals which emigrate find suitable habitats, which they otherwise would not have reached (KALELA, 1949). On the other hand, these emigrations do not start until the density in the favourable habitat has already reached a very dangerous level. Hence this regulation is less effective than that in the Great Tit, where the limit of density in the mixed wood seems rather safe.

V. SUMMARY

1. We have studied densities of the spring populations of Great Tit, Blue Tit and Coal Tit (*Parus major, P. coeruleus,* and *P. ater*) in two woodland districts in Holland. Both districts contained attractive mixed woods (high densities) and unattractive pinewoods (low densities). Fluctuations from year to year were small in the former and considerable in the latter. As the number of tits increased, density in the mixed wood rose much more slowly than density in the pinewoods. Apparently some mechanism buffers density in the mixed woods.

2. The nature of this buffer mechanism is examined in the Great Tit. Reproduction and mortality can be excluded as possible causes;

apparently the mechanism is a component of habitat selection. As a limiting influence of nesting holes, roosting holes and food can be rejected, we assume the following explanation. Great Tits have an aversion for localities which bear a dense population of the same species. The birds seem to prefer mixed woods to pinewoods, but as the mixed woods become more densely populated, excess birds settle in the pinewoods. Thus the attractiveness of the mixed wood per se is counterbalanced by the repelling influence of the population already present. In one case this resulted in the mixed wood being always filled up till a constant level.

3. The limit of tolerable crowding is not the same in all habitats but higher in the more attractive ones. During our observations this limit probably never was reached in the pinewoods.

4. Dispersal of Great Tits over different habitats takes place in autumn and in spring. We conclude that the buffer mechanism works at least in spring. It is unknown whether it acts also in autumn. A partial redistribution of birds during spring must be assumed.

5. Local differences in the census data are discussed. They are most easily explained by the assumption that the buffer mechanism is the more effective, the greater the difference in attractiveness between rich and poor habitat.

6. Innate standards of tolerable crowding and displacements of individuals seem to be important factors in the regulation of density of Great Tits. Besides the principle discussed in this paper, emigration of young birds in autumn (probably density dependent) much influences the number of birds which settle in a given area (KLUYVER, 1951).

7. Great Tits are territorial birds. In spring the mixed wood is completely divided into territories. Since the density effect works during this season, we assume that it is a consequence of territorial behaviour. A complete proof for this assumption cannot be given, but it seems to us the most reasonable hypothesis.

8. The census data for Blue Tit and Coal Tit agree with the above explanation, but it is not possible to exclude alternative hypotheses.

9. The biological significance of the buffer mechanism in the Great Tit is discussed and some comparisons are made.

VI. REFERENCES.

BRIAN, A. D., 1949: Dominance in the Great Tit. Scott. Naturalist, **61**, 144.

CROMBIE, A. C., 1944: On intraspecific and interspecific competition between larvae of graminivorous insects. J. exper. Biol., **20**, 135.

ERRINGTON, P. L., 1934: Vulnerability of Bob-white Populations to Predation. Ecology, **15**, 110.

ERRINGTON P. L., 1943: An Analysis of Mink Predation upon Muskrats in North Central United States. Iowa State Coll. Agr. Exper. Sta. Res. Bull., **320**.

ERRINGTON, P. L., 1946: Predation and Vertebrate Populations. Quart. Rev. Biol., **21**, 144.

ERRINGTON, P. L. & F. N. HAMERSTROM, 1936: The Northern Bob-white's Winter Territory. Iowa State Coll., Agr. Exper. Sta. Res. Bull., **201**.

HOWARD, H. E., 1920: Territory in Bird Life, London.

HUXLEY, J. S., 1934: A natural experiment on the territorial instinct. Brit. Birds, **27**, 270.

KALELA, O., 1949: Über Fjeldlemming-Invasionen und andere irreguläre Tierwanderungen. Ann. Zoöl. Soc. Zoöl. Bot Fenn., **13**, no 5.

KLUYVER, H. N., 1951: The population ecology of the Great Tit. Ardea, **39**, 1.

KRÄTZIG, H., 1939: Untersuchungen zur Siedlungsbiologie waldbewohnender Höhlenbrüter. Ornithol. Abh., **1**, Berlin.

LACK, D., 1946: The life of the robin. 2nd. Ed., London.

PLATTNER, J., 1946: Ergebnisse der Meisen- und Kleiberberingung in der Schweiz (1929–1941). Ornithol. Beob., **43**, 156, and **44**, 1. With postscript by E. SUTTER.

SIIVONEN, L., 1941: Über die Kausal-Zusammenhänge der Wanderungen beim Seidenschwanz. Ann. Zoöl. Soc. Zoöl. Bot. Fenn., **8**, no. 6.

SOUTHERN, H. N. & A. MORLEY, 1950: Marsh tit territories over six years. Brit. Birds, **43**, 33.

TINBERGEN, L., 1946: De Sperwer als roofvijand van zangvogels. Ardea, **34**, 1.

TINBERGEN, N., 1939: The Behavior of the Snow Bunting in Spring. Trans. Linn. Soc. New York, **5**, 1.

8

Reprinted by permission of the publisher from *J. Animal Ecol.*, **33**, (Suppl.), 159–173 (1964)

A LONG-TERM STUDY OF THE GREAT TIT
(*PARUS MAJOR*)

By DAVID LACK

Edward Grey Institute of Field Ornithology, Oxford

INTRODUCTION

This population study of the great tit *Parus major* L. was started in 1947 in the 63 acres (26 ha) of Marley Wood on the Wytham estate near Oxford as a result of a visit to Dr H. N. Kluijver, who informed me of the Dutch work on the same species begun in 1912 and still in progress (Kluijver 1951, etc.). I am extremely grateful to Drs J. A. Gibb, D. F. Owen and C. M. Perrins who carried out the main field work in Marley from 1947 to 1951, 1952 to 1957 and 1958 to 1962 respectively on which this paper is based, also to the other members of the Edward Grey Institute team, to Dr B. C. Campbell who provided census data for the Forest of Dean and to Dr R. F. Scott for a statistical test discussed later, while H. N. Kluijver, R. E. Moreau, C. M. Perrins and G. C. Varley kindly discussed the paper in manuscript.

In western Europe the great tit is primarily adapted to broad-leaved woods, and Marley, a mixed broad-leaved wood, seems highly suited to it, except for a paucity of old trees; but the resulting deficiency of natural holes for nesting was made good by the provision of nesting boxes far in excess of the birds' needs. Virtually all the great tits nested in boxes. The blue tit *P. caeruleus* was also studied, but this species is commonest among oaks, which are sparse in Marley, while many pairs breed in small natural holes, so the number in boxes is not the total population. Perrins (1963) showed that movements in or out of Wytham are negligible in the great tit, but slightly more extensive in the blue tit.

ANNUAL CENSUSES

Annual censuses of the breeding pairs of three tit species are set out in Table 1, both for Marley and for various other woods, the nature and area of each wood being shown in a footnote. Since the numbers are solely of breeding pairs, it should be noted that lines joining the points for consecutive years, as in Figs. 1–3, are misleading in that, between successive points, there has been a big (usually four-fold) increase due to reproduction, followed by a corresponding decrease due to mortality (Gibb 1954a).

Fig. 1 shows that the main, though not all, changes in the great tit population have been synchronous in Marley, the Dean (pure oak 50 miles west), the Breck (pure pine 90 miles north-east) and the Hogh Veluwe (poor mixed woodland with much conifer 300 miles east in Holland). Fig. 2 shows that the main changes in the blue tit have also been synchronous with each other in these woods. Fig. 3 shows that the changes in the great and blue tits have also been mainly synchronous in Marley, and Table 1 shows that this likewise held in the other woods. But the coal tit *Parus ater*, which is adapted to conifers, has in the main fluctuated independently of the other two species (though it increased with the others in 1957, see Table 3, also Lack 1955, 1958).

That the annual fluctuations in breeding pairs should be mainly synchronous in two

A long-term study of the great tit

species over a wide area and in very different types of woods strongly suggests a direct or indirect link with climatic factors. This recalls the account of *Thrips imaginis* in Australia by Andrewartha & Birch (1954), who therefore claimed that density-dependent control could be rejected, whereas Nicholson (1958), Smith (1961), Klomp (1962) and Varley (1963) postulated that, for climatic factors to have acted in this way, some governing (density-dependent) factor must also have played a critical part. Further, Figs. 1 and 2

Table 1. *Number of breeding pairs of tits in different woods*

Year	Great tit (*Parus major*)				Blue tit (*P. caerulerus*)			Coal tit (*P. ater*)	
	Marley	Dean	Breck	Veluwe	Marley	Dean	Breck	Mousehall	Cranwich
1946	–	32	–	34	–	18	–	–	–
1947	7	17	–	44	9	6	–	–	–
1948	21	30	–	46	19	30	–	–	–
1949	30	44	*c.* 50	77	26	61	*c.* 19	15	–
1950	31	37	58	47	34	37	21	25	33
1951	32	31	33	45	34	54	6	16	24
1952	20	22	15	67	17	41	10	14	12
1953	21	17	10	47	14	38	5	11	9
1954	31	27	14	47	18	48	7	25	29
1955	27	17	5	40	13	35	3	7	17
1956	24	25	10	36	15	50	7	13	28
1957	49	75	31	107	32	79	22	20	41
1958	27	28	–	65	17	49	–	–	–
1959	41	36	–	107	20	55	–	–	–
1960	51	20	–	78	25	49	–	–	–
1961	86	45	–	148	44	72	–	–	–
1962	43	21	–	60	21	46	–	–	–
Mean number	34	31	25	64	22	45	11	16	24
Mean density per hectare	1·3	1·4	0·2	0·2	0·9	2·1	0·1	0·3	0·3

The figures give the number of pairs breeding in nesting boxes. In Marley and the Forest of Dean some of the blue tits bred in natural holes, but this applied to a negligible extent in the other censuses.

The area of each wood in hectares is Marley 26, Dean 22, Breck 134, Veluwe 265. Marley is fairly rich mixed broad-leaved woodland with rather few oaks, the Dean is pure oak woodland, the Breck is pure 20–30-year-old conifer, Mousehall is Scots pine, Cranwich is Corsican pine, the Veluwe is poor mixed woodland with much conifer.

show that the synchronous fluctuations have operated around very different levels in different woods, which also requires explanation.

As the Marley study is only in its seventeenth year, there are many points on which it is not yet possible to draw firm conclusions. In the fifteenth year, for instance, the breeding population was 60% higher than in any other, and but for this occasion, which might well have needed many more years for its manifestation, we might have drawn false conclusions as to the greatest number of pairs that the area could hold. Again, either beech-mast or severe weather, or both, might influence the winter mortality, but by chance, in the years so far studied, there was little or no beechmast in most cold winters, while five out of six winters when beechmast was plentiful were mild, so the possible influence of these two factors cannot yet be separated. Because of such considerations, the conclusions in this paper are highly provisional, and statistical tests are not being applied until data are available for further years.

REPRODUCTIVE RATE

The factors determining the reproductive rate of the great tit are here discussed only briefly, as they have been treated in detail before and the earlier conclusions still stand

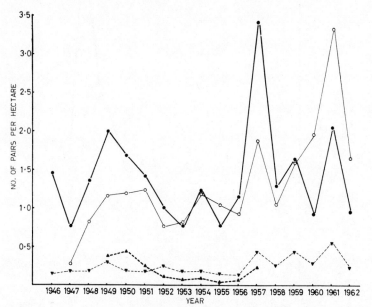

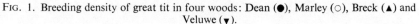

FIG. 1. Breeding density of great tit in four woods: Dean (●), Marley (○), Breck (▲) and Veluwe (▼).

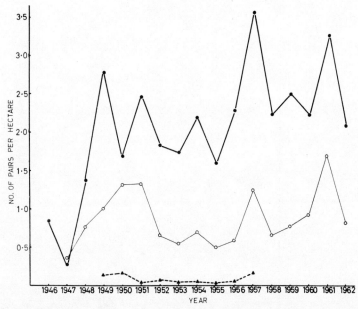

FIG. 2. Breeding density of blue tit in three woods: Dean (●), Marley (○) and Breck (▲).

(Lack 1958 modifying Lack 1955; Lack, Gibb & Owen 1957; Perrins 1963). The basic data are set out in Table 2. The mean clutch in Marley has varied in different years between 7·8 and 12·3 eggs. The two phenotypic variations responsible for this are (a) a

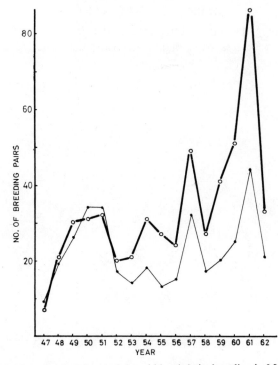

FIG. 3. Number of pairs of great (○) and blue (●) tits breeding in Marley Wood.

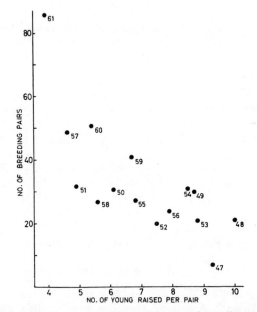

FIG. 4. Production rate of young in relation to density of breeding pairs in great tit in Marley Wood. Each figure beside a point on the graph represents a year, *e.g.* 47=1947.

small reduction in clutch-size at higher population densities, first shown by Kluijver (1951), and (b) a reduction in years when breeding starts later. The significance of (b) is not known, but (a) is presumably adaptive, as it is probably slightly harder, on average, to collect food for the nestlings when the breeding pairs are denser than when they are sparser. It should be added that the annual variations in mean clutch-size are not correlated with the annual variations in the density of the caterpillars which comprise the main food for the young (Lack 1958 correcting Lack 1955; Perrins, in preparation).

Table 2. *Population changes of great tit in Marley Wood*

	(1)	(2)	(3)	(4)	(5)	(6)	(7)	(8)	(9)	(10)
				No. of		Proportion of juveniles		Estimated		
	No. of		Most	young		to each adult		maximum		
	breeding	Mean	productive	leaving	Per cent loss			winter	Beech-	Hard
Year	pairs	clutch	brood-size	nest	summer–spring	Summer	Late autumn	pop.	mast crop	winters
1947	7	11·1	(13)	65	47	4·6	–	–	2	
1948	21	12·3	14, 15	209	76	5·0	(0·9)	–	8	
1949	30	10·1	9	260	81	4·3	0·9	115	1	
1950	31	9·1	9	190	75	3·1	–	–	8	
1951	32	7·8	5	157	82	2·5	0·3	83	0	·
1952	20	9·9	9, 10	150	78	3·8	–	–	3	
1953	21	9·7	11, 12	184	73	4·4	1·3	97	5	+
1954	31	9·7	9	262	83	4·2	0·5	93	0	+
1955	27	10·0	8	183	80	3·4	–	–	0	+
1956	24	10·0	9	189	59	3·9	–	–	6	
1957	49	9·2	8	225	83	2·3	0·2	118	0	
1958	27	9·2	9	150	60	2·8	1·5	135	4	
1959	41	8·8	11	277	72	3·4	1·0	164	0	
1960	51	8·3	11	274	53	2·7	2·0	304	5	
1961	86	8·0	5	365	83	2·1	0·2	206	0	+
1962	43	8·8	14	226	75	2·6	c. 2·6	312	4	+

In 1947–49, 1959, 1960 and 1962 nests in other parts of Wytham were included in the estimates of the mean clutch and the most productive brood-size, but not for other variables. *The most productive brood-size was that which, on average, gave rise to most survivors over 3 months old per brood; the figure for 1947 was based on very few broods, but was certainly higher than usual. The percentage mortality between each summer and the following spring* (column 5) was the percentage reduction between twice each figure in column 1 added to each figure in column 4, as compared with twice each figure in column 1 for the following year, *e.g.* in 1947,

$$\frac{(2 \times 7) + 65 - (2 \times 21)}{(2 \times 7) + 65} \times 100$$

The proportion of juveniles to each adult in summer (column 6) is found by dividing the figures in column 4 by twice those in column 1. Double the figures in column 6 gives the number of young raised to fledging per pair (plotted in Fig. 4). *The juvenile: adult ratio in late autumn* (column 7) was, up to 1957 inclusive, based on the small total of birds previously ringed, but thereafter on the large totals of all birds trapped. The figure was almost certainly too low in 1948, while it is not known why in 1960 and 1962 the ratio should have been higher in the late autumn than the summer, which suggests that juveniles moved into the wood from outside. *The estimated maximum winter population* (column 8) was assessed from the juvenile: adult ratio each winter on the assumption that none of the adults breeding in the previous spring had yet died. The true figures will have been lower, but probably not much lower. *Beechmast* (column 9) is shown on an arbitrary scale in which 0 is crop failure, 2 is poor, 4 is moderate, 6 is good and 8 is abundant, with the odd numbers intermediate in each case. In column 10 *the winters with extended cold weather* are marked with a cross, the winter of 1962–63 being more severe than the others.

SURVIVAL OF YOUNG

In most years in Marley, hardly any young great tits died in the nest (Lack 1955, 1958). But in a few years there was heavy predation through weasels *Mustela nivalis*, and Perrins (1963) showed that this was heavier among the larger and the later broods, evidently because such nestlings were hungrier and called more than those in the smaller and earlier broods respectively. In 1961, when caterpillars were much sparser and breeding great tits were much more numerous than in any other year, many young also starved in the nest, especially among the larger broods, as shown in the footnote to Table 3.

Fig. 4 shows that, on average, rather fewer young great tits were produced per pair when the population was high than when it was low. This is mainly due to the variation of clutch-size with density discussed in the previous section, but it is slightly increased by the factors discussed in the last paragraph. A similar correlation was found by Kluijver (1951; also summary in Lack 1954, p. 68), and it was more marked in Holland because the proportion of second broods (always low in England) was much affected by density.

The influence of breeding density on the production rate was particularly clear in 1961, when the Wytham population rose much higher than ever before. In Marley nesting boxes were still well in excess of the number of pairs, but this did not hold in Wytham

Table 3. *Relation of great tit reproduction to density in* 1961 *at Wytham*

	Marley Wood	Great Wood
Density of breeding pairs per hectare	3·3	1·1
Mean clutch	8·0	7·8
Mean weight of nestling (g)	18·0	19·0
Per cent broods predated	29·2	8·1
Per cent young dying in nest (excluding predation)	16·5	4·9
Per cent fledged young recovered after 3 months	5·2	11·3
Per cent hatched young recovered after 3 months	3·3	9·3
Most productive brood-size	5	10
No. of young recovered per brood after 3 months	0·26	0·66

Figures obtained by C. M. Perrins. The nestling mortality from starvation in Marley Wood was higher in the larger broods, 3% from broods of 2–4, 8% from broods of 5–7, 17% from broods of 8–10 and 24% from broods of 11–13 (broods of 11–13 being artificially enlarged). As the caterpillars disperse aerially, there is every reason to think that they were equally dense in Marley and Great Wood, but their density was not measured in Marley.

Great Wood, where as a result, the breeding density was only one-third of that in Marley. Presumably correlated with this, in Marley as compared with Great Wood, as set out in Table 3, the nestlings on the fifteenth day were a gramme lighter, the predation rate (which is correlated with food shortage as explained) was over three times as high, so was the proportion of young dying of starvation in the nest, while of the fledged young which safely left the nest, proportionately less than half as many were later recovered. The most productive brood-size (discussed later) was only five in Marley but ten (close to average) in Great Wood. Finally the number of young per brood recovered when at least 3 months old was only 0·26 in Marley but 0·66 in Great Wood, *i.e.* each pair raised, on average, two and a half times as many young in Great Wood as in Marley.

While in most years few young great tits died in the nest, they usually did so in large numbers between leaving the nest and the start of the winter. Analysis of ringed birds recovered at least 3 months after they left the nest has shown that, in this period, the young from the smaller broods survived, on average proportionately better than the young from the larger broods (Lack, Gibb & Owen 1957; Lack 1958). This differential mortality was such that, in most years, broods of 9 or 10 gave rise to the largest number of survivors per brood, and as nine is the commonest clutch, this indicates that clutch-size has been evolved through natural selection to correspond with the most productive brood-size. This is not, however, the whole story. In particular, young from late-hatched broods survive less well than those from early-hatched broods of the same size. Further, Perrins (1963) showed that the variations in survival-rate correlated with nestling weight, brood-size and date of hatching respectively differed quite markedly in different years.

In column 3 of Table 2, I have calculated the most productive brood-size in each year. While this has most often been nine, it has in several years been much higher and in two years much lower. These annual variations are presumably due to differences in the feeding conditions in different years, either when the young are in the nest or soon after they leave it. Conditions in the nest are probably of minor importance, since the mean weight of broods of the same size did not vary greatly in most years (though it was lower than usual in Marley in 1961), whereas the chances of subsequent survival of poorly nourished young of the same weight when in the nest varied greatly in different years. Hence most of the annual differences were presumably due to differences in the feeding conditions after the young left the nest, and probably to conditions within a month of their leaving, since any undernourished nestlings presumably either make up the deficit, or succumb to it, within a month of becoming independent.

In the poor mixed woodland with much conifer at Oranje Nassau's Oord, Holland, studied by Kluijver, the commonest clutch, as in England, was nine, but that which gave rise to most survivors per brood was only eight (Lack et al. 1957). This is presumably because the west European great tit is not adapted to conifer plantations.

ANNUAL CHANGES IN BREEDING POPULATION

On the basis of the highest average clutch in any year, namely 12·3 eggs, the greatest possible increase of the great tit population between one year and the next, i.e. without any mortality, would be over 600%. Fig. 3 shows that, in comparison with this, the annual increases have usually been very small; the decreases have also been restricted. This suggests a density-dependent relationship, but Fig. 5 shows that there has been no simple variation with density. It may be particularly noted that the second highest population, in 1960, was followed by an increase which in terms of the number of pairs (as distinct from the percentage change) was the largest recorded. In contrast, the Dutch population studied by Kluijver (1951) appeared to fluctuate strongly with density, normally rising after being low and falling after being high, but as mentioned later, density may not have been the actual cause of this.

Kluijver (1951) considered that in his population the number of young raised to fledging per pair was positively correlated with the population fluctuations at low, though not at high, densities, but as shown later, this apparent correlation disappears when other factors are taken into account. Fig. 6 suggests that the annual variations in the production-rate in Marley Wood have had little if any influence on the population fluctuations. It is true that some years with a high production rate were followed by an increase and others with a low production rate by a decrease, but the low production rates of 1958 and 1960 were followed by fairly large increases, while the high production rate of 1954 was followed by a decrease and that of 1949 by a very small increase.

The number of great tits present in Marley just after breeding is the number of breeding pairs together with the number of young leaving the nest (ignoring any early-fledged young or adults which died before the latest young left). The difference between this

A long-term study of the great tit

summer population and twice the number of breeding pairs in the following year gives the loss between each summer and the following spring and this has been expressed as a percentage of the summer population in column 5 of Table 2. The figures in column 5 seem strongly correlated with the population fluctuations. Thus all four years in which the loss was at most 60% were followed by an increase, while five out of the six years in

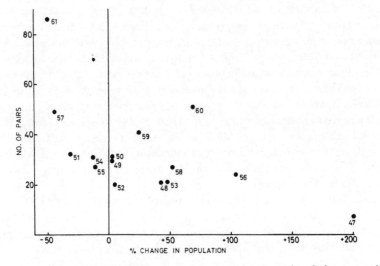

FIG. 5. Change in breeding population between one year and next in relation to previous breeding density in great tit in Marley Wood.

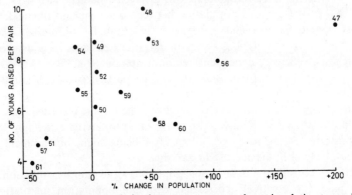
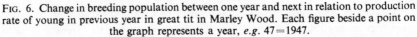

FIG. 6. Change in breeding population between one year and next in relation to production rate of young in previous year in great tit in Marley Wood. Each figure beside a point on the graph represents a year, *e.g.* 47=1947.

which the loss was at least 80% were followed by a decrease and the other by a very small increase. This apparent correlation is, of course, merely another way of stating that the population fluctuations are not correlated with either the initial density of breeding pairs or with the production rate of young, since the subsequent loss is the remaining variable. Since the adult great tits almost never move out of Marley Wood, while the interchange of juveniles between Marley and the rest of Wytham presumably occurs equally in both directions, nearly if not quite all of this loss was presumably due to mortality.

THE RATIO OF JUVENILES TO ADULTS

In 1958 Perrins (1963) found that juvenile great tits can be distinguished from older birds by their greenish instead of greyish primary coverts, and he was able to estimate the juvenile : adult ratio for most earlier winters from the recoveries of ringed birds, though these totals were unfortunately small. He then found that the juvenile : adult ratio each winter, set out in column 7 of Table 2, is strongly correlated with the population fluctuations, as shown in Fig. 7.

Column 6 of Table 2 shows the juvenile : adult ratio in summer at the time when the young leave the nest. Comparison with the ratio in winter shows first that it is much higher in summer, and secondly that the marked annual differences in the winter ratio

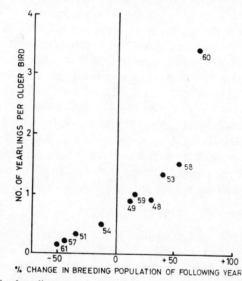

FIG. 7. Change in breeding population between one year and next in relation to juvenile : adult ratio in winter in great tit in Marley Wood.

are little if at all correlated with those in the summer. Between the summer and the early winter, the overall population of great tits decreases markedly (Gibb 1954a), while adults do not enter the wood, hence the marked annual differences in the winter ratio must be due to corresponding marked differences in different years in the rate of loss of the juveniles between the summer and the start of the winter. This indicates that the annual fluctuations in the breeding population are correlated with the loss of juveniles between summer and winter.

As already explained, the young from broods of larger size survived after leaving the nest proportionately much better in some years than others, so that the most productive brood-size differed in different years. It is reasonable to suppose that the survival-rate from the largest broods would be correlated with that of the young as a whole, and comparison between the figures in columns 3 and 7 in Table 2 suggests that this is so. Thus the three years when the most productive brood-size in Marley was smallest (1951, 1957 and 1961) were those when the juvenile : adult ratio in winter was lowest, while three of the five years when the most productive brood-size was largest (1953, 1960 and 1962) were those when the juvenile : adult ratio was highest. However, some years fitted less well, notably 1959 and 1960. (1948 was probably not a genuine exception, since all other evidence strongly suggests that the juveniles survived unusually well in 1948, and as the juvenile : adult ratio was based on a small sample, it may have been erroneous.)

Since the most productive brood-size tends to be correlated with the juvenile : adult ratio in winter, while the latter is correlated with the population fluctuations, the most productive brood-size is also correlated with the population fluctuations. This is not illustrated here, as in the main it reinforces Fig. 7, but it is of value first because the measurement of the most productive brood-size is independent of that of the juvenile : adult ratio, and secondly because data are available for additional years.

Furthermore, as already noted, the most productive brood-size is probably determined by the survival of the young in the first month after they leave the nest, usually in June. Hence the correlation between the most productive brood-size, the juvenile : adult ratio in winter, and the annual fluctuations in the breeding population, indicates that these fluctuations are determined mainly by what happens to the young within a month of their leaving the nest. This view is supported by another point, that the breeding population of the blue tit tends to fluctuate in parallel with that of the great tit, which suggests a common cause. During most of the year these two species are ecologically isolated, but they feed on mainly the same foods at three seasons, with young in the nest in May on caterpillars, with young just out of the nest in June on aphids, and in winter on beech-mast in years when it is available (Hartley 1953; Gibb 1954b; Betts 1955). But as already discussed, the period in the nest in May has at most only a subsidiary influence on the winter ratio of juveniles to adults, while this ratio has been determined prior to the fall of the beechmast. Hence the parallel between the fluctuations of the great and blue tits supports other evidence that June is the critical period for survival.

BEECHMAST

Perrins (in preparation) noticed further that, in those winters with a rich crop of beech-mast in Wytham, tits seemed unusually numerous and the juvenile : adult ratio was unusually high. The Forest Research Station (Alice Holt Lodge) of the Forestry Commission kindly supplied a record of the beechmast crop each year, which is in general similar over a wide area in Britain and continental Europe. When the figures in column 9 of Table 2 are compared with the juvenile : adult ratio in column 7, there is seen to be an obvious correlation. As shown in Table 4, in all four winters with a low ratio there was no beechmast, while in all four with a high ratio there was at least a moderate crop. (A comparison between columns 9 and 3 shows likewise that there is a strong correlation between the beechmast crop and the most productive brood-size, as was to be expected, since the latter is correlated with the juvenile : adult ratio.)

Perrins then found a strong correlation between the beechmast crop and the population fluctuations, as summarized in Table 5, above for Marley between 1947 and 1962, and below for Oranje Nassau's Oord, Holland, between 1912 and 1942. Both great and blue tits feed heavily on beechmast when it is available, which suggests that beechmast would have a direct effect on their winter survival and hence on their population fluctuations. But as already stressed, these fluctuations and the beechmast crop are also correlated both with the juvenile : adult ratio at the start of the winter, before the beechmast falls, and with the most productive brood-size, which evidently relates primarily to feeding conditions in June. Presumably, therefore, the correlation is primarily due to certain conditions which affect both the production of beechmast and the survival of juvenile great tits in summer. The formation of beechmast is partly determined by the weather 18 months earlier (Matthews 1955), but it is also influenced by various factors in the summer in question, one or more of which might well influence the insects, including aphids, eaten by young tits shortly after they leave the nest, but this problem has not yet been studied further.

Table 4. *Relation of beechmast to juvenile : adult ratio in great tit at Wytham*

(Number of years in each category)

Beechmast	No. of juveniles per adult in late autumn		
	0·2–0·5	0·9–1·0	1·3 +
None	4	1	0
Poor	0	1	0
Moderate or larger	0	(1)	4

Table 5. *Relation of beechmast to changes in great tit population*

(Number of years in each category)

Beechmast	Population change		
	Decrease	Negligible	Increase
MARLEY 1947–61			
None	5	0	1
Poor	0	2	1
Moderate or larger	0	1	5
ORANJE NASSAU'S OORD 1912–42			
None	6	1	0
Poor	6	3	4
Moderate or larger	1	1	9

Based on an analysis by C. M. Perrins. The decrease was of at least 7%, the increase of at least 20%, and the negligible change varied between a decrease of 2% and an increase of 5%. The Marley figures are from Table 2, the O.N.O. populations are from Kluijver (1951, Table 48, p. 110) and the beechmast figures were supplied by the Forestry Commission.

WINTER MORTALITY

No actual counts are available of the population each year at the start of the winter, but the maximum possible population at that time can be estimated from the juvenile : adult ratio on the assumption that all the breeding adults of the previous summer were still alive. The figures calculated in this way are shown in column 8 of Table 2 (that for 1948 being excluded because, as already discussed, the juvenile : adult ratio for that winter was almost certainly too low). Should any adults have died between the summer and winter, the number of adults and juveniles present will, of course, have been lower than shown, but since the main mortality in this period is of juveniles, the figures in column 8 are probably only a little higher than the true figures.

If these figures are compared with those for the subsequent breeding populations, it is seen that a high population at the start of the winter was usually followed by a high breeding population and a low population at the start of the winter by a low breeding population. The reduction between each estimated winter population and the subsequent breeding population varied between 36% (in 1953–54) and 58% (in 1961–62); Kluijver's (1951) figures for mortality based on ringed individuals at Oranje Nassau's Oord, Holland, in nearly the same period (December–May) were very similar. These estimated mortality rates for the winter and early spring in Marley do not appear to have been density-dependent, or to have been correlated with the fluctuations in the breeding population, or to have been affected by the beechmast crop. Nor did they appear to be related to the coldness of the winter, since one of the three winters with mortality rates of less than 40% (1953–54, 1958–59 and 1959–60) was cold, while two of the four winters with mortality rates of over 50% (1951–52, 1957–58, 1961–62 and 1962–63) were mild. However, the estimated mortality rate was as high as 75% in the cold winter of 1962–63, which was far colder than any other, and the only one of the Continental type. It should be emphasized that these conclusions are based on estimated, not actual, winter populations. But so far as they go, they support the view that the population fluctuations in Marley are due primarily to the loss of juveniles before the winter, and though one would have expected the subsequent winter mortality to have seriously modified this relationship at times, there is no evidence that it did so in the years under review.

At Oranje Nassau's Oord, Holland, Kluijver (1951) found a marked decrease after some of the abnormally cold winters, but not after others, and established a low but significant correlation between winter cold and the population fluctuations, which was higher when restricted to the years of high density. In Marley, as can be seen from Table 2, the population usually declined after a hard winter, but as already mentioned, the cold winters were also those with little or no beechmast, and since both great and blue tits feed extensively on beechmast when it is present, it seems likely that its presence would reduce their winter mortality, though there is as yet no evidence for this.

To separate these interrelated variables, further years of observation are needed, and an analysis for Marley Wood in terms of 'key factors' (Morris 1959; Varley & Gradwell 1960) is being postponed until data for further years are available. A re-analysis by R. F. Scott of Kluijver's (1951) data for Oranje Nassau's Oord 1912–42 showed a significant correlation with beechmast regardless of other variables, while a correlation with winter cold was probably significant but was so interrelated with the population density as such that the two could not be separated; there was no significant correlation with the production rate of young.

AVERAGE DENSITY AND FOOD SUPPLY

While the population fluctuations in Marley Wood seem correlated chiefly with the juvenile mortality in summer, this leaves unexplained the general level around which they occur, a level which, as shown in Table 1, has differed markedly in different woods. The four potentially density-dependent factors are predation, disease, food supply and the birds' own territorial or other behaviour. Of these, the first two can be virtually excluded, since there is very little predation on the tits after they leave the nest and no evidence that they die in appreciable numbers from disease, though this is hard to exclude with certainty. This suggests that the food supply might be critical, and there is no evidence pointing against this. Further, both the great tit and its food supplies seem to be much sparser in conifer plantations than in Marley Wood. That, however, is as far as the argument can yet be taken, for Marley includes such varied vegetation, and the great tit feeds so diversely, that it has not been possible to measure there either the density of its foods or the bird's impact upon them.

For these reasons, the study of food in relation to population density was transferred to the coal tit in the much simpler and more uniform habitat of the Breckland pine plantations. Gibb (1960) found that in the coal tit, as in the great tit, the breeding population each spring was not correlated with the production rate of young in the previous summer. Nor is it correlated with the previous density of breeding adults. He found further that, unlike what we now know happens in the great tit, the breeding population each spring was not correlated with survival between the previous summer and September. This difference between the two species may well be due to the fact that, in late summer, caterpillars are very sparse in broad-leaved woods but comparatively plentiful in conifers.

Gibb found, however, that the breeding population of the coal tit each spring was correlated with the losses occurring between the preceding October and March inclusive. This mortality, in turn, was correlated with the amount of insect food available in winter in the pines, a substantial proportion of which was removed by the birds in the course of the winter. The relationship between the percentage reduction in numbers of the coal

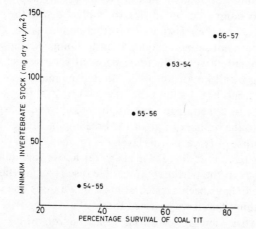

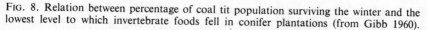

FIG. 8. Relation between percentage of coal tit population surviving the winter and the lowest level to which invertebrate foods fell in conifer plantations (from Gibb 1960).

tit between October and April and the lowest level to which their available foods were reduced is shown in Fig. 8. Hence the evidence points to the coal tit being critically limited in numbers by the density of its winter food, and to the density of its winter food being critically influenced by the birds' feeding operations, so that the relationship is density-dependent, while no evidence points against this view. As, however, the study was continued for only four seasons, and as there was also a correlation between the availability of winter food and the mean temperature in winter, data for further years would have been desirable, particularly as this seems to be the first case in which any population of wild birds has been shown to be up against the food limit outside the breeding season.

TERRITORIAL BEHAVIOUR

The ultimate density-dependent control of the great tit population may likewise prove to be the food supply at some critical period or periods. But even if this is so, there is a possible complication, that the bird may adjust its own density by territorial behaviour in relation to the food situation. That territorial behaviour results in a local spacing of the available breeding pairs can hardly be doubted. The critical questions are first whether such behaviour limits the number of available pairs, and secondly whether this limit is related to the food resources.

In Holland, Kluijver & Tinbergen (1953) found that the breeding density of great tits was both much higher and much more stable in strips of broad-leaved woodland than in poor conifers near at hand, and claimed that this was due to territorial limitation in the broad-leaved areas, a conclusion widely accepted (*e.g.* Gibb 1956; Wynne-Edwards 1962). However, the fluctuations of the great tit in poor conifers on the Breckland were of similar extent to, not smaller than, those in favoured broad-leaved woods in England (Lack 1958), while within Marley Wood, Perrins (in preparation) found stronger fluctuations in the more favoured areas with canopy trees than in the less favoured areas with fewer canopy trees. I think, therefore, that judgment should be suspended for the present on Kluijver & Tinbergen's claim.

Different workers differ in respect to how territorial behaviour limits population density, but the prevailing view today is that of Huxley (1934) (supported, for example, by Tinbergen 1957), who compared territories to rubber discs; the more they are compressed, the greater becomes the resistance to further compression. On this view, though territorial behaviour does not set an absolute limit to numbers, one would expect that at high densities the population would fluctuate between very restricted limits. Further, if territorial behaviour is at all important, there should be many years in which numbers are near to this upper limit. Fig. 1 shows that the picture in Marley and other woods has been very different. Until 2 years ago, one might possibly have argued that territorial behaviour set a limit to the number of great tits breeding in Marley at around fifty pairs, even though, in the course of 14 years, this level was reached only twice, in 1957 and 1960. But in 1961 eighty-six pairs bred there. This is so far above the number in any previous year that, if the analogy of the partly compressible disc is valid, effective pressure from territorial behaviour can have been exerted at most only in this one year; and there is, of course, no reason to think that it was exerted then.

There is an alternative view of territorial behaviour on which, it is claimed, birds modify the size of their territories in accordance with the food situation. On this view, one would expect the breeding density of the great tit in Marley each year to be closely correlated with the density of the caterpillars on which the young are fed. No such correlation is found (see data in Lack 1958 for earlier years, Perrins (in preparation) for later years). On this view, in particular, it should have followed that in 1961 the great tits defended much smaller territories than usual (or even that territorial behaviour temporarily broke down) because of an unusually favourable food situation. But in fact defoliating caterpillars were sparser in 1961 than in any other summer of this study, while the average number of young raised per pair was also the lowest yet recorded (see column 6 in Table 2). In Wytham Great Wood that year, as already mentioned, the density of breeding pairs was much lower than in Marley owing to a shortage of nesting boxes, and the figures in Table 3 strongly suggest that, if the birds breeding in Marley had been able to restrict their density to that in Great Wood, each pair would have raised about two and a half times as many surviving young. Hence territorial limitation would have been highly advantageous for the reason claimed for it by its advocates, but in fact it did not operate.

Summing up, the evidence indicates that, whatever view of territorial behaviour is adopted, it has played no effective part in the regulation of the number of pairs of great tits breeding in Marley. As predation and disease also seem ruled out, the critical factor is presumably the food supply acting directly, in the way that it evidently does on the coal tit in the Breckland pine plantations, but this cannot yet be established for the great tit in Marley.

SUMMARY

1. The breeding population of the great tit in Marley Wood between 1947 and 1962 fluctuated between seven and eighty-six pairs. The fluctuations were mainly synchronous with those of the blue tit, and also with those of the great tit in other woods in England and Holland.

2. The commonest clutch was nine, and in most years the brood-size which on average resulted in most surviving juveniles was 9–10, though it varied in different years between five and fourteen.

3. Clutch-size and the production-rate of nestlings were slightly lower at higher breeding densities. They had no apparent influence on the fluctuations in the breeding population.

4. The critical factor determining the population fluctuations was the mortality of the juveniles between leaving the nest and the beginning of winter, probably mainly in June. This mortality was correlated with the ensuing beechmast crop, presumably because both were affected by certain conditions in summer.

5. Predation and disease seem to have been unimportant, and the evidence suggests that the breeding density was not limited by territorial behaviour. Hence the chief density-dependent factor was probably food shortage, but this could not be studied in the great tit in Marley; the coal tit was probably limited by its winter food in conifer plantations.

REFERENCES

Andrewartha, H. G. & Birch, L. C. (1954). *The Distribution and Abundance of Animals.* Chicago.

Betts, M. M. (1955). The food of titmice in oak woodland. *J. Anim. Ecol.* **24**, 282–323.

Gibb, J. (1954a). Population changes of titmice, 1947–1951. *Bird Study,* **1**, 40–8.

Gibb, J. (1954b). Feeding ecology of tits, with notes on treecreeper and goldcrest. *Ibis,* **96**, 513–43.

Gibb, J. (1956). Territory in the genus *Parus. Ibis,* **98**, 420–9.

Gibb, J. (1960). Populations of tits and goldcrests and their food supply in pine plantations. *Ibis,* **102**, 163–208.

Hartley, P. H. T. (1953). An ecological study of the feeding habits of the English titmice. *J. Anim. Ecol.* **22**, 261–88.

Huxley, J. S. (1934). A natural experiment on the territorial instinct. *Brit. Birds,* **27**, 270–7.

Klomp, H. (1962). The influence of climate and weather on the mean density level, the fluctuations and the regulation of animal populations. *Arch. néerl. Zool.* **15**, 68–109.

Kluijver, H. N. (1951). The population ecology of the great tit, *Parus m. major* L. *Ardea,* **39**, 1–135.

Kluijver, H. N. & Tinbergen, L. (1953). Territory and the regulation of density in titmice. *Arch. néerl. Zool.* **10**, 265–89.

Lack, D. (1954). *The Natural Regulation of Animal Numbers.* Oxford.

Lack, D. (1955). British tits (*Parus* spp.) in nesting boxes. *Ardea,* **43**, 50–84.

Lack, D. (1958). A quantitative breeding study of British tits. *Ardea,* **46**, 91–124.

Lack, D., Gibb, J. & Owen, D. F. (1957). Survival in relation to brood-size in tits. *Proc. Zool. Soc. Lond.* **128**, 313–26.

Matthews, J. D. (1955). The influence of weather on the frequency of beechmast years in England. *Forestry,* **28**, 107–16.

Morris, R. F. (1959). Single-factor analysis in population dynamics. *Ecology,* **40**, 580–8.

Nicholson, A. J. (1958). Dynamics of insect populations. *Annu. Rev. Ent.* **3**, 107–36.

Perrins, C. M. (1963). Survival in the great tit *Parus major. Proc. XIVth Int. Orn. Congr.* (In press).

Smith, F. E. (1961). Density dependence in the Australian *Thrips. Ecology,* **42**, 403–7.

Tinbergen, N. (1957). The functions of territory. *Bird Study,* **4**, 14–27.

Varley, G. C. (1963). The interpretation of insect population changes. *Proc. R. Ent. Soc. Lond.* (C), **27**, (In press).

Varley, G. C. & Gradwell, G. R. (1960). Key factors in population study. *J. Anim. Ecol.* **29**, 399–401.

Wynne-Edwards, V. C. (1962). *Animal Dispersion in Relation to Social Behaviour.* Edinburgh.

M J.E.

9

Reprinted from *Acta Biotheoret.*, **19**, 16–36 (1969)

ON TERRITORIAL BEHAVIOR AND OTHER FACTORS INFLUENCING HABITAT DISTRIBUTION IN BIRDS

I. THEORETICAL DEVELOPMENT [1])

by

STEPHEN DEWITT FRETWELL [2])
and
HENRY L. LUCAS, JR.

(Biomathematics Program, Department of Experimental Statistics, North Carolina State University, Raleigh, N.C., 27607, U.S.A.)

(Received 17.X.1968)

TABLE OF CONTENTS

I. GENERAL INTRODUCTION

HOWARD (1920) suggested several roles for territorial behavior in birds, some of which involve the dispersal of the species over available habitats. Since this classic work, there have been a large number of sometimes conflicting statements regarding the relationship of territorial behavior to the habitat distribution of a species (*e.g.*, STEWART & ALDRICH, 1951, KLUYVER & TINBERGEN, 1953, TINBERGEN, 1957, and WYNNE-EDWARDS, 1962, have all argued that territorial behavior is involved in the dispersal of birds; LACK,

---

1) Work done while senior author was on fellowship support from Public Health Service Grant no. GM-678 from the National Institute of General Medical Sciences.
2) Present address: Department of Biology, Princeton University, Princeton, New Jersey, 08540, U.S.A.

1954, 1964 and JOHNSTON, 1961, have rejected this view, arguing instead that the behavior only isolates individuals or pairs). LACK in particular has dealt with this problem and recently (1966, p. 136) has emphasized the absence of a clear understanding of it. The purpose of this paper is to present an interpretation of the different hypotheses considered by different authors, with the aim of clarifying and resolving the different points of view. The re-defined hypotheses will yield predictions which can be compared with field observations in order to ascertain the probable role of territorial behavior in determining the distribution of a given species.

This study will be presented in three papers. The first will provide a theoretical statement of the general problem of habitat distribution. Hypotheses of the role of territorial behavior will be defined within the context of this theory of habitat distribution, and predictions derived from each. The problem of testing these predictions will be discussed in general. The second and third papers of the series will provide sample studies which demonstrate the application of the theory.

II. INTRODUCTION TO THEORETICAL DEVELOPMENT

In offering different hypotheses about the role of territorial behavior in habitat distribution, it will be necessary to first present some ideas about the factors which lead to birds being present in one place or another, at one level of abundance or another. By considering concepts related to overcrowding and evolutionary optima, we will develop a theory to describe a particular way in which bird populations might distribute themselves over the available living places. This distribution will be called the ideal free distribution. We then will define three hypotheses for the role of territorial behavior. One hypothesis describes how territorial behavior might be part of the mechanism by which the ideal free distribution is achieved. Another describes how territorial behavior might modify the distribution from the ideal free form. A third hypothesis contends that territorial behavior has a role that is unrelated to the ideal free distribution.

III. THEORY OF HABITAT DISTRIBUTION

In order to describe these hypotheses clearly, an agreed upon statement of the problem of habitat distribution in general must be developed. Only with such agreement can the possible relationships of territorial behavior to habitat distribution be differentiated. We proceed with the following definitions.

2

Definitions

Habitat. A habitat of a species is any portion of the surface of the earth [3]) where the species is able to colonize and live. The total area available to a species can be divided into different habitats. The area of any one habitat can be large or small, and different habitats of the same species may be of different sizes. A given habitat can consist of several subdivisions which are not contiguous. We will define habitats so that all of the area within each habitat is, at zero density of the species, essentially homogeneous with respect to the physical and biological features which we believe to be most relevant to the behavior and survival of the species. We will also frame our definitions so that different habitats are not identical with respect to those same physical and biological features.

This definition does not imply that all measurable variables within a single habitat must take constant values over all of that habitat. Some variables may be irrelevant. Others, such as temperature and humidity may compensate for one another. In the latter case, the "relevant feature" which is "homogeneous" is some function of the compensating variables. Thus evaporative heat loss rate which depends on temperature and humidity may be the relevant feature which is homogeneous, in which case temperature and humidity may still vary over a single habitat. Because of compensating variables, all of the area within a habitat does not need to *appear* homogeneous to our measuring devices. However, if an area is uniform in all measurable variables, it is in a single habitat.

Habitat distribution of a species. Suppose the total area available to a species is divided into different habitats and that the area of each habitat is known. The habitat distribution of the species is the set of numbers which state the number of individuals resident in each of the habitats. It can also be expressed as the proportions of the total population resident in the different habitats, or as the density in each habitat.

Factors affecting habitat distribution

Habitat selection. We can now consider how the habitat distribution is achieved. Habitat distribution in birds is usually based on habitat selection, at least some individual birds being exposed to a variety of habitats of which just one is chosen for residence. Therefore, the distribution may be considered as a behavioral phenomenon, involving stimuli and responses. This means that an understanding of the habitat selection responses in given en-

---

3) In a number of cases, a habitat may be restricted to a layer parallel to the surface of the earth.

vironmental circumstances will lead to an understanding of the habitat distribution. In order to understand behavioral responses, we should consider the environmental factors (excluding direct within-species individual interactions) which caused the natural selection leading to the evolution of the behavior.

Suitability. In the case of habitat choice, these factors include differences in goodness or suitability of habitat because individuals which choose relatively poor habitats are selected against. Although the stimuli directly influencing the choice of habitat may be no more than correlated with habitat goodness, it is the goodness itself which is a basic (or ultimate — see below) determinate of the behavior. To summarize, the relative suitabilities of the different habitats give rise to habitat selection which in turn determines the habitat distribution. The habitat distribution then depends on the relative goodness of the habitats.

In developing an agreed upon statement of habitat distribution, the next matter is to examine the relative suitability of the various habitats. Suppose the habitats are indexed by i, $i = 1, 2. ..., N$, where N is the total number of habitats. The goodness of each occupied habitat is related to the average potential contribution from that habitat to the gene pool of succeeding generations of the species. We are interested in some measure of that goodness, which may be called the suitability, and denoted for the ith habitat as S_i. The suitability of the habitat cannot here be precisely defined, but may be thought of as the average success rate in the context of evolution (and/or "adaptedness") of adults resident in the habitat. Stated formally, if s_{iq} is the expected success rate of the qth individual ($q = 1, 2, ..., n_i$, $n_i =$ number of birds resident in the ith habitat), then

$$S_i = \frac{1}{n_i} \sum_{q=1}^{n_i} s_{iq}. \tag{1}$$

The habitat suitability will be determined by several factors such as food supply and predators. The influence of some of these factors is density dependent so that the suitability in a habitat is affected by the density of birds there. Let us assume for the moment that the effect of density is always a decrease in suitability with an increase in density. This assumption would imply that ALLEE's principle does not operate (ALLEE, *et al.*, 1949), and it may be truly valid only when densities are not close to zero. ALLEE's principle states that survival and reproductive rates increase with population size up to some maximum. Further increase in population size leads to a decrease in survival and reproduction, as assumed here. ALLEE's principle certainly holds at very low densities. A solitary male, for example, cannot have as high a reproductive rate as a male-female pair. At moderate densities, the assumption that suitability decreases with increased density is reasonable since predators may become more active at higher densities and competition for food more severe (LACK, 1966).

We can now define a habitat distribution which will provide a reference for the discussion of the role of territorial behavior. This is the ideal free distribution noted previously, and rests on assumptions about habitat suitability and the adaptive state of birds.

Ideal free distribution

Assumptions on suitability. Ignoring ALLEE's principle, if we assume that suitability always decreases with density, then it would follow that the maximum suitability occurs when the density approaches zero. Let us call this maximum value the *basic suitability*, denoted for the *i*th habitat as B_i. The basic suitability of the *i*th habitat is affected by such factors as potential predators, food density, and cover.

These considerations lead to an equation expressing the suitability of the *i*th habitat as a function of the basic suitability there, and the density (denoted d_i). We write

$$S_i = B_i - f_i(d_i), \quad i = 1, 2, ..., N. \tag{2}$$

The term $f_i(d_i)$ expresses the lowering effect on suitability of an increase

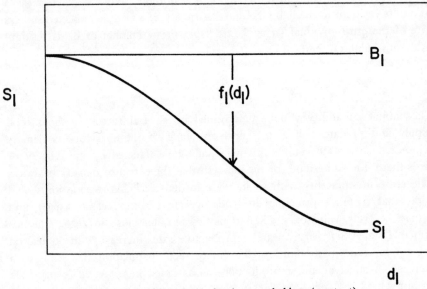

Figure 1. Suitability versus density: 1st habitat (see text).

in habitat density. Since $f_i(d_i)$ always increases with density, S_i always decreases. The Equations (2) will here be assumed to be the same through time. A possible example of Equation (2) for some value of i is plotted in Figure 1.

Before going any further, let us order the habitats in terms of their basic suitabilities, so that $B_1 > B_2 > \ldots > B_n$. By definition, no two habitats have equal basic suitabilities. This is consistent with our restriction on habitat definition (stated above), that no two habitats are identical with respect to relevant features.

Assumptions on birds. A description of the suitabilities of the various habitats has been considered in order to understand the habitat selection behavior and the habitat distribution. In applying this description, we make two additional assumptions. These are (1) all individuals settle in the habitat most suitable to them, and (2) all individuals within a habitat have identical expected success rates.

The first assumption demands that the birds have habitat selection behavior which is ideal, in the sense that each bird selects the habitat best suited to its survival and reproduction. Such birds will be referred to as *ideal* individuals. It is not an unreasonable assumption since individuals which are closest to being ideal will be selected for in the evolution of the species. Therefore, if the environment has sustained the same selective pressures for a large number of generations, the behavior of actual individuals should be approximately ideal.

The second assumption demands, first of all, that the birds be *free* to enter any habitat on an equal basis with residents, socially or otherwise. For example, if a population in a habitat is limited by nest holes and if all these nest holes are occupied by residents which neither share nor are displaced, then a newly settling individual may expect to be totally unsuccessful, although the average of all residents is rather high. In this case, newly settling individuals are not free to enter the habitat on an equal basis with residents. If, when a new bird arrived, all of the occupants of the habitat came together to draw lots for the nesting holes, and those losing remained in the habitat, then the individuals would be free. This, of course, is unrealistic.

The second assumption demands also that individuals are alike, genetically and otherwise. This aspect of the assumption may heavily restrict the application of this theory, at best to local populations.

A particular difficulty which arises from this assumption concerns habitat accessibility. According to this assumption, if accessibility is relevant, every habitat must be equally accessible to all members of the species. This is absurd when we regard widely distributed birds species and only local habitats.

However, if we restrict attention to habitats as widely distributed as the bird species, or to local populations as narrowly distributed as the habitats, then this difficulty is bypassed.

Note that the second assumption, and Equation (1), imply

$$S_i = s_i.$$

Each bird expects success equal to the habitat average.

The ideal free distribution. With these assumptions we now use the suitability Equations (2) to determine the habitat selection of the individuals in the population. The ideal assumption states that each individual will go where his chance of success is highest. The assumption of homogeneity states that each individual's chance of success is highest in the habitat of highest suitability. The two assumptions together then assert that each individual will go to the habitat of highest suitability. Thus, a description of the relative habitat suitabilities to some degree determines the choices of ideal free individuals. These choices in turn determine a distribution, which may be called the *ideal free* distribution. This distribution, which is formally described below, will form a convenient basis for discussing the hypothesized relationships of territoriality to distribution.

If all individuals choose the habitat of highest suitability, then from the point of view of unsettled individuals, the suitability in all occupied habitats must be approximately equal and not less than the suitability in all occupied habitats. This is true because if some habitats had a clearly lower suitability, then some of the birds in that habitat could improve their chance of success by moving to the habitats of higher suitability. If they did not make that move, they would not have ideally adapted habitat selection behavior, contradicting the ideality assumption. The distribution is stable only when suitabilities are equal in all habitats. With Equations (2), a fixed set of habitat areas, and given population size, the condition of equal suitabilities in all occupied habitats completely determines the ideal free distribution. To prove this, let a_i be the area of the ith habitat, and M the population size. If exactly l habitats are occupied, then the equal suitabilities condition says that

$$S_1 = S_2 = ... = S_l. \tag{3}$$

Note that the first l habitats are the l occupied habitats. These are the l habitats with the highest basic suitabilities, B_i. In fact, if the $(l + p)^{th}$ habitat were occupied, some $(l—q)^{th}$ habitat must be unoccupied (p, q positive integers such that $0 < p \leq N—l$ and $0 < q < l$), since only l habitats are occupied. But, because d_{l-q} equals zero, the suitability in habitat $l—q$ is B_{l-q} and

$$S_{l-q} = B_{l-q} > B_{l+p} \geq S_{l+p},$$

or

$$S_{l-q} > S_{l+p}. \tag{4}$$

Because under our assumptions individuals settle where the habitat suitability is highest, the birds in habitat $l + p$ would move to $l-q$ where by (4) the suitability is higher. Thus, if there are l occupied habitats, they are the first l.

It is also true that M, the total population size of a given species over all its occupied habitat, is given by

$$M = a_1 d_1 + a_2 d_2 + \dots + a_l d_l, \tag{5}$$

since
$a_i d_i$ is the number of birds in the ith habitat, and the total population is the sum of all the birds in all occupied habitats.

From Equations (2),

$$S_i = B_i - f_i(d_i), \; i = 1, \dots, l$$

so that $S_i = S_{i+1}$ in (3) implies

$$B_i - f_i(d_i) = B_{i+1} - f_{i+1}(d_{i+1}), \; i = 1, \dots, l-1. \tag{6}$$

There are $l-1$ of these equations in l unknowns; $d_1, d_2, \dots, d_l$. With Equation (5), there are l equations. These l equations can be solved uniquely [4]) for the d_i ($i = 1, \dots, l$) in terms of the constants M and the a_i. The distribution can be expressed as the proportion of birds in each habitat. Denote the proportion in the ith habitat as P_i. Then clearly

$$P_i = \frac{d_i a_i}{M},$$

and the distribution is seen to be a function of the density in each habitat. Since the densities are determined by the condition of equal suitabilities, so is the distribution.

An example of the solution of the equations is given in Figure 2 for $l = 1$, 2, and 3. The suitability curves are drawn for three habitats, 1, 2, and 3. When no birds are present, the suitability is highest in 1 and equals B_1. Therefore, if a small number of birds now settle in the habitats, they will all go to 1, because they settle where the suitability is highest. Then the density in 1 will increase from zero, and the suitability will decrease, following the

---

4) Because the $f_i(d_i)$ are always increasing.

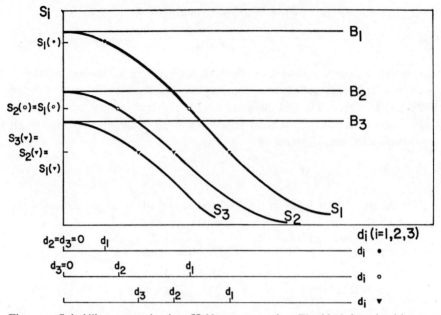

Figure 2. Suitability versus density: Habitats 1, 2, and 3. The ideal free densities are shown on the extra density coordinates at three values for the total population size, $M' < M'' < M'''$. The situation at each population size is denoted by ($\bullet$) for M', ($\bigcirc$) for M'', ($\blacktriangledown$) for M'''. At M', the lowest population size, all the population is in 1; the densities in 2 and 3 are zero. At M''', the largest population size, all three habitats are occupied.

curve labelled S_1. As the population size increases, more and more individuals will settle in 1, untill the density there is so high that the suitability is equal to the basic suitability in 2 (B_2). Now, any additional birds have a choice of habitats; 1 and 2 are equally suitable. However, these additional birds must increase the density in both habitats, and in such proportions that the suitabilities in both remain equal. Further increases in population size will raise the density in both habitats. If the population increases enough, the densities in 1 and 2 will be so high that the suitabilities in both habitats are reduced to B_3, the basic suitability in habitat 3. Any additional birds must increase the density in all three habitats in such proportions that the suitabilities in all three remain equal.

Allee-type ideal free distribution

Let us now briefly consider the effect on the distribution of ALLEE's principle which we have heretofore assumed does not apply. In this case, the S_i curves first increase with density up to a maximum then decrease.

These curves do not always have unique inverses; there are sometimes two densities corresponding to a single suitability. Therefore, the Equations (3) and (5) do not necessarily have a unique solution. Consideration of ALLEE's principle suitability curves is best done graphically as in Figure 3. At low population sizes, the birds will presumably go to habitat 1, and as the population increases, will enjoy an increasing suitability up to some maximum. Further increases in population size with all birds settling in 1, will cause a decrease in suitability in habitat 1 until at some higher population size (A), the density (d_i) in 1 is such that the suitability there equals the suitability in 2 at density 0. Now a remarkable event may occur. With a further slight increase in population size (A+), some birds will settle in 2 and perhaps some in 1. But the suitability in 2 *increases* with an increase in density, while the suitability in 1 decreases; therefore, $S_1 < S_2$, and suddenly it be-

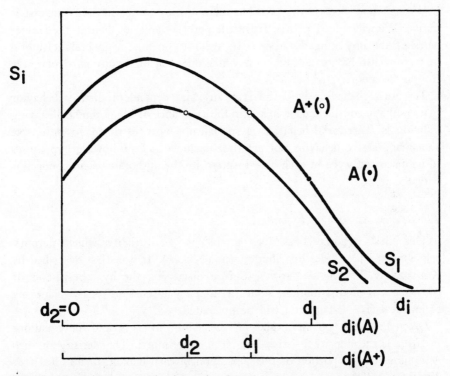

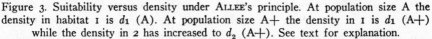

Figure 3. Suitability versus density under ALLEE's principle. At population size A the density in habitat 1 is d_1 (A). At population size A+ the density in 1 is d_1 (A+) while the density in 2 has increased to d_2 (A+). See text for explanation.

comes advantageous for birds in 1 to go to 2. Being ideal, they will so move, and may continue to move until the suitability in 2 is maximal. They would then fill the two habitats in such a way that $S_1 = S_2$, and may well become common in both (open circles in Figure 3). Thus, a very small increase in population size may result in a very large change in the distribution. By manipulating the S_1 curves, many such changes can be produced, and one can generally conclude that under the conditions of this theory, species following Allee's principle may demonstrate erratic changes in distribution with small changes in population. One can imagine curves which lead to complete shifts in population while other curves may lead to no erratic behavior at all.

Unless otherwise noted, the Allee-type ideal free distribution will not be considered in the following section.

IV. TERRITORIAL HYPOTHESES AND EFFECTS ON DISTRIBUTION

We can now discuss the possible relationships of territoriality to distribution. We will consider three distinct hypotheses and put them in the framework of the preceding discussion. In order to do this, a general definition of territorial behavior will be made which does not imply any particular role for the behavior.

Territorial behavior is defined as any site dependent display behavior that results in conspicuousness, and in the avoidance of other similar behaving individuals. Territorial behavior is specifically not restricted to defensive and/or aggressive behavior nor are they excluded. The following hypotheses to be discussed were in each case inspired by the authors cited in connection with them.

The density assessment hypothesis

This first hypothesis ascribes a role to territoriality which permits achieving the ideal free distribution (non-ALLEE). It was first described by KLUYVER & TINBERGEN (1953). Before discussing the hypothesis and its consequences in distribution, some preliminary discussion on the achieving of an ideal free distribution will be presented.

Given the total population size M, the habitat areas a_i and the equations in S_i (2), the density in each habitat is determined. The density in turn determines the proportion of the population in each habitat, or an ideal free distribution. The actual values will be expressed in terms of M and the a_i which for a given year are constants for the purpose of the present discussion. However, M, and perhaps the a_i vary with time in a somewhat irregular fashion. Therefore, the proportion of the population in a given habitat may also vary between different years. This is demonstrated in Figure 2. The suitability curves for three habitats are drawn, and the habitat densities at three population sizes shown. The relative densities in each of the three habitats are shown on the extra abscissas, and are markedly different at the three population sizes.

If the total population size does not vary, then a species could consistently achieve an ideal free distribution by being composed of individuals, a fixed proportion of which prefer each habitat at all times, or by being composed of individuals which prefer each habitat a fixed proportion of the time. However, if the variation in population size does occur and leads to considerable variation in the ideal free distribution, then such a fixed habitat selection scheme will usually not work. In this case, the habitat selection of the individuals, if it is to be ideal, must adjust to the changing conditions.

In order for individuals to be able to modify their habitat selection in accordance with changes in the ideal free distribution, there must be some cue or cues which reflect these changes. The changes come about due to variation in population size and possibly habitat areas. These variables affect the densities in the various habitats and therefore the relative suitabilities. Any cue, such as foot-print abundance, which would reflect habitat densities, or any cues which would reflect population size and perhaps habitat areas could be used by ideal individuals in achieving the ideal free distribution. The individuals would have their preference for a given habitat depend on the state of the cue.

KLUYVER & TINBERGEN (1953) suggested that the territorial behavior of resident individuals is used as a density cue by unsettled individuals so that they can avoid highly populated habitats where the chance of breeding success is presumably lower than elsewhere. These authors observed that the habitat distribution of some *Paridae* (tits) was dependent on population size. At low population levels, most of the individuals were found in a single habitat type, while at higher levels many individuals occupied another habitat type, but at a lower density (compare this with population sizes M' and M'' in Figure 2 of this paper). They emphasized that breeding success (and therefore, suitability) was not noticeably different in the two habitats. Thus, the distribution of the tits was apparently nearly ideal free, despite changing population size. No appropriate cue for density other than territoriality was observed, nor was there evidence for a cue for population size.

There is no evolutionary difficulty in supposing that territorial behavior serves as a density index. It is obviously to the residents' advantage to provide such a cue, since they suffer if their habitat is crowded to the extent that its suitability is lower than that in other habitats. It is also the advantage of the settling individuals to respond to such a cue, for by so doing they avoid habitats where high density makes their chance of success lower than elsewhere. Since the population size of tits varies considerably, the development and use of some density cue, such as territorial behavior, might be expected.

The density limiting hypothesis

Introduction. The second territorial hypothesis is based on a model from HUXLEY (1934), who described a territory as a rubber disk. The disk can be compressed, but with an increased amount of force necessary as it gets smaller. This hypothesis is relevant to the free aspect of the homogeneity assumption about birds in the ideal free model. That assumption states that any individual is free and may therefore enter any habitat on an equal basis with the birds already resident there. With the rest of the homogeneity assumption, this means that the average success of the occupants of a habitat is also the suitability that an unsettled bird will have (on the average) on settling in that habitat. The free assumption, as already mentioned, fails if the species is limited by nest holes which, once occupied by a resident, are not relinquished or shared. The second territorial hypothesis describes another possible way in which this assumption might fail. Suppose the residents of the habitat, by their territorial behavior, made it dangerous for unsettled individuals to enter the habitat. Then the average success of newly settling individuals will be lower than the habitat average, and the assumption fails. If so, ideal individuals maximizing their own success would not necessarily settle where the habitat suitability is highest, and the habitat suitabilities no longer must be equal. Since Equation (6) no longer holds, the distribution is not determined as before. There exists a new, different distribution.

The supposition that the territorial behavior of residents restricts non-residents from settling is reasonable. In evolution such behavior effectively prevents the density in the habitat from increasing, maintaining the suitability (see Figure 4). This would give the aggressive residents a selective advantage and the behavior, as suggested by BROWN (1964), would spread throughout the population. However, if all the individuals, settled and unsettled alike, became equally aggressive, it might not then be possible for the settled birds to make a habitat less suitable to unestablished birds. This is perhaps unlikely; for example, dominance seems to depend on experience (NICE, 1936; SABINE, 1955) which for a given area should vary considerably from bird to bird. In the following discussions we will assume that all individuals are not equally aggressive, so that social dominance hierarchies are established and the free assumption does not hold.

Consequence of hypothesis on the distribution. We consider in more detail the altered ideal distribution which would arise from this hypothesis. By definition the ideal bird always goes to the habitat where his potential success is highest. In the ideal free distribution, the potential of a new individual settling in a given habitat is equal to the average of all individuals resident

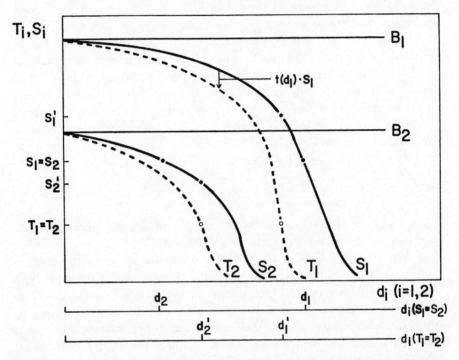

Figure 4. A comparison between ideal free and ideal dominance distributions: Habitats 1 and 2. Population size and habitat areas are constant. S_i is the suitability for established birds. T_i, the apparent suitability to unestablished birds. The filled circles represent equilibrium conditions $S_1 = S_2$ for the ideal free distribution, and densities for that distribution are given below as (d_1, d_2) on the abscissa labelled $(S_1 = S_2)$. The open circles represent equilibrium conditions $T_1 = T_2$ for the territorial distribution, and densities for that distribution are given as (d_1', d_2') on the abscissa $(T_1 = T_2)$. At $T_1 = T_2$ the actual suitabilities are given as (S_1', S_2'), marked on the S_i curves with a star. Note that in going from the ideal free to the ideal dominance distribution, at constant population size and habitat areas, the density in 1 decreases $(d_1' < d_1)$ while the density in 2 increases $(d_2' > d_2)$. Some individuals have in principle been forced from habitat 1 into habitat 2. See text for further discussion.

there, including the new one. This is the habitat suitability described by Equation (2). If the unsettled birds are restricted by the territorial behavior of residents, then their potential success is less than the average of the habitat. This suggests that we define some quantity t, $0 \leq t \leq 1$, which can be subtracted from one and multiplied into the S_i, to yield the apparent habitat suitability from the point of view of the unsettled bird. This apparent habitat suitability may be denoted by T_i, defined symbolically in Equation (7):

$$T_i = S_i \,(1{-}t) \qquad\qquad (7)$$

Let us now consider some of the properties of t. This quantity will depend on density, and can be reasonably assumed to always increase as the density in a habitat increases. This assumption is justified because t is related to the resistance of the established birds. The resistance from each male evidently increases as the territory size diminishes (HUXLEY, 1934); that is, as the density increases. Thus, the per bird resistance always increases with density. Also, the number of birds increases with density. Therefore, the overall resistance, and t, must increase with density. We will assume that t does not vary with habitat, since it is unlikely that the territorial resistance is dependent on habitat, except as density varies. Thus, Equation (7) may be rewritten

$$T_i = S_i \,(1{-}t(d_i)). \qquad\qquad (8)$$

The function T_i, like S_i, always decreases with density, since S_i always decreases and $t(d_i)$ always increases with density. Also, T_i is always less than or equal to S_i. The actual suitability of the habitat remains S_i; T_i is just the apparent suitability from the point of view of the unsettled bird.

The T_i in (8) are defined such that an ideally adapted individual will always go to the habitat where T is highest, assuming as before that all (unsettled) individuals are alike in their adaptation to the habitats. This leads, as with the S_i, to an equilibrium condition where the T_i are equal in all occupied habitats. The resultant equations, with (5), completely define a set of habitat densities and a new distribution. This distribution may be called the *ideal dominance* distribution. Like the ideal free distribution, the ideal dominance distribution is a useful basis for discussion, but because of the underlying assumptions, can only approximate any real life situation.

Interpretation. Some remarks about the T_i in (8) may make them easier to understand. If within an area some habitats are better than others then the territorial restrictions may just restrict newly settling individuals to the less good habitats. Then T_i represents the average success expected in the area for individuals in the less good areas. However, in a uniform habitat where no less favorable regions exist, the territories of all occupants will be, on the average, about equally suitable. In this case, $t(d_i)$ represents only an entering risk to unsettled individuals. If they can successfully settle in the habitats, they may expect to be as successful as the habitat average.

But the act of settling may involve a serious risk of physical harm. There may be a high probability of failure with accompanying mortality. Thus, the T_i may not be a measure of the success of some members of the habitat, but a measure of the average success of a hypothetical group of individuals which tried to enter a habitat until either successful or dead.

These two territorial hypotheses exemplify the distinction between ultimate and proximate determinants of behavior. These are defined as follows. Ultimate determinants of behavior are the environmental factors which produce the natural selection that leads to the genetic basis for the behavior. Proximate determinants are the stimuli that prompt the behavior. The behavior that we are considering is the habitat selection of individuals. The first hypothesis is that territorial behavior is only a proximate factor providing information about density. The second hypothesis is that territorial behavior is an ultimate factor which by directly influencing the survival of past individuals which selected certain habitats to breed in has actually caused the population-genetic basis for the habitat selection to change in some way. And since the distribution is determined by the habitat selection, the territorial behavior is an ultimate cause of the phenomenon as well. However, in the second hypothesis territorial behavior is probably also a proximate factor. It could easily provide information about itself and provide a stimulus to which unsettled individuals could respond. This is quite likely because territorial behavior generally involves display and vocal announcement as a substantial part of the "defense" of boundaries.

The spacing hypothesis

LACK (1964) and JOHNSTON (1961) have supported a third territorial hypothesis, that the territorial behavior has been evolved only to space individuals within a habitat. This means that the density in a habitat is determined independently of the behavior, but that given a certain density, the individuals separate as much as possible and have non-overlapping home ranges. For example, suppose a population of a territorial species can achieve distribution without the territorial behavior either cueing density variation or restricting habitat occupancy. Then the role of the territorial behavior may be purely behavioral, isolating mating adults to strengthen pair bonds (LACK, 1954). Or the behavior could prevent the spread of disease by a quarantine effect. Whatever its role, if the territorial behavior only spaces individuals, it has nothing to do with whether or not an individual will settle in one habitat or another. It only affects the movements of the individuals *within* the habitat. The effect of spacing is to keep the *instantaneous* distribution of individuals within the habitat fairly uniform. It does not alter the number of occupants of the habitats, nor does it affect the average number of individuals using any portion of the area within the habitat (the average being taken over some time period). Thus, the average density over any piece of land (*i.e.*, any habitat) is not influenced by territorial behavior under this hypothesis. If territorial behavior only spaces individuals, it has no effect on the habitat distribution, either ultimately or proximately.

Application of the theoretical development

Approach. The next matter is to consider the problem of identifying the actual role of the behavior relative to habitat distribution in a given species. We have defined two theoretical distributions under certain assumptions and three different territory hypotheses. We may, as a first approximation, assume that if territorial behavior has one of the hypothesized roles in a given species the actual habitat distribution will be similar to the theoretical distribution described for that hypothesis. Thus, if we can show that the actual species distribution is approximately ideal dominance, this may be taken as evidence in the support of the hypothesis that the role of the behavior is to limit density. Or if we can show that the distribution is consistently ideal free in spite of variation in population size, and can find no other evident cues for density and/or population size variation being used by the species, then this may be taken as evidence in the support of the hypothesis that the behavior is used for density assessment. If we can show that the distribution is not consistently ideal free in the face of variation in population size, nor ideal dominance, then this may be taken as evidence for a lack of any density assessment or density limiting mechanism, and spacing remains as the most tenable of the hypotheses offered. (Other hypotheses may exist, but we will not pursue them here). The approach to the problem rests on identifying the distribution as (approximately) ideal free, ideal dominance, or neither.

Density and habitat suitability. If a population has an ideal free distribution, then by definition the suitability of all habitats is equal. We used this to show that the ideal free assumptions determine a distribution. However, the densities in the different habitats are not necessarily equal (see Figure 2). Thus, if a species has a nearly ideal free distribution, habitats with different densities will show similar success rates.

If a population has an ideal dominance distribution, then the suitability of all habitats is not necessarily equal. Only the apparent suitabilities (the T_i) will be everywhere equal, assuming ideally adapted individuals. In this case, it will now be shown that, for two occupied habitats, p and q, if $d_p > d_q$ then

$S_p > S_q$: the suitability is higher in habitats of higher density. From the equilibrium condition, for occupied habitats q and p,

$$T_q = T_p. \tag{9}$$

Since, by (8)

$$T_q = S_q \ (1 - t(d_q)),$$
$$T_p = S_p \ (1 - t(d_p)),$$

then substituting in (9) we obtain

$$S_p \ (1 - t(d_p)) = S_q \ (1 - t(d_q))$$

and

$$S_p/S_q = (1 - t(d_q))/(1 - t(d_p)). \tag{10}$$

If $d_p > d_q$, then $t(d_p) > t(d_q)$, since the t function always increases with density. Then

$$(1 - t(d_q)) > (1 - t(d_p)),$$
$$(1 - t(d_q))/(1 - t(d_p)) > 1,$$

and therefore, by (10),

$$S_p/S_q > 1.$$

From this it follows that $S_p > S_q$ as asserted. Thus, if a species has an ideal territorial distribution, then the success rate in habitats with higher densities of residents will be higher. This assumes that the t function does not depend on habitat, and always increases with density.

It is possible using another approach to show that this relationship between density and suitability exists. Assume that territory size is inversely related to the number of birds against which it is defended (HUXLEY, 1934). The better a territory is, the more there are that are less good, and there are more birds potentially trying to overtake it. Since the territory must be defended against more birds, it must be smaller. Hence, more suitable territories are smaller, and density, which is inversely related to territory size, is seen to be higher in high suitability habitats.

Figure 4 shows examples of the S_i and T_i curves for $N = 2$, and a sample population distributed freely (filled circles) and territorially (open circles). Note that the suitability in the ideal free distribution is equal in both habitats $(S_1(.) = S_2(.))$. In the territorial distribution, the suitability and the density are higher in habitat 1 than in habitat 2 $(S_1(*) > S_2(*), d_1 > d_2)$. [5]

---

[5] We should note the discussion of GIBB (1961) in which attention is drawn to the relationship between the role of territorial behavior and differences in habitat suitability. Gibb's remarks are not formally developed, and do not distinguish the density assessment and density limiting hypotheses. But they do foreshadow several of the ideas presented above.

These conclusions suggest that the role of territorial behavior in the habitat distribution of a territorial species can be ascertained as follows: If the high density habitats show consistently higher success rates, and if no density limiting mechanism other than territorial behavior is evident, then the role of the territorial behavior is evidently to limit density. If suitabilities in all habitats are equal, even though densities are not, if the distribution changes with changing population size, and if no alternative density or population size cue is apparent, then the role of the territorial behavior is evidently to serve as a density assessment mechanism. If neither of the above two criteria are met, then the role of territorial behavior seems to be only to space individuals. Underlying this approach is the assumption that the birds are approximately ideally adapted. This assumption may well fail, and this possibility should always be considered. There is also the possibility, ever present, of hypotheses different from those considered here.

Uncertainties in application. There are a number of uncertainties involved in these conclusions which should be given careful consideration. Some of this uncertainty is inherent in the theory, which describes the relationships of expected values or population means. Any given realization of the theory (*e.g.*, the observed habitat suitabilities in a given year) is expected to deviate from the average values, even if the assumptions of the theory are met. This deviation will alter the succeeding year's distribution somewhat (by accidental selection), and so some fluctuation in the distribution may also be expected.

Another source of uncertainty lies in inherent failure of the assumptions, particularly the ideality assumption. For example, the sensory reception of the bird cannot be perfect, and so the bird can be expected to misread whatever environmental cues it uses to assess the suitability of a habitat. Also, the correlation between these cues and suitability is probably never perfect. Finally, the birds may reasonably be expected to be always evolving towards the ideal state without ever achieving it. These failures of the ideal assumption will lead to errors in the individual judgments of habitat suitability.

Hopefully the uncertainties in our predictions will be generally independent in different years, or even in different regions in the same year. Probably most of the error-causing factors are rather local in effect. A major exception is weather, but even this factor changes from year to year. Thus, in most cases, sampling over several years, or perhaps over widely separated regions in the same year should provide a way of estimating or controlling these errors. If the errors are not independent even over years, then the species may be considered to be evolving and changes over time should be detected. Thus, results which are consistently obtained over a number of years may be reasonably considered free from these errors.

Measurement of suitability. The problem of measuring suitability remains. The suitability of a habitat is a reflection of the average genetic contribution of resident adults to the next generation, and must be closely related to the average lifetime production of reproducing offspring in the habitat. Therefore, it must depend on several components, including reproductive rate, and survival of adults and immatures. Since territoriality is normally associated with breeding behavior and habitat distribution during the breeding season, we will usually associate suitability with such things as nesting success, feeding rates, and clutch size.

V. SUMMARY OF THE THEORETICAL DEVELOPMENT

The purpose of this report is to define precisely the problem of the role of territorial behavior in habitat distribution. Models for the habitat distribution of an ideally adapted species are developed under certain hypotheses for the role of territorial behavior. These models provide well-defined statements of what each hypothesis means, and also provide under certain assumptions a way of ascertaining, for a given territorial species, what the role of territorial behavior is. The hypotheses considered were that territorial behavior (1) is part of a density assessment mechanism, (2) limits density, or (3) only spaces individuals.

VI. ACKNOWLEDGEMENTS

The theoretical developments of this paper were stimulated by correspondence with R. L. Haines and H. A. Hespenheide. T. L. Quay provided encouragement and advice relative to the field aspects of the study as reported in the next two papers; H. R. van der Vaart, C. Proctor and W. Standaert read parts or all of the manuscripts and made many valuable suggestions.

Glinda Goodwin and my wife, Armeda, prepared the manuscript.

We are extremely grateful to all of the above individuals for their interest and help.

VII. REFERENCES

ALLEE, W. C., A. E. EMERSON, O. PARK, T. PARK & K. P. SCHMIDT (1949). Principles of Animal Ecology. W. B. Saunders, Philadelphia. xii + 837 pp.

BROWN, J. L. (1964). The evolution of diversity in avian territorial systems. — Wilson Bull., 76, 160-169.

GIBB, J. (1961). Bird populations, pp. 413-446. In: A. J. MARSHALL (ed.), Biology and Comparative Physiology of Birds. II. Academic Press, New York.

HOWARD, F. (1920). Territory in Bird Life. Atheneum, London, 239 pp.

HUXLEY, J. S. (1934). A natural experiment on the territorial instinct. — Brit. Birds, 27, 270-277.

JOHNSTON, R. F. (1961). Population movements of birds. — Condor, 63, 386-388.

3*

KLUYVER, H. N. & L. TINBERGEN (1953). Territory and the regulation of density in titmice. — Arch. Neerl. Zool., 10, 265-289.

LACK, D. (1954). The natural regulation of animal numbers. Clarendon Press, Oxford, viii + 343 pp.

——. (1964). A long term study of the great tit (*Parus major*). — J. Anim. Ecol., 33 (Suppl.), 159-173.

——. (1966). Population Studies of Birds. Clarendon Press, Oxford. 341 pp.

NICE, MARGARET M. (1937). Studies in the life history of the Song Sparrow I. — Trans. Linn. Soc., N. Y., viii + 246 pp.

SABINE, WINIFRED (1955). The winter society of the Oregon Junco: the flock. — Condor, 57, 88-110.

STEWART, R. E. & J. W. ALDRICH (1951). Removal and repopulation of breeding birds in a spruce-fir forest community. — Auk, 68, 471-482.

TINBERGEN, N. (1957). The functions of territory. — Bird Study, 4, 14-27.

WYNNE EDWARDS, V. C. (1962). Animal Dispersion in Relation to Social Behavior. Hafner Publishing Company, New York. xi + 653 pp.

10

Reprinted from *Auk*, **68**, 471–482 (1951)

REMOVAL AND REPOPULATION OF BREEDING BIRDS IN A SPRUCE-FIR FOREST COMMUNITY

BY ROBERT E. STEWART AND JOHN W. ALDRICH

In 1949, while engaged in population studies of birds in northern Maine, the authors accumulated considerable information concerning population dynamics of birds inhabiting the Spruce-Fir forest community. This information was obtained in connection with investigations of the effective control by breeding birds of an infestation of the spruce budworm, *Choristoneura fumiferana* Clem.

Field studies were conducted during June and July on two separate areas; a square 40-acre tract which was used as an experimental area, and a rectangular 30-acre tract that served as a control area. The two areas were 1.25 miles apart and were both within one-half mile of the shore of Cross Lake, located about 14 miles southeast of Fort Kent, Aroostook County, Maine. In the experimental area an attempt was made to eliminate or to reduce drastically the bird population by use of firearms, while in the control area the natural bird population was allowed to remain unmolested. Both skins and stomachs of the birds killed were preserved for the Fish and Wildlife Service collections. The spruce budworm populations were studied in both areas before, during, and after the removal of the birds in order to see if any differer :al developed between the two areas in their total populations. The entomological portion of the field study was conducted by Philip B. Dowden and V. M. Carolin of the Bureau of Entomology and Plant Quarantine of the U. S. Department of Agriculture. Stomach contents of the birds collected were analyzed by Robert T. Mitchell of the Fish and Wildlife Service. The results of the various phases of this work concerning the effective control by birds of spruce budworms will appear in a separate report. The present paper includes an analysis only of the data on removal and repopulation of birds in the experimental area.

The Spruce-Fir forest in the experimental area was somewhat varied. The canopy in certain sections of the forest had been partially opened by selective lumbering, resulting in a fairly dense understory growth of conifers. The other portions of the forest with a closed canopy contained little or no understory. Balsam fir, *Abies balsamea*, and black spruce, *Picea mariana*, were the primary dominant trees throughout the forest. The majority of the trees ranged between 40 and 60 feet in height. Small sections of mixed coniferous-deciduous brushy undergrowth were present and the area was crossed by a state highway.

The population of territorial males in the area was determined prior to the removal of birds, by using the spot-mapping method (see Breeding-Bird Census, Audubon Field Notes, 4 (2): 185–187). Twenty-five census trips totaling 55 man-hours were conducted on the area from June 6 to June 14, inclusive. Most of the census trips were taken during early morning hours between 6:00 and 10:30 a.m. A few trips were also made in the evening for the determination of thrush populations.

Beginning on June 15 and extending through July 8, an attempt was made to remove all birds on the experimental area by shooting with 16-gauge shotguns. Shells used were loaded with very fine (No. 12) shot so that birds would not be too mutilated for specimens. By using this method the total population on the area was greatly reduced, although the actual degree of reduction could not be readily appraised. The chief difficulties were that the breeding territories were completely disrupted during the period when the original occupants were being removed and at the same time new adult males were constantly invading the area. Most of the adult females were so inconspicuous and secretive that it was not possible to census them accurately. In all, a total of about 130 man-hours were spent on the area collecting birds. The number of birds removed from this 40-acre tract during the entire collecting period was 455 (420 adults and 35 birds in juvenal or immature plumage).

Nearly all of the birds in the Spruce-Fir forests appeared to be at or near the peak of their breeding cycle during this study. With the exception of the American Crow, Golden-crowned Kinglet, and Red-breasted Nuthatch, it is doubtful if any species, at the onset of collecting, had produced nestlings. Nestlings, as well as three or four groups of fledglings, of the Golden-crowned Kinglet and Red-breasted Nuthatch were observed, but they were practically all removed from the area soon after they were first encountered during the first few days of collecting. The constant pressure of collecting and the resultant turn-over in adult birds completely prevented the development of other nestling populations in the study area during the collecting period.

Small, roving groups of adult non-breeding birds were found in the area from time to time. The most common of these were the American Robin and Cedar Waxwing, both species being occasionally represented by from one to nine individuals. These birds were noted with increasing frequency during the latter part of the collecting period, from about July 2 to July 8. Other non-breeding birds occurring more rarely included the American Goldfinch, Pine Grosbeak, and Chipping Sparrow. A few wandering immature birds were also seen during the last few days of collecting. These had undoubtedly moved in from outside areas after having become independent of their parents. An attempt was made to collect all of these birds whenever they were encountered.

After the first nine days of collecting (June 15–June 23) the resident males of all species except the Myrtle Warbler and Olive-backed Thrush had been reduced to 15 per cent or less of the original numbers. Continued collecting coupled with a steady influx of new males maintained these species at about this level for the remainder of the collecting period. The number of male Magnolia Warblers was reduced to the 15 per cent level by June 23 even though only 16 males (67 per cent of original population) had been taken by that time. Possibly some of the males had deserted owing to the disturbance of walking and shooting in the area; the Magnolia Warbler, being an understory species, would probably be affected by such disturbance more than tree-top birds would be. The number of male Myrtle Warblers was reduced to about 25 per cent and maintained this approximate level thereafter. Apparently a proportionately larger number of non-territorial males of this species was present to fill evacuated territories. The Olive-backed Thrush was found to be a very difficult species to collect. It was much more wary and secretive than the other species and was seldom seen, although its songs and call-notes were evidence of its continual presence. Although a total of 18 adult Olive-backed Thrushes was taken during the collecting period, the rate of collection was so low and repopulation so fast that the resultant voids in the population were filled by new males almost as soon as the voids were produced. As a result, the population level for males of this species was reduced only slightly below (probably about 80 per cent of) the pre-collecting population.

The approximate reduction in total number of adult males of all species was from an average of 148 to 28 or, in other words, from a density of 370 males per 100 acres to 70 males per 100 acres (reduction of 81 per cent). This density of 70 males per 100 acres is the approximate level that was maintained from June 24 to July 8. With the exception of wandering non-breeding birds, including the Arctic Three-toed Woodpecker, Cedar Waxwing, Black-throated Green Warbler, Chestnut-sided Warbler, Pine Grosbeak, American Goldfinch, Chipping Sparrow, and most of the Robins, practically all of the adult males taken appeared to be on their breeding territories at the time they were collected.

Most of the new males which took possession of the evacuated territories probably arrived during the night or early morning, as practically all of them when first observed seemed to be actively establishing territory, and no evidence of any late arrivals during the day was noted. The males of many species appeared to be completely eradicated time after time, only to be replaced by other males on following days. The characteristic behavior pattern of most of the new arrivals was quite distinct from the behavior displayed by males with territories that had been maintained over a longer period. The songs of these newcomers were louder and uttered more frequently, while at the same time the birds were much more active in covering and inspecting the areas within their territories. It is probable that most of these new arrivals were unmated. Lack (1946) in his study of the English Robin found that "In the late spring nearly all the best robin song comes from cocks which are unmated." ". . . almost immediately after the cock has obtained a mate, its song declines to a rather moderate intensity and remains so, except during fights, unless the mate is lost, in which case the cock again comes into loud song."

Due to the variable nature and uneven growth of the Spruce-Fir forests in the area, the required habitat niches for the various resident species were not uniformly distributed. Because of this, the pre-collecting distributional pattern of breeding territories was quite different for each species. It was therefore of interest to note that subsequent to the collection of the original territorial males, the new males of each species almost invariably established territories in the same places that were occupied by their predecessors. Since an abundant food supply was available for birds throughout the area, these observations substantiate the importance of "the niche" in explaining local distribution of breeding birds.

Prior to the collecting period, the number of territorial males on the shooting area was 148, as determined by the spot-mapping method from June 6 to June 14. The number of adult males collected on the area during the period June 15 to July 8 was 302 + (Table 1). Thus, over twice as many males were removed from the area as were present before the collecting was started. The rapid influx and establishment of new territorial males, following the removal of the former occupants, account for the large number of males collected and are indicative of the amount of population pressure that was present in this community. It would appear that this pressure was due to severe competition, between individual males of the same species, for suitable areas to establish their territories. Apparently, the appropriate habitat

TABLE 1

PRE-COLLECTING POPULATION AND COLLECTION TOTALS OF ADULT MALES

Species	Pre-collecting population, territorial males June 6–14	Total males collected June 15–July 8	Collecting periods		
			June 15–23	June 24–July 1	July 2–8
Tetraonidae					
Ruffed Grouse, *Bonasa umbellus*	+	1	1		
Picidae					
Yellow-shafted Flicker, *Colaptes auratus*	1	2	1		1
Arctic Three-toed Woodpecker, *Picoides arcticus*	V	1		1	
Tyrannidae					
Yellow-bellied Flycatcher, *Empidonax flaviventris*	4	3 (4)	2		1
Corvidae					
Blue Jay, *Cyanocitta cristata*	1.5	2	2		
American Crow, *Corvus brachyrhynchos*	+	1 (1)		1	
Paridae					
Black-capped Chickadee, *Parus atricapillus*	2	3 (4)	3		
Brown-capped Chickadee, *Parus hudsonicus*	2.5	3 (5)	3		
Sittidae					
Red-breasted Nuthatch, *Sitta canadensis*	3.5	4 (1)	4		
Troglodytidae					
Winter Wren, *Troglodytes troglodytes*	2	1	1		
Turdidae					
American Robin, *Turdus migratorius*	2	14 (4)	4	4	6
Hermit Thrush, *Hylocichla guttata*	2	4	2	2	
Olive-backed Thrush, *Hylocichla ustulata*	7.5	11 (2)	4	2	5
Regulidae					
Golden-crowned Kinglet, *Regulus satrapa*	4	4	3	1	
Ruby-crowned Kinglet, *Regulus calendula*	2	2	2		
Bombycillidae					
Cedar Waxwing, *Bombycilla cedrorum*	V	6 (2)			6
Vireonidae					
Solitary Vireo, *Vireo solitarius*	3	4 (1)	3	1	
Red-eyed Vireo, *Vireo olivaceus*	1.5	7	2	3	2
Parulidae					
Tennessee Warbler, *Vermivora peregrina*	3	1	1		
Nashville Warbler, *Vermivora ruficapilla*	4	6	5		1
Parula Warbler, *Parula americana*	1	4	2		2
Magnolia Warbler, *Dendroica magnolia*	24	36	16	10	10
Cape May Warbler, *Dendroica tigrina*	12	23	13	10	
Myrtle Warbler, *Dendroica coronata*	8.5	29	11	9	9
Black-throated Green Warbler, *Dendroica virens*	V	1		1	
Blackburnian Warbler, *Dendroica fusca*	9	15	7	2	6
Chestnut-sided Warbler, *Dendroica pensylvanica*	V	1		1	
Bay-breasted Warbler, *Dendroica castanea*	35	81	34	21	26
Oven-bird, *Seiurus aurocapillus*	1.5	1	1		
Canada Warbler, *Wilsonia canadensis*	.5	1	1		
Fringillidae					
Purple Finch, *Carpodacus purpureus*	2.5	6	3	1	2
Pine Grosbeak, *Pinicola enucleator*	V	3			3
American Goldfinch, *Spinus tristis*	V	2			2
Slate-colored Junco, *Junco hyemalis*	5	13 (6)	6	2	5
Chipping Sparrow, *Spizella passerina*	V	1 (1)			1
White-throated Sparrow, *Zonotrichia albicollis*	3.5	5 (3)	5		
TOTAL	148	302 (34)	142	72	88

( ) undetermined sex   V visitor in area

+ indicates that less than .25 of one territory was present on area.

niches for most species were not numerous enough to satisfy the territorial requirements of all the males present.

The indications are that a large, surplus, "floating" population of unmated males must have been present in the Spruce-Fir forests of this region. Since there was no evidence that spring migration was still continuing at the time the collecting was initiated, it was assumed that the birds which invaded the area were largely unmated wandering males that had been in the general region for some time. Presumably these males were searching for suitable sites to establish territories, and whenever unoccupied habitat of the right type was found they lost little time in taking over and proclaiming their ownership by song. This is in line with the belief of Nice (1941) that "birds which fail to obtain territory, form a reserve supply from which replacements come in case of death of owners of territory." In a study of the territories of Marsh Tits, Southern (1950) stated that "the speed with which deaths among the breeding population were replaced suggests that there was definitely a non-breeding population throughout the spring."

It is probable that at least some of the new males that were collected were those that had established territories in areas adjacent to the collecting area. Upon the removal of the original territorial males in the collecting area, some of the males in outside areas could have taken advantage of the reduction in population pressure and expanded the boundaries of their territories into the collecting area. In a study of banded English Robins, Lack (1946) found that "when the owner of a territory disappears, the owners of the neighboring territories expand into the vacant site almost at once, and often the ground is fully occupied within twenty-four hours." In a comprehensive study of Ruffed Grouse, Bump, *et al.* (1947) discovered that the birds "tend quite rapidly to move into a covert whose population has been depleted out of proportion to those surrounding it." In the present study, however, the fact that more new males first appeared in the center of the collecting area than on the periphery would indicate that most of them were wandering, unmated birds rather than neighbors expanding their territories.

The ratio of males collected to the number of territorial males on the area prior to the collecting period was found to vary considerably with different species. In the case of birds belonging to Old World families such as the kinglets, chickadees, and nuthatches, there was very little difference between the two figures, indicating that their populations were comparatively stable. However, for most of the thrushes, vireos, warblers, and fringillids, the number of males collected was from one and one-half to three times as great as the number present in the area before collecting. The Red-eyed Vireo showed the greatest difference of all with nearly five times as many males collected as were formerly present in the area (the relatively high numbers of American Robins and Cedar Waxwings that were collected were due to the presence of wandering non-territorial birds). These striking differences in ratios of individuals present before collection to those collected for the different species would seem to indicate that the surplus of males was much greater for some than for others. In the case of Brown-capped Chickadees, American Robins, Golden-crowned Kinglets, and Red-breasted Nuthatches, it is possible that because of their earlier period of nesting they might have passed that stage in their breeding cycle at which they were interested in establishing territories and, therefore, the number of individuals present before collecting was relatively similar to the number collected.

Differences in the ratio of males collected to number of males present in the area before collecting were quite pronounced even among species that were closely related and that had similar habits. The five most abundant species in the area were warblers, all belonging to the genus *Dendroica*. The number of males collected for every male that was present in the area before collecting is indicated for each of these species, as follows: Magnolia Warbler, 1.5; Blackburnian Warbler, 1.7; Cape May Warbler, 1.9; Bay-breasted Warbler, 2.3; and Myrtle Warbler, 3.4. Since these species are about equally conspicuous in the field, it is unlikely that such noticeable differences in the ratios can be attributed to variations in the ease with which they may be collected.

Those species that showed a definite surplus of males could be considered as having *supersaturated* populations (see Kendeigh, 1947). In other words for these species, the carrying capacity of the area, insofar as breeding territories were concerned, was insufficient to take care of the needs of all males present. Possibly, the presence of the infestation of spruce budworms could have accounted for this. The great abundance of food resulting from the budworm infestation might have attracted such a large number of males to the area that the territorial requirements of all of them could not be satisfied. The importance of territorial behavior in limiting bird population in a given habitat is stressed by Nice (1937) who stated, however, that "climate and other factors may keep the numbers of a species in a region so low that territorial behavior has no chance to limit populations." Thus, while various factors may affect the numbers of breeding birds occurring in an area, the true balancing agent between bird populations and environment is the competition for territories among individual males of the same species. This is in accord with the findings of Nicholson (1933) who wrote "for balance, it is essential that the action of a controlling factor should be governed by the density of the population controlled, and competition seems to be the only factor that can be governed in this way."

It would seem probable that when certain elements, such as food supply, in the habitat vary, the saturation point for each species will vary also. For example, the saturation point for most insectivorous birds in a Spruce-Fir forest without a budworm infestation would probably be much lower than it was in the area studied. This belief is based on the assumption that the minimum territorial requirements are affected by the relative ease with which food can be obtained. Kendeigh (1947) has pointed out that in the case of the Tennessee, Cape May, and Bay-breasted warblers there appears to be a direct correlation between the abundance or occurrence of these species and the severity of spruce budworm infestations. The principle of balance between changing animal populations and environment was expounded by Nicholson (1933) who stated that "if a population is in a state of balance with the environment, its density must necessarily change in relation to any changes in the environment." This principle is further clarified by Elton (1936) who maintained that "the chief cause of fluctuations in animal populations is the instability of the environment."

The number of adult breeding females was also greatly reduced during the collecting period, although the actual degree of reduction could not be determined from the data available. The small number of females recorded at any time was somewhat puzzling. Although female birds are generally much more difficult to observe in the field than are males, the exceedingly small number actually seen and collected over a three-week period would, assuming that most males had been mated, indicate that a preponderance of the females deserted the area shortly after their mates had disappeared. The close attachment to the nests and general inconspicuousness of most females undoubtedly explains in part why such a small number was seen and collected during the first few days. The presence of large numbers of conspicuous territorial males at that time would naturally tend to overshadow the count of the females. However, the continued recording of low numbers of females following the initial period of collecting is thought to be due, in large part, to an extensive withdrawal of females from the area. After the majority of the males had been collected, any type of bird activity, slight though it might be, became much more noticeable; thus the presence of any appreciable number of females would surely have been noted.   At least 130 man-hours were spent in the area during the collecting period and practically all of the females seen were collected (Table 2).   The females collected had well developed brood-patches on their breasts, indicating that they either were or had been sitting on eggs.   If females had remained undetected in the area and continued to sit on eggs throughout the period, many of them would have been feeding young before the period was over and, therefore, would have become much more conspicuous at that time.   The females of the later-breeding species did not become conspicuous at any time, and there was no evidence of young produced on the shooting area.

Certain species showed a much greater disparity in the proportions of the sexes seen and collected than others (Table 2).   The greatest differences in the ratio of the sexes were found in the warblers (Parulidae) and vireos (Vireonidae).   In the case of the warblers, the males of which comprised about 67 per cent of the total adult male population of all species, only 29 (13 per cent) of 228 adults collected proved to be females.   For most birds, other than the warblers and vireos, the females were observed nearly as often as the males and were almost as readily collected from the start.   It would seem to be significant that several species, of which almost equal numbers of males and females were collected, were earlier breeders which were caring for young at the time collecting began.   This fact could be responsible for the females, being more conspicuous and more readily collected.   It is well known also that females desert their nests and territories more readily before incubation becomes well advanced.

Variations in the sex ratios of the birds collected may be indicative of actual differences in sex ratios of certain of the species represented. The sex ratios, of birds collected, for the seven most abundant species are indicated as follows (ratio of males to females): Blackburnian Warbler, 15–1; Bay-breasted Warbler, 9–1; Magnolia Warbler, 6–1; Cape May Warbler, 6–1; Myrtle Warbler, 6–1; Slate-colored Junco, 2.5–1; Olive-backed Thrush, 2–1.   It is unlikely that such marked differences as these could be due entirely to differences in comparative conspicuousness between males and females of the species listed.

Summary

The number of territorial male birds in a 40-acre tract of Spruce-Fir forest in northern Maine was 148 during the period June 6 to June 14, 1949.   Birds were removed from this area by shooting from June 15 to July 8, inclusive, with the intention of reducing the population to and keeping it at as low a level as possible.   By this means the number

TABLE 2
Numbers of Adult Males and Females Collected

Species	Males	Females	Undeter-mined sex[1]
Tetraonidae			
Ruffed Grouse	1	1	
Picidae			
Yellow-shafted Flicker	2		
Arctic Three-toed Woodpecker	1		
Tyrannidae			
Yellow-bellied Flycatcher	3		4
Corvidae			
Blue Jay	2	1	
American Crow	1		1
Paridae			
Black-capped Chickadee	3	1	4
Brown-capped Chickadee	3	4	5
Sittidae			
Red-breasted Nuthatch	4	6	1
Troglodytidae			
Winter Wren	1	1	
Turdidae			
American Robin	14	11	4
Hermit Thrush	4		
Olive-backed Thrush	1	5	2
Regulidae			
Golden-crowned Kinglet	4	4	
Ruby-crowned Kinglet	2	1	
Bombycillidae			
Cedar Waxwing	6	2	2
Vireonidae			
Solitary Vireo	4	1	1
Red-eyed Vireo	7		
Parulidae			
Tennessee Warbler	1	1	
Nashville Warbler	6		
Parula Warbler	4		
Magnolia Warbler	36	6	
Cape May Warbler	23	4	
Myrtle Warbler	29	5	
Black-throated Green Warbler	1		
Blackburnian Warbler	15	1	
Chestnut-sided Warbler	1		
Bay-breasted Warbler	81	9	
Oven-bird	1	3	
Canada Warbler	1		
Fringillidae			
Purple Finch	6	3	
Pine Grosbeak	3	1	
American Goldfinch	2	3	
Slate-colored Junco	13	5	6
Chipping Sparrow	1		1
White-throated Sparrow	5	5	3
Total	302	84	34

[1] Many birds lodged in tree-tops at time of collecting and were not examined closely.

of territorial males was reduced to approximately 19 per cent of the original by June 24. Continued collecting coupled with a steady influx of new birds maintained this low level until July 8. A total of 455 birds (420 adults and 35 young) were removed from the area during the entire collecting period.

Following the collection of the resident males, new males entered the area either during the night or early morning. The behavior of the new arrivals differed from the older residents in that they were much more active and vocal, singing more vigorously and more frequently. The importance of habitat niches in controlling the distribution of most species was indicated by the fact that most of the new arrivals established themselves in the same places that had been occupied by former residents of the same species.

For most species, over twice as many adult males were collected on the area as were present before the collecting started. This was due to the rapid influx and establishment of new territorial males following the removal of the original occupants. The carrying capacity of the forest from the standpoint of suitable sites to establish territories was not sufficient to accommodate all the males present, resulting in a surplus population. These surplus birds served as a reserve supply, replenishing areas that had been depleted. Some of the additional males collected were probably those that had territories adjacent to the shooting area, since the reduction in population pressure in the area due to collecting would permit an expansion of their territories into the shooting area. However, it is believed that the majority of evacuated territories were filled by entirely new birds. A few species were represented in the area only by wandering non-breeding birds.

The ratio of males collected to the number of territorial males on the area prior to collecting varied considerably with different species. The competition for territories among individual males of the same species appears to be the balancing agent between the size of bird populations and environment.

The number of adult females of later-breeding species collected was much less than the number of males. Although the general inconspicuousness of most breeding females undoubtedly explains in part why such small numbers were collected, there is evidence that a majority of the females in these species deserted the area after their mates had disappeared. The greatest differences in the ratios of the sexes collected occurred in the warblers (Parulidae) and vireos (Vireonidae).

In general, the later-breeding species, such as warblers, vireos, and thrushes, showed a greater disparity in the sexes of collected specimens than did the earlier breeders, such as the chickadees, nuthatches and kinglets. This condition seemed to be correlated with the fact that these earlier breeders did not refill evacuated territories to nearly the extent that the later breeders did. This could indicate a smaller reservoir of non-territorial birds and thus a lower population pressure in these species; or it could be the result of a more advanced condition of the breeding cycle during which the instinct to establish territory is absent or diminished in intensity.

LITERATURE CITED

BUMP, GARDINER, ROBERT W. DARROW, FRANK C. EDMINSTER, WALTER F. CRISSEY. 1947. The ruffed grouse. (New York State Cons. Dept.), pp. xxxvi + 915, 4 col. pls., figs. 1–171, tables 1–186, 127 sketches.

ELTON, CHARLES. 1936. Animal ecology. (Macmillan Co., New York), pp. xxx + 209, 8 pls., 13 figs.

KENDEIGH, S. CHARLES. 1947. Bird population studies in the coniferous forest biome during a spruce budworm outbreak. Dept. Lands and Forests, Ontario, Canada, Biol. Bull. 1: 1–100, figs. 1–32.

LACK, DAVID. 1933. Habitat selection in birds with special reference to the effects of afforestation on the Breckland avifauna. Journ. Animal Ecol., 2 (2): 239–262.

LACK, DAVID. 1946. The life of the robin. (H. F. & G. Witherby Ltd., London), pp. xvi + 224, pls. 1–8 (1 col.), figs. 1–6, rev. ed.

NICE, MARGARET MORSE. 1937. Studies in the life history of the song sparrow. I. Trans. Linn. Soc. New York, 4: vi + 247, 3 pls. (1 col.).

NICE, MARGARET M. 1941. The role of territory in bird life. Amer. Midl. Nat., 26 (3): 441–487.

NICHOLSON, A. J. 1933. The balance of animal populations. Journ. Animal Ecol., 2 (1): 132–178.

SOUTHERN, H. N. 1950. Marsh-tit territories over six years. Brit. Birds, 43 (2): 33–47.

U. S. Fish and Wildlife Service, Patuxent Research Refuge, Laurel, Md., and Washington, D. C., August 24, 1950.

11

Reprinted from *Auk*, **79**, 687–697 (1962)

TERRITORIAL BEHAVIOR: THE MAIN CONTROLLING FACTOR OF A LOCAL SONG SPARROW POPULATION

Frank S. Tompa

It has been shown frequently that territories are compressible. It is also known that the average territory size for a species in a given area may be significantly influenced by local conditions. However, it has been doubted whether territories can ever shrink to a point beyond which they can become no smaller and thus set an upper limit to the numbers of that local population.

This paper describes a situation in which a local passerine population, affected by favorable environmental conditions, could reach high densities for several consecutive years, and in which the upper limit was set by territorial behavior. It throws some light on how territorial behavior, combined with other factors, may regulate a given population. Conclusions, are based in part on data already available; nevertheless, references are made to factors—food and nestling mortality—that have not yet been fully investigated.

Song Sparrows (*Melospiza melodia*), the subjects of this study, defend a territory that normally includes mating, nesting, and feeding grounds, and thus falls into the territory category "A" of Nice (1941). This study, undertaken in order to find out the regulatory mechanism in a local Song Sparrow population, was started in the spring of 1960 and is still in progress. Some data concerning this population were also available for the period between 1957 and 1960 (R. Drent and G. van Tets, pers. comm.).

Study Area and Environment

This study was made on Mandarte Island in the Gulf Islands archipelago, on the southern coast of British Columbia; appropriate additional small-scale habitat and population surveys were carried out also on neighboring islands (Figure 1). Mandarte Island, with an area more than five hectares, rises abruptly from sea level to an average elevation of 15 meters. A longitudinal groove in the limestone block divides the island into a northeastern and southwestern half. The SW half is a grassy plateau, and is bordered by 20- to 25-meter-high cliffs along the shoreline, while the other half slopes gradually from the groove toward three- to five-meter cliffs of the NE shore.

The vegetation of Mandarte can be divided into three main zones: barren rocks and cliffs (including rocky beaches), grassy meadows, and shrubbery. The shrubbery is of primary importance in this study; it follows the longitudinal axis of the island and is supported by the relatively thick soil

687 The Auk, **79**: 687–697. October, 1962

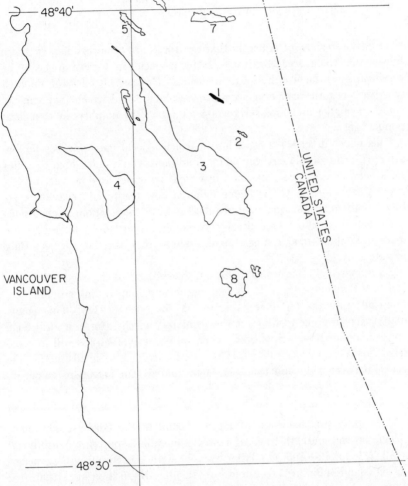

Figure 1.   Mandarte Island (in black) and neighboring islands. 1. Mandarte Island; 2. Halibut Island; 3. Sidney Island; 4. James Island; 5. Forrest Island; 6. Damville Island; 7. Gooch Island; 8. D'Arcy Island.

accumulated in the groove. It reaches the NE shore at some points and altogether covers somewhat more than one hectare, while grassland covers more than 50 per cent of the island.

The composition of shrubbery is given in order of abundance: Waxberry (*Symphoricarpos albus*), Wildroses (*Rosa* spp.), Saskatoon berry (*Amelanchier florida*), Blackberries (*Rubus* spp.), Ocean Spray (*Holodiscus dis-*

color), and Fireweed (*Epilobium angustifolium*), one or another dominating at different localities. At some places the shrubbery is heavily interwoven with stems of Northern Bedstraw (*Galium boreale*). Although a
few small cherry trees (*Prunus virginiana* and *P. emarginata*) are scattered through the shrubbery, taller trees can be found at one point only,
where the groove has expanded and allowed more soil to accumulate. Here
a group of trees—Douglas-fir (*Pseudotsuga menziesii*), Grand Fir (*Abies
grandia*), and Pacific Madrone (*Arbutus menziesii*)—has established itself
over an area of some 500 square meters, together with rich undergrowth
of willows (*Salix* spp.), Ocean Spray (*Holodiscus discolor*), and English
Hawthorn (*Crataegus oxyacantha*). It is important to note that at least
during the last 50 years no substantial change has occurred in the vegetation cover of the island (Anderson, 1916).

In the summer, grasslands are occupied by breeding colonies of Glau-
cous-winged Gulls (*Larus glaucescens*). On the steep SW cliffs two species
of cormorants (*Phalacrocorax auritus* and *Ph. pelagicus*) breed regularly,
while crevices along the shoreline offer nest sites to Pigeon guillemots
(*Cepphus columba*). Regular breeders in the shrubbery are the crows
(*Corvus brachyrhynchos*) and Song Sparrows, while Rufous Hummingbirds (*Selasphorus rufus*) are suspected to breed there. The number of
breeding crows was approximately 50 in each of the last four seasons. In
1961 two pairs of Red-winged Blackbirds (*Aegelaius phoeniceus*) raised
single broods after failures in previous years. There are neither amphibians
nor reptiles on Mandarte, and deermice (*Peromyscus maniculatus*) are the
only mammals.

SONG SPARROWS

Dense vegetation cover, absence of predators and nest parasites, sufficient moisture, and favorable climate, as well as other factors, offer exceptionally good conditions to Song Sparrows, when compared with adjacent areas. Because of the usually mild winters in this region, these
birds, belonging to the race *Melospiza melodia morphna* Oberholser, are
residents the year around. The length of the breeding season normally
extends from the second half of March to late July.

Field work was carried out during the last two summers and involved
the color banding of adult and juvenile birds, and recording features of
the breeding season. A total of 401 (93 adults and 308 young) birds were
banded. The sexes of the birds were determined by behavioral characteristics. Population counts were made approximately every fortnight during
the summers, and nearly every month through the winters 1960–1961 and
1961–1962. The number of breeding pairs remained relatively constant
(46–48 in 1960 and 47 in 1961), but the number of nonbreeding adults

was unknown for the first summer, because at that time they were not all banded. In 1960 the total adult population was 98–100. In 1961, 47 pairs of Song Sparrows started breeding on Mandarte. Five additional territories were occupied by unmated males throughout the season, while two unmated males composed the floating population. Thus the total number of adults at the onset of breeding was 101.

Survey of territories. Every available place in the shrubbery was utilized by Song Sparrows during the course of the study. Helped by obvious perching trees and singing posts, territory-owner males kept sharply defined boundaries throughout the breeding seasons, although the aggressiveness of the birds gradually decreased toward the end of the summers. Territorial activity reached its minimum in late July and early August, when adults entered their postnuptial molt. Suthers (1960) distinguished between utilized area and maximum territory occupied by Song Sparrows inhabiting a lakeshore environment; on Mandarte territories the shrubbery were too small and tightly packed for this distinction to be made.

When they were feeding nestlings, adults mainly searched for food in the shrubbery. However, when foraging for themselves, they frequently entered the grassland adjacent to their territories, and occasionally the tidal zone also—the latter, to a certain extent, serving as a common feeding ground. It was only at the end of the breeding season, when caterpillars became scarce, that parents collected food items, mainly lacewings and other insects, from the grassland.

Territory boundaries in the meadows, unlike those in the shrubbery, were difficult to define. Areas in the former used by adults often overlapped; occasional fights occurred, although never with such vigor as in the shrubbery. These observations, and the obvious insignificance of the area of utilized grassland compared with the area of shrubbery defended by a pair (as will be shown later), suggest that these grassy areas should not be considered as part of territories, but rather as a constituent of home ranges.

In 1961, before the territories were measured, daily observations were made throughout the season to determine the boundaries. During that period the size of individual territories proved to be constant. Measurements were made with a 33-meter (100-foot) tape. Certain errors are due to the very irregular shape of some territories, especially at places where the dense vegetation precluded accurate measurements. However, estimations showed that these errors were not more than 5 per cent. Measurements of grassy areas for the calculating of home ranges were rather approximate, because the boundaries here were not sharply defined, there were no obvious landmarks, and the ranges overlapped.

Territory and home range measurements are shown on Table 1. Considering the shrubbery part of home range as real territory, the average

TABLE 1

TERRITORY AND HOME RANGE MEASUREMENTS FOR BREEDING AND
UNMATED MALES IN 1961
(Measurements given in square meters)

Status of males	Number	Area of territories			Area of home ranges		
		Min.	Max.	Av.	Min.	Max.	Av.
Breeding	47	110	400	288	167	822	473
Unmated	5	65	105	82	(98	135	120)[1]

[1] Home range measurements for four unmated males. The fifth, which possessed 65 m² of shrubbery, utilized some 300 m² of grassland. However, the home range boundaries were too loose to take accurate measurements.

territory size for 47 breeding pairs was one tenth of the minimum for Ohio as reported by Nice (1943). Including grassy areas, the average home range size was still one fifth of the size of the Ohio minimum. Territories of the five unmated males were without exception smaller than the minimum for breeding pairs. Four of these males defended their territories throughout the season; in fact, they were still singing in late June and early July, when breeding males were very rarely heard. However, the fifth male did not show any sign of aggressiveness toward neighboring males. Two additional males were unable to establish territories; they stayed in the same general area of the island throughout the summer, apparently tolerated by territory-owning males.

As mentioned before, there was no correlation between the area of utilized grassland and that of defended shrubbery. Even when the latter was well below average, breeding pairs were able to rear two or three broods without any sizable grassland. On the other hand, one male was unable to obtain a mate with only 65 square meters of shrubbery, even though he also used more than 300 square meters of grassland with a home range area of ca. 365–375 square meters. Although features of the shrubbery might influence the size of individual breeding territories, the five unmated males, each defending an area of shrubbery of 105 square meters or less, possessed territories distributed over the island in a way that covered the range of all vegetation types. This suggests that the determining factor of successful mating was the amount of shrubbery defended by a male rather than the total area utilized.

Site tenacity and emigration. During the first summer 55 adult birds were banded—somewhat more than 50 per cent of the total adult population. Of these birds 29 survived the winter and started to breed in 1961. With one exception they all kept their old territories, apart from minor changes in boundary lines. One male, which stayed unmated through the summer of 1960, in the next summer moved to a neighboring territory where the owner had perished during the winter.

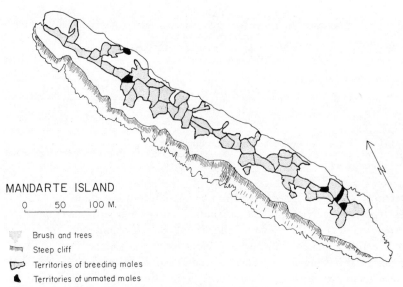

MANDARTE ISLAND

0   50   100 M.

⩗   Brush and trees

⩗   Steep cliff

⊂⊐   Territories of breeding males

▲   Territories of unmated males

Figure 2. **Map shows the arrangement of Song Sparrow territories.**

Of the 113 young banded in 1960, 21 were recovered in 1961. Nineteen of these stayed on Mandarte Island, while two had emigrated to adjacent islands. Those young that stayed can be further divided in the following way: six breeding males, 11 breeding females, and two unmated males. all of them occupied places left vacant through adult mortality. In one area of the island, following a heavy loss of adults and their replacement by yearlings in 1961, the territories were completely rearranged, although the number of territories for this area remained unchanged. (Figure 2 shows the subdivision of shrubbery into individual territories in 1961.)

During the summer of 1961 there was only one example of territory desertion by a female, whose mate had disappeared in July. This female settled down in September on a territory ca. 150 meters away, where the owner had lost his mate in August. From the beginning of breeding in 1961 to the end of January 1962, the loss in adult population was 22 per cent. Vacant places caused by adult mortality during the summer were filled by young birds in the fall, which once their postjuvenile molt had been completed, showed signs of territoriality.

Emigration of young birds first occurred in late summer and lasted until October. In late August two Mandarte young, both from the first brood, were observed on Halibut Island, ca. 1,300 meters to the south. They showed definite signs of territoriality. These birds were followed shortly by one young from the second and three from the third brood, which also established themselves on Halibut Island. The males obtained territories, and stayed there through the winter. One young of unknown sex from the

196

second brood was recovered on 1 September six km to the west of Mandarte on James Island. Both of these islands are forested and support local Song Sparrow populations with lower densities when compared with Mandarte Island.

This first phase of emigration was markedly correlated with the revival of territoriality in autumn. It affected those young that already showed signs of increasing aggressiveness, even though they were unable to establish themselves on Mandarte, since previously vacant places had been filled by other young males. Following this period, territoriality gradually decreased, reaching its minimum during November and December. During this time territory owners, both old and young, usually stayed on their home ranges, although the surplus young population gathered in loose feeding groups of 5–10, frequenting the meadows and the abandoned nesting grounds of cormorants.

At the end of January 1962 definite signs of the revival of territoriality were observed. Singing was often heard, and frequent chasing occurred. Vacant places caused by winter mortality were already filled by first-year birds. There had been no further sign of emigration since the fall. There were approximately 55 young in excess, when compared with the breeding population in 1960 and 1961. As indicated by data from 1961, the second phase of emigration occurs in February and mainly in March, which is the period of spring territorialism and includes the final spacing and mating, before the onset of breeding. This spring emigration concerns young males left without a territory, as well as first-year females that could not settle down on an already established territory. It is worthwhile to note that there was no detectable immigration of Song Sparrows to Mandarte Island during 1961, and no indication that this had happened in 1960.

DISCUSSION AND CONCLUSIONS

Howard's original theory of the functions of territory has been continuously argued and modified since 1920, partly because of different ways of interpreting it, partly because of the very complex nature of territorialism. The object of territory defense—nest site, mating and feeding ground, etc.—varies from species to species, and even within a given species it is under the influence of seasonal changes and features of the habitat. Nevertheless, most contemporary authors agree on two main functions of the territory, *i.e.*, behavioral and ecological. The former mainly concerns pair formation and maintenance of a pair, while the latter may include the assurance of adequate food supply during certain parts of the annual cycle and/or the regulation of population densities. However, the behavioral and ecological functions are often so closely related that distinction between the two becomes very difficult.

Based upon these considerations the following questions suggest themselves:

1. Does territorial behavior play any significant role in the control of
   the Mandarte Island Song Sparrow population?
2. If so, is the minimum required territory size for successful mating
   and/or breeding determined by:
   a. the amount of food available;
   b. other features of the habitat, such as type and density of the vege-
      tation, number of perching posts, exposure to prevailing wind, etc.
3. Or is the regulation mainly behavioral, the size of the territories depending upon the aggressiveness or tolerance of the individuals, thus
   assuring the owners of permanent mates, sufficient nesting sites, and
   the avoidance of interference during the breeding season.

On Mandarte Island the Song Sparrow population has remained at a
high density for at least the last two years, presumably as a result of the
suitable habitat, and the absence of severe mortality factors during the
annual cycle, especially in the breeding season. Although quantitative data
are not available for the years previous to 1960, observations carried out
by other students do not indicate any significant changes in the population
since 1957 (Drent and van Tets, pers. comm.). The unchanged environmental conditions on the island during the past 50 years would also favor
relative stability in population numbers.

The territories, as shown above, have been remarkably smaller than
those reported for this species on the continent. The fledging success during the last two years was higher (more than 60 per cent of the eggs laid)
than in Ohio (ca. 36 per cent, Nice, 1937) and San Francisco Bay (49.3
per cent, Johnston, 1956). Nest destruction, nest parasitism, and predation did not play any important role in nestling mortality. The abundance
and availability of food might be influenced by the vegetation, which
varies from one territory to another; thorough investigation is necessary
to find out whether starvation plays any substantial role in the less than
40 per cent egg and nestling losses, especially as the numbers of caterpillars, the main food item of nestling Song Sparrows, show a gradual decrease toward the end of the breeding season. Although there is no indication that the size of breeding territories is affected directly by the availability of food supply, the data are not quantitative enough to support
final conclusions on the food value of territories on Mandarte Island.

On the other hand, observations, population counts, and the survey of
territories suggest that territorial behavior plays an important, if not the
main, role in the control of this local population. Under different conditions this function of the territory is not always obvious enough to be
recognized. In an area with yearly changing physical and biological con-

ditions, with less suitable habitat, and with a significant annual immigration, the fluctuations in the breeding population may be substantial. In one year, because of favorable conditions, the habitat may become overcrowded, the territories will be relatively small and tightly packed, and hostility of the individuals will increase. On the other hand, following a bad year with low reproductive success, heavy mortality, and/or decreased rate of immigration, the population will be scarce, the territories loosely attached to each other with vacant places in the habitat, and the chance for encounter between neighboring birds relatively low.

Very rarely are conditions favorable for a length of time in one area for the same species, thus permitting high numbers for consecutive years with little fluctuations. But when this happens, the so far latent or less obvious function of territorial behavior in population control becomes operative. Where the suitability of habitats has been artificially raised, *e.g.*, by the provision of nest boxes, significant increase in breeding populations has occurred. This phenomenon was well demonstrated by experiments of Kluyver and Tinbergen (1953) in European titmice, and of von Haartman (1956) in Pied Flycatchers (*Muscicapa hypoleuca*). However, such an increase cannot be indefinite. With higher densities, under favorable conditions, the size of territories will decrease. This shrinkage in size is accompanied by an increasing resistance on the part of individuals, which in time will reach a point, beyond which no further decrease in territory size can occur. This procedure has been described in detail by Tinbergen (1957) in his discussion of the role of hostility (including both aggressiveness and avoidance) in the mechanism of dispersion.

That the same phenomenon can occur also under natural circumstances was shown again by Kluyver and Tinbergen (1953). In their study of titmice they found differential regulation of densities in neighboring, but basically different, habitats. In mixed woods, described as desirable for titmice, territories were relatively small, boundaries well defended, and the population showed small-scale yearly fluctuations. On the other hand, in adjacent, less favorable pine woods, territories were larger, with loose boundaries and less fighting, and the number of breeding birds changed significantly from one year to the other. Also, there was a detectable emigration from the mixed woods to the less favorable habitats with lower densities.

A similar situation due to habitat selection was described by Glass (1960) in the European Chaffinch (*Fringilla coelebs*). In this study, the difference in population densities between the stable populations in desirable habitats and the unstable ones exhibiting yearly fluctuations in less favorable habitats (and one-way emigration as a result of population pressure) was still more pronounced than in the study of Kluyver and Tinbergen.

Apparently the same is true on Mandarte Island, where favorable conditions allowed a stable population with high density during at least the last two but presumably five or more years. The territories have become extremely small. They are tightly packed, and all available space in the shrubbery is being utilized. Because of the increased pressure, the yearly population surplus emigrates to adjacent islands with less desirable habitats for Song Sparrows, in a manner similar to that shown with the titmice and chaffinch. The emigration of Song Sparrows occurs in two steps as shown by population surveys and field observations. The first phase begins in late summer and reaches its peak during the autumnal territoriality. This involves mainly young males, which already show signs of territorality, and cannot settle down since places vacant as a result of adult mortality have been filled. The second phase of emigration occurs during late winter and early spring, when territory establishment and mating takes place prior to the onset of breeding, and affects those yearlings that are still in excess and have a lower tolerance threshold toward crowdedness than the others. Similar two-phase emigration is characteristic of several territorial passerine species (Kalela, 1958).

This emigration of yearly surplus from Mandarte into less suitable habitats, with no detectable immigration from those areas, reveals the real importance of territorial behavior in the mechanism of population control at high density levels. While the size of individual territories is determined by the aggressiveness and tolerance of the neighboring males, and to a certain extent is under the influence of the vegetation cover, the success of obtaining a mate for the breeding season is dependent upon the amount of shrubbery defended by a male. Thereby the females, when rejecting or accepting a particular part of the habitat, may also play an important role in population control. By the acceptance of an appropriate amount of vegetation cover, the chance for successful breeding is increased, by preventing interference and possibly by assuring shelter and adequate food supply for the young. However, the role of the food on Mandarte Island is still to be investigated.

Acknowledgments

I am indebted to Drs. J. F. Bendell, D. H. Chitty, I. McT. Cowan, and M. D. F. Udvardy for advice and criticism throughout the course of the study, as well as to Messrs. R. Drent, P. R. Grant, and K. Vermeer for their friendly help in the field.

Summary

In 1960 a study was undertaken to reveal the controlling mechanism of a local Song Sparrow population on Mandarte Island, British Columbia. The present paper discusses the role of territorial behavior in the regulation of breeding numbers.

This population is at a high density, and has showed relative stability over the past years. The survey of territories, observations on behavioral aspects, population counts, and the emigration of yearly surplus into neighboring, less attractive habitats indicate that territorial behavior plays the most important role in the regulation of Song Sparrows on this island. It has been concluded that the size of individual territories is determined by the aggressiveness and tolerance of the individuals, while the chance for successful mating is apparently dependent upon the size of shrubby area defended by a male.

LITERATURE CITED

ANDERSON, W. B. 1916. Bare Island Indian Reserve. B.C. Proc. Mus. Nat. Hist. Annual Reports, 1915. Pp. 14–16.

GLASS, P. 1960. Factors governing density in the chaffinch (Fringilla coelebs) in different types of wood. Arch. neerl. Zool., **13**: 466–472.

HAARTMAN, L. VON. 1956. Territory in the Pied Flycatcher *Muscicapa hypoleuca*. Ibis, **98**: 460–475.

JOHNSTON, R. F. 1956. Population structure in salt marsh inhabiting Song Sparrows. Part II. Density, age structure, and maintenance. Condor, **58**: 254–272.

KALELA, O. 1958. Über ausserbrutzeitliches Territorialverhalten bei Vögeln. Ann. Acad. Sci. Fenn. Ser. A. IV. Biol., **42**: 42 pp.

KLUYVER, H. N., and L. TINBERGEN. 1953. Territory and the regulation of density in titmice. Arch. neerl. Zool., **10**: 265–289.

NICE, M. M. 1937. Studies in the life history of the Song Sparrow I. Trans. Linn. Soc. N.Y., IV. 247 pp.

NICE, M. M. 1941. The role of territory in bird life. Amer. Midl. Nat., **26**: 441–487.

NICE, M. M. 1943. Studies in the life history of the Song Sparrow II. Trans. Linn. Soc. N.Y., VI. 329 pp.

SUTHERS, R. A. 1960. Measurements of some lake-shore territories of the Song Sparrow. Wils. Bull., **72**: 232–237.

TINBERGEN, N. 1957. The functions of territory. Bird Study, **4**: 14–27.

Department of Zoology, University of British Columbia, Vancouver, B.C., Canada.

12

Reprinted from *Proc. 13th Intern. Ornithol. Congr.*, 740–753 (1963)

Ecological Significance of Territory in the Australian Magpie,

Gymnorhina tibicen

ROBERT CARRICK

Division of Wildlife Research, C.S.I.R.O., Canberra, Australia

Much has been written on the possible significance of territorialism in birds and other animals; but, while some of the functions of territory appear self-evident enough, actual proof of their operation in nature has been difficult to obtain. In a comprehensive review of this subject, which cites the relevant literature to that time, Hinde (1956) was still able to write: "There is no direct evidence that territory limits the total breeding population in all habitats. . . . Territorial behaviour may reduce disease, but this is unlikely to be a significant consequence except in some colonial species. . . . The functions of territorial behaviour are extremely diverse, and the quality of the evidence available for assessing them is little different from that available to Howard." This last point is still substantially true, half a century after Howard. Wynne-Edwards (1962) has given a fully documented account of the territory habit, which he rightly interprets as no different in purpose from the other forms of social behavior that constitute the homeostatic machinery whereby populations of animals are widely dispersed and excessive increase of numbers, with consequent depletion of food and other resources, is prevented.

The two main questions arise from each side of the population equation, and each contains several others. *Firstly*, does territorialism reduce productivity (fecundity) significantly below the biotic potential of the species? To what extent does it do so, and how is the reduction achieved? Are adult females unable to breed through denial of suitable nest sites, mates, or food supply? Or is maturity prevented by lack of the necessary proximate stimuli, or even by inhibitory factors? *Secondly*, does territorialism buffer its adherents from important causes of mortality? Does it prevent or reduce the risk of starvation, i.e. what is the relation between territory and food supply? Does it confer safety from predators or protection from disease?

A main difficulty of research on this problem is that the effects of territorialism usually have to be inferred from the study of the territorial individuals alone; there is no nonterritorial element in the same species, or at least it is barely visible, to serve as a control and provide comparative data on natality and mortality under the two systems. This stems from the fact that those individuals that fail to attain territorial status are either excluded from the habitat that the species requires for food, shelter, and reproduction, and so they succumb, or else they live cryptically in and around the margins of the preferred habitat. In the case of the strongly territorial Australian Magpie (*Gymnorhina tibicen*), however, there is a large and obvious overflow

Proc. XIII Intern. Ornithol. Congr.: 740–753. 1963

population outside the wooded breeding territories that is not territorial, at least in the same sense as the breeding birds, and that can maintain its numbers without recourse to migration. This stems mainly from the fact that this species is primarily an insectivorous ground-surface feeder, but is versatile enough to explore other food sources and even resorts to carrion and pasture foliage when necessary. The controlled experiment that we would like to set up in so many other species exists naturally.

Fig. 1. Adult cock Australian Magpie (*Gymnorhina tibicen*) giving the aggressive carol at the boundary of its territory. Photo by E. Slater.

G. tibicen (Fig. 1) is a member of the Australo-Papuan family Cracticidae, allied to the Corvidae. It stands about 9 inches high; and its adult plumage pattern, with jet black underparts and white nape, rump, and wing-flash, advertises the fact that predation on it is unlikely to be important; so exposed habitats can be used. A reasonable solution to the problem of diurnal shelter from the elements, especially heat and wind, can usually be found even in open country, and these individuals resort to communal night roosts some distance from their feeding grounds. This magpie is a sedentary species, with conspicuous behavior; it is not shy and it is readily trapped. Its aggressive carol, energetic defense of the territorial boundary, and readiness to attack intruders, including ornithologists, further assist field study (Fig. 1). The immature first-year birds are distinguished by their grayish, not black, plumage, and the sexes by the grayish lower nape and rump and the shorter bill of the female. It nests typically in trees, but shows considerable adaptability.

Fig. 2. The central part of the study area at Gungahlin, Canberra. Open savannah woodland (*Eucalyptus* spp., exotic conifers, and deciduous trees) and adjacent pasture are the habitat for permanent breeding territories. Photo by C. Totterdell.

The scene of the present study is 5 sq miles of open savannah, woodland, and pasture (Fig. 2 and 3) around Gungahlin, the headquarters of the Division of Wildlife Research, C.S.I.R.O., outside Canberra, Australia. The native gums, among which *Eucalyptus blakeleyi* predominates, form sparse cover with ground feeding places throughout and around them; exotic trees, including conifers and elms, are planted more compactly, and offer equally acceptable nest sites to the magpie. The study area was chosen to include samples of breeding habitat with intervening open ground. The basis of this study is individual color banding of territorial birds and group banding of others; over 650 of the former, and 2,500 of the latter, have been banded during 1955–62. Adults and young in territories on the study area have been banded annually, and in the four winters 1957, 1958, 1960, and 1961, about

80 percent of the nonterritorial birds living in the treeless pasture habitat have been banded. Since 1955, some 220 territorial groups have been studied. The area of woodland cover around Gungahlin (Fig. 2), which contains two-thirds of the territorial breeding groups in the study area, has been most intensively studied; every magpie there is color banded, i.e. about 150 birds in 40–45 territories. Over 1,000 specimens for dissection have been taken from comparable open and wooded terrain several miles from the study area, and experiments involving manipulation of internal or external environment

Fig. 3. A marginal part of the study area at Gungahlin, Canberra. The open pasture is populated by nonbreeding flock birds and groups in open territories; marginal territories form around isolated trees or bushes. Photo by C. Totterdell.

have been made mainly outside the study area. Counts of the territorial birds in the area, with identification of color-banded individuals, are made every 3 months, and the free-flying juveniles still present in February receive their color combination then.

This is a preliminary account of the main findings that relate to the ecological significance of territorialism. These results are based on extensive data from birds of known identity and history, and a full account of this study will be published in a future issue of *C.S.I.R.O. Wildlife Research*.

SOCIAL ORGANIZATION AND USE OF HABITAT

The Australian Magpie forms social territorial groups of 2–10 birds. Most territories fall within the 5- to 20-acre range, with an average of about 10

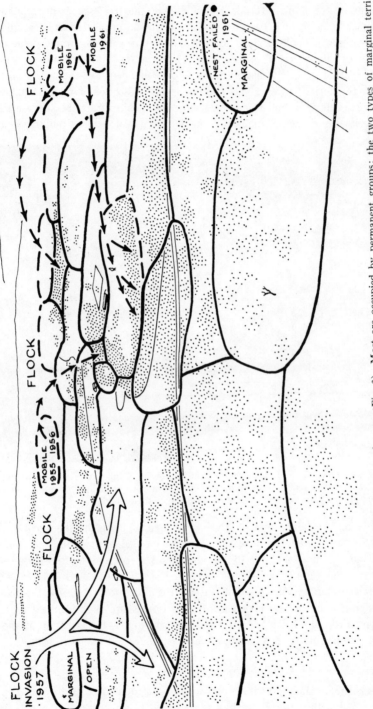

Fig. 4. The territories in the central part of the study area (Fig. 2). Most are occupied by permanent groups; the two types of marginal territory are shown at top left (inadequate cover) and bottom right (inadequate pasture); small arrows show where mobile groups attempted to nest, and the large arrow shows where flock birds invaded breeding territories in the hard winter and spring of 1957.

acres, but smaller areas are held where surrounding pressure is strong, and larger ones at the margins of the territorial area where there is no neighboring group. A group of two birds is always adult cock and hen; there may be six adults in a group, with any combination of sexes but a maximum of three breeding individuals of either sex in one group. The average number of adults per group is three, and males seldom outnumber females. Bigamy is common, and trigamy occurs. There is no relation between the size and quality of the territory and the number of birds that occupy it. At any time a large territory may have a small group, and vice versa; groups can fluctuate in time within the range of 2–10 birds without change of boundary.

The upper limit of territory size is, by observation, the largest area that the group can obtain and hold effectively; the better territories contain far more nest habitat and shelter than the group can use, and may well contain a food surplus also, although this requires to be tested by experimental alteration of food level. In a few instances the constant lateral pressure at territory boundaries enabled a group to increase its area when a neighboring group departed, but this gain was later surrendered, presumably through inability to defend the larger area. The lower limit of territory size is set by the amount of feeding pasture required to sustain the group, for a much smaller area than this can contain superabundant cover. Thus, territory size is largely determined by group size, although it is difficult to see what determines the level of the latter, which is similar throughout the range of the black-backed and white-backed forms of *G. tibicen* in eastern Australia. It is tempting to suggest that the group is limited by the number of birds that the dominant member can control, but the Western Australian Magpie (*G. dorsalis*) differs in having groups of up to 26 birds, with as many as six adult males in some groups, that occupy territories of 30–150 acres, and there are apparently no flocks (Robinson, 1956).

In many changes of territory ownership, no healthy reigning group has been dispossessed, regardless of the size of its territory or the relative strength of opponent groups. The members of defending and attacking groups fight as a team, with the advantage strongly in favor of the former.

It is convenient to recognize five social categories based on the quality of habitat occupied by each (Fig. 4 and 5), but these form a graded series and the system is anything but static, for birds and groups in the poorer environments are continually striving to improve their position in the habitat scale. Groups compete for tree cover with adjoining pasture feeding areas, which results in the open and marginal woodland, and also some open pasture, becoming subdivided into territories that are held for periods and defended with a tenacity proportionate to their quality as places to breed and feed.

1) *Permanent* groups hold territories that provide an adequate or surplus amount of all requirements all the year round. There are many more trees than the small number of birds requires for shelter, roosting, or nesting, and seasonal weights give no indication of food shortage at any time. Birds

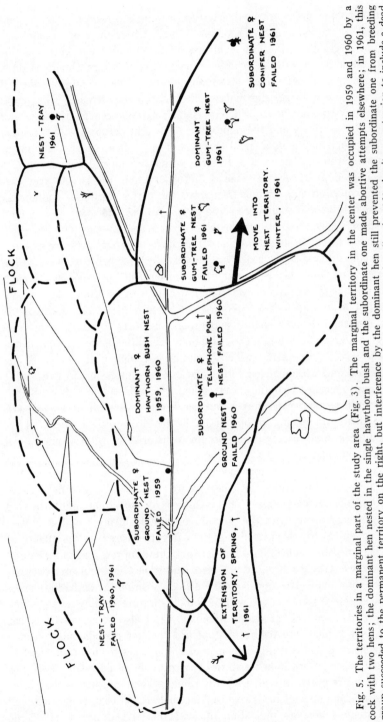

Fig. 5. The territories in a marginal part of the study area (Fig. 3). The marginal territory in the center was occupied in 1959 and 1960 by a cock with two hens; the dominant hen nested in the single hawthorn bush and the subordinate one made abortive attempts elsewhere; in 1961, this cock with two hens; the dominant hen nested in the single hawthorn bush and the subordinate one made abortive attempts elsewhere; in 1961, this group succeeded to the permanent territory on the right, but interference by the dominant hen still prevented the subordinate one from breeding successfully. A group of two cocks and three hens found the center territory inadequate for breeding in 1961, despite an extension to include a dead tree. A similar variability of response was shown by the open groups at top left and top right, and only the latter nested on the artificial tray provided.

remain in these optimal territories all day and all year, and make no attempt to move. Virtually all successful breeding is done by these birds. A permanent group may contain birds of all ages.

2) *Marginal* groups occupy territories with an inadequate amount of either cover or feeding area. They form around one or two small trees or bushes (or even artificial nest sites such as telegraph poles or tall wireless masts) on the outskirts of better cover; also in open woodland poorly defended by the surrounding groups but with inadequate pasture for feeding at all seasons. Attempts to breed rarely succeed to the point of fledging. Marginal groups usually consist of adult birds only.

3) *Mobile* groups commute between a separate feeding area in the open and a nesting–roosting area among trees. The latter is held against the strong opposition of neighboring groups, and mobile groups exist mainly during the breeding season. Breeding always fails, usually at an early stage, and mobile groups do not contain first-year birds.

4) *Open* groups form in areas of treeless pasture that provide adequate feeding all year, except possibly during severe drought or hard frost. They roost in the denser woodland that is not otherwise used by this species; the daily flight is usually within a mile, and members of the same open group may go to different roosts. An open group can last several years, and some become mobile groups in spring. Open groups contain only adult birds, but they make no attempt to nest.

5) *Flock* birds are nonterritorial. They are birds of all ages and both sexes, and some may have bred as members of territorial groups now disbanded. They form loose flocks of a few up to several hundred individuals that feed in open pasture and roost in woods. More intensive study might well reveal that most flock adults are in fact in open groups with varying degrees of attachment to feeding area or constancy of membership, but more stable during the breeding season. The flocks show slight mobility throughout the year, and about half of the open terrain in the study area is occupied and the other half left untenanted by them at any one time. They do not attempt to nest.

CHANGES OF STATUS

In this territorial system, there is considerable temporal and spatial stability of groups and individuals; in May 1962, 16 of the original 38 permanent groups present in 1955 still occupied the same territory, and 20 percent of males and 18 percent of females were in the territories where they were first banded as adults in 1955 or early 1956. These figures are rather low because they include losses from human activities that would not occur over most of the magpie's range. This stability is maintained by constant vigilance in a dynamic situation in which there is continual daily effort all the year round, with an upsurge of activity in July to October, i.e. before and during breeding, on the part of both groups and individuals to improve their social status.

Individual changes may result in increase, replacement, or decrease in the members of a group. Most birds leave their natal territory during their first year, some in the second, and a few in the third, but a small proportion continues to live and breed there. More females than males do this, and the oldest hen that has remained in the parental territory is now almost 7 years old, while the oldest cock is 3½ years old. It is exceptional for an adult to be added to an existing group, but this has occurred when a sick hen was unable to repel a flock hen that became established in the group before the resident hen recovered. Members of a group repel their own sex, but each sex supports the other once a contest is under way. When a vacancy is created, as by mortality, for either sex, a replacement by one or more birds of the same sex may occur; this is more usual in the case of females than males, presumably because of the greater ability of the latter to hold on to their territory. Mortality causes decrease in a group, and either sex may emigrate from a group in an inadequate territory to one in a better territory where there is a sex vacancy for it.

A group preformed in the flock, or one in occupation of a poor habitat, may succeed in forcing its way into a better habitat, thus creating a new territory. Loss of the dominant adult, usually the male, often leads to break-up and displacement of a group by a new one. A group seldom becomes too large, but in one case a group of 3 males, 4 females, and 3 immatures became subdivided and the adult male and female, which separated from the others, continued to occupy part of the original territory. Expansion of territory with change in the composition of the group can occur when a neighboring group goes out and the territory is not immediately claimed by an incoming group; unless the first territory is small, the group is not usually able to defend the expanded area effectively and has to surrender all or part of it eventually.

THE GONAD CYCLE AND BREEDING

In the Canberra region, egg laying extends from August through October, and some seasons start earlier than others. This is preceded by increased epigamic activity of adult males from July onward, when many immatures are evicted from territorial groups and many open and marginal groups make determined attempts to secure adequate tree cover for nesting. The female alone selects the nest site and builds the nest; copulation occurs only at her invitation. The clutch contains 1–6, average 3, eggs, and some re-laying after failure occurs in earlier seasons, but little in a normal one. The cock does not brood, but may feed the hen on the nest and may play a variable part in feeding the young.

Males of all ages and all environments and social positions have motile sperm in the breeding season. Testes are largest in adult males in permanent territories, and much smaller in territory or flock 1-year-olds. Where nutrition is adequate, as it always seems to be, physical environmental stimuli alone appear capable of bringing the testis to maturity, after which age and social status in that order determine how far development will proceed. Testis size and sperm production do not appear to be affected by antagonistic relations between groups, or between a group and trespassing individuals, or within the group.

No 1-year-old female has been known to ovulate. Some 2-year-old hens breed, but even in permanent groups some females of this age or even older may not breed. In spring the ovary undergoes partial development in every case, and the final rapid increase in size of oocytes, with associated nest-building behavior, depends on the presence of certain critical stimuli as well as the absence of inhibiting factors. To attain ovulation, the hen must be a member of a social group in a territory that offers an acceptable nest site, but the threshold value of the latter in different individuals varies from a high tree to a low bush or post, and in one exceptional case the ground. Tradition is probably important, and preliminary experiments in open territories with artificial sites in the form of wooden trays on poles, bare or decorated with foliage, and with small trees, suggest that foliage as such sometimes has valency. Male stimulation of the female does not occur, for hens whose adult cocks were caponized with oestrogen implants and made effeminate to the point of building nests and soliciting, continued to build and lay (infertile eggs) normally on the same dates as control groups.

Even in the presence of adequate proximate stimuli, oocyte development and nest building can be inhibited by emotional factors, such as intrusion of a strange magpie of either sex into the territory, an undue amount of boundary fighting, or domination by another female of the same group. The psychosomatic effect of alien individuals, even on hens in first-grade permanent territories, has been observed as it occurred naturally in several situations, notably in 1957 when flock birds, which overran some territories (Fig. 4) during the frosty winter and were not evicted by spring, caused inhibition of nesting among the resident hens; this effect has been confirmed experimentally. The response of individual hens to similar stimulatory and inhibitory factors in the environment varies widely.

No open group has reached the stage of nest building. The breeding performance of mobile groups varies from failure to commence building to an occasional successful hatching, but, because of predation while the adult is absent at the feeding ground, no hen in a mobile group has been able to fledge its young. Those mobile hens that become sufficiently established to build and lay often lose their eggs from the direct attack of neighboring magpies; or else the eggs become addled, or eggs or nestlings fall to predators when the hens are engaged in boundary fights or are absent in the feeding area. The most common predator of eggs and nestlings is the Australian Raven (*Corvus coronoides*). Marginal groups often fail to nest, but a small percentage rear young to the free-living stage. It is the permanent groups that produce the annual increment to the population, but even their breeding

TABLE 1.—NUMBERS OF *Gymnorhina tibicen* IN THE GUNGAHLIN STUDY AREA

Status When Censused	TERRITORY[a] AUG.–SEPT.		FLOCK[b] MAY–JULY		TOTAL IN POPULATION	
	No.	Percent	No.	Percent	No.	Percent
ADULT FEMALES						
1957[c]	92	36	161	64	253	100
1958	103	35	189	65	292	100
1960	112	39	178	61	290	100
1961	111	40	168	60	279	100
AVERAGE DURING 1957–61						
Males	79	21	296	79	375	100
Females	107	38	174	62	281	100
Total adults	186	28	470	72	656	100
(Sex ratio ♂ : ♀)	(43 : 57)		(63 : 37)		(57 : 43)	
1st-year birds[d]	ca. 15		ca. 70		ca. 85	
Total population	ca. 200		ca. 540		ca. 740	

[a] Territory birds include all hens that have an opportunity to breed, i.e. permanent, marginal, and mobile groups.
[b] Flock birds are nonterritorial individuals plus open groups.
[c] Owing to improved methods of observation and trapping in subsequent years, the numbers for 1957 may be rather low.
[d] Counts taken in midwinter.

rate is reduced by aggression between groups and by sex dominance within groups, as well as by the usual factors not directly associated with social territorialism. The extent to which this system reduces breeding to one-quarter of the potential it would have in the absence of territorial capitalism of breeding sites and sociosexual aggression and dominance is shown in Tables 1 and 2.

ANNUAL PRODUCTIVITY AND MORTALITY

In a good breeding season about one juvenile magpie per adult territorial female reaches the free-flying stage in January, and the number is much

TABLE 2.—DEGREE OF BREEDING FAILURE CAUSED BY TERRITORIALISM

Year	1958		1960	
	No.	Percent	No.	Percent
Total number adult females in study area	292	100	290	100
Nonbreeding females in flocks	189	65	178	61
Failing females in territories, due to intergroup aggression	16	6	21	7
and intragroup dominance	13	5	14	5
Total reduction of nests	218	76	213	73

lower in poor seasons. The high survival of adult birds during the course of this study indicates that a low annual death rate, especially in the permanent territorial groups, is adequate to cancel the normal increase from natality.

Starvation has not been evident in this study, although birds of all ages in the wide range of habitats, but each with good feeding pasture, have shown significant differences of body weight throughout the year. The same is true of stomach contents and fat reserves.

Predation by crows and hawks occurs up to the free-flying stage, and is more severe in poorer cover, but natural losses among older birds, even including those due to feral cats, are not considered serious. The Peregrine Falcon (*Falco peregrinus*) has occasionally hunted the area and taken adult flock birds in the open. Immature *Homo sapiens* of all ages take a steady but small toll of nestlings, which are popular if illegal pets, and of adults, which afford target practice in the absence of other game and, in the case of the cock during the breeding season, engender retaliation for their unprovoked attacks on people. Territories on main roads consistently lose their juveniles, and an occasional adult, in traffic accidents, and rabbit traps and other human agencies also account for a small number of birds, mainly in territories.

Diseases of many kinds have been identified during this study, some of them lethal, and at least one is considered to be an important primary cause of death. This is *Pasteurella pseudotuberculosis*, which killed large numbers of flock birds during the cold wet winter of 1956, but, being contact-spread, did not cause a single death among the territory individuals, all of which were banded, in woodland closely adjacent to open pasture where dead and dying birds could be picked up daily at the height of the epidemic. The flock birds exposed to this infection showed no sign of debility, as compared with territory birds, that might have predisposed them to mortality, and the conditions that favor the disease appear to occur too infrequently for development of resistance to it. During the harder winter of 1957, the food on the more exposed pastures became unavailable, and many flock birds concentrated on softer ground and haystacks, where they picked up the spores of the fungus *Aspergillus*, which became a secondary cause of death of some importance that year. Both of these diseases and several others take a constant low-level toll of magpies.

CONCLUSIONS

Territorialism and associated sociosexual interactions limit breeding to about one-quarter of the adult population of *G. tibicen*.

Territorialism buffers that element of the population against important mortality from disease, and probably also protects it from predation.

The completeness of the territorial habit in *G. tibicen*, in which the social group lives permanently within its territory, indicates that the food supply of this area is always adequate; nor is there evidence that flock birds come up against the food limit.

ACKNOWLEDGMENTS

I am much indebted to F. N. Ratcliffe and the Executive of the Commonwealth Scientific and Industrial Research Organisation, Australia, for enabling this study to be done. I am also grateful, among many others, to my wife, who has been mainly responsible for identifications of banded birds, to W. J. M. Vestjens, who has borne the brunt of the trapping and collection of specimens and breeding data, to Mrs. Amy Bernie for maintenance of records and analysis of data, to R. Mykytowycz for pathological examinations, and to I. C. Rowley and K. Keith for assistance during the earlier years.

During the discussion after this paper was read at the Congress, significant questions and comments were made by David Lack of Oxford and Richard F. Johnston of Kansas.

SUMMARY

A study of population ecology in the Australian Magpie (*Gymnorhina tibicen*) during 1955–62 at Gungahlin, Canberra, is based on 650 birds individually color-banded and 2,500 others banded; 220 territorial groups have been studied, and three field experiments done on proximate factors in the stimulation and inhibition of breeding. A preliminary account of the evidence on the ecological significance of territorialism is given in this paper.

Open savannah woodland and adjacent pasture are permanently occupied by breeding groups of 2–10 birds, with a maximum of 3 adults of each sex; an average territory is 10 acres. Similar social groups also hold marginal territories that are deficient in feeding pasture or tree cover for shelter and nesting. Open treeless pasture supports a large nonbreeding element, including former breeders, in the form of slightly nomadic flocks and some territorial groups that may become mobile in spring between separate feeding and breeding areas. Successful breeding is virtually confined to the permanent groups, and productivity is low; territorialism reduces breeding to one-quarter of its potential. Groups and individuals in the poorer habitats make constant efforts to improve their status; the causes of changes in the composition or status of groups are described.

Testes mature in all situations, but the response of individual females to the environmental situation is more variable. The adult ovary requires adequate stimuli from group status and suitable nest site. It can be inhibited by sociosexual factors, such as conflict with neighboring groups or intruders in the territory, or by dominance within the group, which cause a psychosomatic reaction involving ovarian repression.

In this species, territorialism has led to a high degree of numerical and spatial stability. It buffers the permanent occupants of the preferred habitat from important disease mortality, and probably also from predation.

LITERATURE CITED

HINDE, R. A. 1956. The biological significance of the territories of birds. Ibis 98: 340–369.

ROBINSON, A. 1956. The annual reproductory cycle of the magpie, *Gymnorhina dorsalis* Campbell, in south-western Australia. Emu 56:233–336.

WYNNE-EDWARDS, V. C. 1962. Animal dispersion in relation to social behaviour. Oliver and Boyd, Edinburgh. 653 p.

IV

Colonial and Group Territories

Editor's Comments on Papers 13, 14, and 15

13 **Buechner:** Territorial Behavior in Uganda Kob
Science, **133,** 698–699 (1961)

14 **Robel:** Booming Territory Size and Mating Success of the Greater Prairie Chicken
(*Tympanuchus cupido pinnatus*)
Animal Behaviour, **14,** 328–331 (1966)

15 **Bartholomew and Hoel:** Reproductive Behavior of the Alaska Fur Seal, *Callorhinus ursinus*
J. Mammal., **34,** 425–435 (1953)

Certain birds and mammals utilize small, highly clustered breeding territories defended by the male. This is the lek, or arena, form of territorial behavior shown by certain grouse, manakins, hummingbirds, pinnipeds, and ungulates. Recently Paul J. Campanella has described somewhat comparable lek behavior in dragonflies (*Behaviour,* in press). All lek species have a polygynous mating system with strong competition between males. Territories on a lek are small, have precise boundaries, and are contested vigorously.

The discovery of the lek system in the Uganda kob (Paper 13) was the first to be made among ungulates. In kob the breeding season, hence the lek itself, is maintained throughout the year. No male is able to defend his territory for more than a few months (or even days), so the composition of the males on a lek is constantly changing.

Lek behavior is of interest because the phenomena of territory and social hierarchy are inseparable on a lek. It is not sufficient that a male acquire a territory on the lek to be successful in breeding. Certain territories are much more attractive to females than others. And the activity of the individual males is probably also a factor in determining how many females a given male will be able to mate. John W. Scott (1942) had recognized that certain sage cocks did most of the mating. Subsequently, Robert Robel (Paper 14) has reported the same thing in a marked population of prairie chickens and shown how the success of individual males changes from year to year. Robel is Professor of Zoology at Kansas State University. He is best known for his studies of lek-breeding grouse of North America and Europe.

The breeding colonies of certain pinnipeds may contain over a thousand members. One of the most highly developed polygynous mating systems is that of the Alaska fur seal (Paper 15). Among fur seals the factors leading to high mating success seem more complicated than in grouse. They include, in addition to the age and strength of the male, such factors as the degree of physical isolation from competition and proximity to the shoreline.

Bartholomew is best known for his work at the interface between physiology, ecology, and behavior, particularly with regard to desert vertebrates and marine birds. For many years, however, he studied social behavior in marine mammals. He is currently Professor of Zoology at University of California at Los Angeles. He is recipient of the coveted Brewster Award of the American Ornithologists' Union, as well as other awards.

Even in monogamous species, the position of the territory within a breeding colony can affect reproductive success. Although it was not possible to include it here, a paper by Richard Tenaza, "Behavior and nesting success relative to nest location in Adelie penguins" (1971), is important for showing the relation between territorial behavior and selection pressure for colony size. Tenaza found that those birds nesting in the center of a penguin colony performed better than those at the periphery as a result of less predation, less disturbance from other penguins, better nests and larger clutches. Since in a larger colony a higher percentage of all nests would be centrally located, selection should favor large colonies.

Some animals show a higher order of territorial organization in which the territory is held by a group of animals, including adult males and females and their young. The entire group unites in common defense of the territory. A classic study of this kind was that by John A. King: *Social Behavior, Social Organization, and Population Dynamics in a Black-tailed Prairiedog Town in the Black Hills of South Dakota* (1955). This long-out-of-print classic is once again available. King found that the sexes played different roles in defense of the territory. Females sounded the alarm when an intruder appeared, but it was the males who provided the actual defense. All members of the group territory shared a common burrow system, but during the breeding season, each female defended her individual nesting burrow within the group territory. Another example of higher-order group territory is that established by the Australian magpie (Paper 12).

References

King, J. A. 1955. *Social Behavior, Social Organization, and Population Dynamics in a Black-Tailed Prairiedog Town in the Black Hills of South Dakota.* Contr. Lab. Vert. Zool. No. 67. University of Michigan, Ann Arbor, Mich. 123 pp.

Scott, J. W. 1942. Mating behavior of the sage grouse. *Auk,* **59**: 477–498.

Tenaza, R. 1971. Behavior and nesting success relative to nest location in Adelie penguins. *Condor,* **73**: 81–92.

13

Reprinted from *Science*, **133**, 698–699 (1961)

Territorial Behavior
in Uganda Kob

Abstract. Territorial behavior of the Uganda kob, *Adenota kob thomasi* (P. L. Sclater), is largely the defense of small, fixed territories within a central area of concentrated territorial activity. This area is surrounded by a zone of more widely spaced territories. Females enter the territorial ground throughout the year for the purpose of breeding.

The well-defined pattern of year-around territorial behavior of the Uganda kob was discovered (*1*) in March 1957. Although other African antelopes exhibit territoriality, the behavioral pattern of the Uganda kob appears to be unknown among any of the other species. Initial interpretations were verified in several widely separated herds over the following 15 months, and from June through August 1959 an intensive study on marked animals was conducted to ascertain details of this behavior in the Semliki Game Reserve, 20 miles north of Fort Portal, Uganda.

About 10,000 Uganda kob occur in approximately 100 square miles of habitat included by the reserve and vicinity (*2*). The entire population utilized only 13 known territorial breeding grounds, suggesting that certain physiographic requirements limited the number of breeding grounds. Each ground was situated on a ridge, knoll, or slightly raised area characterized by short grass, good visibility, and proximity to a permanent stream. A territorial ground of average size consisted of a central area of intensive activity, about 200 yards in diameter, containing 12 to 15 more or less circular territories varying from 20 to 60 yards in diameter. Some had common boundaries; others were separated by neutral areas. The territories were fixed in position and could be recognized by a central area of closely cropped grasses on heavily trampled ground and boundaries clearly demarcated with longer, less grazed grasses. In the surrounding peripheral area, a zone 100 to 200 yards wide, about twice as many territories were found as in the central area. Territorial breeding grounds remained in the same locations from July 1957 to September 1959. Apparently the grounds are seldom shifted to new locations. Territorial behavior occurred throughout the year, with slight peaks of intensified activity during the two rainy seasons (April to May; October).

Fifty territorial males were captured by the use of paralyzing drugs and marked for identification (*3*) at three territorial grounds separated from each other by distances of 2 to 2½ miles. Only about a third of the tagged individuals were seen again on territories, but almost invariably an individual returned to the same territory it had occupied previously. Some males remained on territory less than a day, others for several days or a few weeks. The longest record was 2½ months. During occupancy of a territory the male left once or twice in daylight to obtain water and forage. Observations on about 50 additional territorial males marked naturally by scars, abscesses, broken horns, and other characteristics showed that individuals not subjected to the experience of capture returned to territories in about the same frequency as marked individuals. A high rate of exchange of individual males within individual territories seems to be a natural feature of the behavior pattern.

Exchange of males on a given territory involved serious fighting between the occupant and a challenging male, the latter running in rapidly, through the peripheral area, into the central area of activity to what appeared to be a predetermined territory, perhaps one that he had occupied previously. This pattern and other aspects of territoriality of Uganda kob have been documented in a 30-minute motion picture film, in sound and color (*4*). Fights for possession of a territory were the longest and most serious of the fights observed. Two deaths positively attributable to fighting were recorded. Successful fights in which the challenging male defeated the occupant of a territory and took possession were observed about a dozen times; more frequently unsuccessful challenges were observed, some of which involved hard fighting. Often males ran into the central area only to be chased out by territorial males, each making short runs or threatening gestures in relay until the invader ran out of the prized area.

Defense of boundaries between occupied territories was accomplished mostly by ritualized display rather than intensive fighting. For example, to maintain the integrity of territories, males frequently walked toward one another with lowered ears and met at the boundary without fighting. Feigning and dodging with lowered heads with slight, if any, clashing of horns was also characteristic of ritualized defense of boundaries. Often brief fights were precipitated by females entering the territorial ground to breed. In attempting to attract the female by driving her toward the territory with prancing display, a male sometimes ran into neutral ground or unoccupied territories. If he approached an occupied territory too closely, vigorous fighting in defense of the boundary ensued. However, without disturbance from outside, serious fights were infrequent, territorial boundaries being maintained through mutual respect and ritualized display. When a female chose to leave one territory (*A*) for an adjacent territory (*B*), male *A* stopped at his boundary and permitted the female to walk into territory *B* without attempting to fight with male *B*. When disturbed by lion, automobile, elephant, or similar influence, the Kob deserted the territorial ground by leaving along established routes. No antagonism between males occurred during such movements. Within 10 to 20 minutes after the disturbance was removed, the kob were again on their individual territories, brief fights occurring frequently as they reestablished themselves.

High population density may be essential for expression of territorial behavior of Uganda kob. In the Rutshuru Plains of the Congo, about 150 miles south of the Semliki Game Reserve, F. Bourlière (*5*) of the Faculty of Medicine in Paris did not observe the behavior in the same subspecies of kob, but the density of the population in the former region was about half of that in the latter.

HELMUT K. BUECHNER
Department of Zoology, Washington State University, Pullman

References and Notes

1. I am indebted to my wife, Jimmie, for her keen observation and unbiased interpretation that led to the discovery of territorial behavior in Uganda kob. Research during 1957 and 1958 was supported by a Fulbright appointment and financial assistance from the Uganda Administration; in 1959 a grant in aid of research from the National Science Foundation permitted concentrated study of the phenomenon. Personnel of the Uganda Department of Game and Fisheries and the Uganda National Parks assisted greatly in field operations.
2. Estimation of the population from aerial counts was made in May 1958 with Dr. William M. Longhurst, University of California, who piloted a Stinson Voyager generously loaned to us by C. D. Margach, Kinyala Estates, Misindi, Uganda.
3. H. K. Buechner, A. M. Harthoorn, J. A. Lock, *Can. J. Comp. Med. Vet. Sci.* **24**, 317 (1960).
4. The film may be rented through the Audio-Visual Center at Washington State University, Pullman.
5. F. Bourlière, verbal communication.

28 October 1960

Reprinted from *Animal Behaviour*, **14**, 328–331 (1966)

BOOMING TERRITORY SIZE AND MATING SUCCESS OF THE GREATER PRAIRIE CHICKEN *(TYMPANUCHUS CUPIDO PINNATUS)**

By ROBERT J. ROBEL

Kansas State University, Manhattan

Greater prairie chickens (*Tympanuchus cupido pinnatus*) are characteristic birds of climax grasslands in North America. Prairie chickens are social birds which exhibit a breeding display during spring months. This display involves primarily males, and the area on which the display is conducted is termed a booming or dancing ground. Such a group display by male birds is similar to leks of blackcocks (*Lyrurus tetrix*), hooting grounds of sharp-tailed grouse (*Pedioecetes phasianellus*), and strutting grounds of sage grouse (*Centrocercus urophasianus*). Territorial behaviour displayed at leks, strutting grounds, hooting grounds, and booming grounds is thought to be an integral force in the natural regulation of tetraonid populations (Tinbergen, 1957; Wynne-Edwards, 1962).

Early descriptive studies of greater prairie chicken booming activity were conducted by Brickenridge (1929) while more complete investigations were reported later by Hamerstrom (1941), Schwartz (1945), and Hamerstrom & Hamerstrom (1955).

The present study was conducted to determine the effects of territoriality on the mating behaviour of greater prairie chickens in a large and stable population of these birds in north-eastern Kansas.

Study Area

This study was conducted about 35 km south of Manhattan on the western edge of the Flint Hills in Kansas. The climax vegetation for this region is true prairie (Herbel & Anderson, 1959) now used primarily as grazing pasture but was cultivated partially until the late 1920's.

The study booming ground was located on the uplands of the region, a ridge 1·6–6·4 km wide extending several kilometres north-west to south-east. Gently rolling ridge tops were covered by grasses 20–35 cm tall, the most conspicuous of which were: tall dropseed (*Sporobolus asper*), buffalo-grass (*Buchloe dactyloides*),

*Contribution Number 351, Department of Zoology, Kansas Agricultural Experiment Station and Serial Number 379, Department of Zoology, Kansas State University, Manhattan.

Japanese brome (*Bromus japonicus*), and side-oats grama (*Bouteloua curtipendula*). Principal forbs were slim-flower scurf-pea (*Psoralea tenuiflora*), Louisiana sagewort (*Artemisia ludoviciana*), and western ragweed (*Ambrosia psilostachya*). The hillsides were dominated by big blue-stem (*Andropogon gerardi*); bottom lands and creeks banks were covered by thick brush and trees.

Methods and Materials

Intensive observations were made three mornings per week at one greater prairie chicken booming ground from 23 March to 19 June, 1964 and 15 March to 4 June, 1965. Observation periods began 1 hr before sunrise and continued for 3 hr. A blind erected at the edge of the booming ground facilitated observations of bird activity.

Six male birds were captured during 1964 by using three Miller cannons which projected a 12·2 × 18·3 m, 5 cm mesh nylon net over the booming birds. Two additional males were captured using the same method in 1965. Each trapped bird received a numbered aluminium leg band plus one or more coloured plastic leg bands. A combination of coloured leg bands permitted identification of individuals at a maximum distance of 150 m.

Territoriality of booming greater prairie chickens was measured by determining the exact location of each banded bird at 15 min intervals during the period of observation. Bird locations were determined from the blind by using an accurate range finder and a surveying compass. Territories of booming greater prairie chickens were determined by constructing diagrammatic maps from these location data. Since birds did not spend an equal amount of time in all areas of their territories, concentric rings were drawn to include approximately 50, 70, and 90 per cent of the location records for each bird. The term territory is used broadly to denote the area in which 90 per cent of the sighting locations were recorded. The area including the central 50 per cent of the location sightings is subjectively termed the *primary* portion of the territory, whereas the area including the next 20 per cent of the sightings is termed the *second-ary* portion and the *tertiary* portion includes the outer 20 per cent. Area determinations of primary, secondary and tertiary portions of the territories were made with the aid of a compensating polar planimeter from the diagrammatic maps.

The number of copulations was used as a measure of mating success for each marked bird.

Results

Observations of booming greater prairie chickens were conducted during twenty-one and thirty-two mornings for the 1964 and 1965 study periods, respectively. During 1964, two flocks of greater prairie chickens utilized the same booming ground, one numbering nine males between March and late May, and the second flock of seven birds from late May to late June. In 1965, the regular greater prairie chicken booming population on the ground was nine from March to mid-May and then decreased to seven by early June. The number of females visiting the booming ground fluctuated during the study period, being greatest during late April and in May. General booming ground

activity was most intense during late April and early May, while the peak of mating occurred during early May in 1964 (Robel, 1964) and mid-April in 1965.

Location recordings of banded males totalled 747 during this study; 205 and 542 for the 1964 and 1965 periods respectively (Table I). Bird M3 (metal band No. 3) was banded on 30 April while birds M4 and M5 were banded on 2 May, 1964. Birds M6, M7, and M8 were banded on 30 May 1964, and were of a different booming flock from birds M3, M4, and M5. Birds M7 and M8 returned to the booming ground and were part of the regular booming flock in 1965. Birds M19 and M20 were banded on 20 April and 30 April 1965 respectively.

When fifty-one location sightings for bird M3 were plotted, a total of 168, 350, and 550·5 m² were included in the first (primary portion), second (secondary portion), and third (tertiary portion) concentric rings, respectively (Fig. 1). Bird M3, compared with other banded birds in 1964, exhibited the largest primary, secondary, and tertiary portions (Table I, Fig. 1). Bird M7 exhibited the smallest primary and secondary portions but the second largest tertiary portion in 1964, while in 1965 it had the smallest primary, secondary and tertiary portions of the four

Table I. Area Included Within Concentric Rings of Territorial Diagrams (Fig. 1) and Distribution of Copulations Observed Involving Marked Birds During 1964 and 1965

Bird number	Total No. of sightings	Area (m²) contained within			Total area of territory	Observed copulations	
		First ring	Second ring	Third ring		No.*	%
1964							
M3	51	168·4	350·3	550·5	1069·2	21	70·0
M4	37	68·3	100·1	163·7	332·1	2	6·6
M5	41	27·3	72·8	136·5	236·6	0	0·0
M6	26	72·8	314·0	268·4	655·2	0	0·0
M7	28	18·3	64·0	451·4	533·7	0	0·0
M8	22	136·5	191·1	150·2	477·8	0	0·0
1965							
M7	174	22·7	41·0	100·1	163·8	0	0·0
M8	207	154·7	232·1	473·1	859·9	29	74·4
M19	91	31·9	77·3	168·3	277·5	0	0·0
M20	70	91·0	190·5	196·4	577·9	2	5·1

*Total copulations observed; 1964 = 30, 1965 = 39,

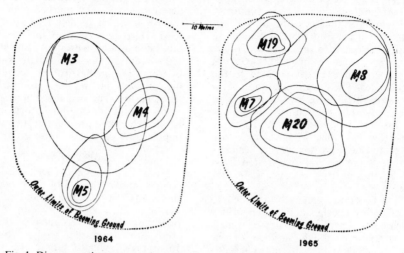

Fig. 1. Diagrammatic representation of primary, secondary, and tertiary portions of territories controlled by greater prairie chickens number M3, M4, and M5 during 1964, and M7, M8, M19 and M20 during 1965.

birds studied. Bird M8 had the second largest primary portion, third largest secondary portion, and fifth largest tertiary portion in 1964. During 1965, bird M8 had the largest primary, secondary, and tertiary portion when compared to the other males for which territories were determined (Table I, Fig. 1). Bird M19 occupied small primary, secondary and tertiary portions during 1965 while bird M20 displayed a moderate sized territory.

Thirty copulations were observed during 1964, of which twenty-three (76·6 per cent) were by marked males. Bird M3 conducted twenty-one (70·0 per cent) of the thirty matings observed, while bird M4 accounted for two (6·6 per cent). An unmarked bird whose territory was adjacent to the territory occupied by M3 was credited with the other seven matings (23·2 per cent).

Thirty-nine matings were recorded for the 1965 study period, of which thirty-one (79·5 per cent) involved banded birds. Bird M8 was responsible for twenty-nine (74·4 per cent) of the thirty-nine matings observed, while bird M20 was involved in two (5·1 per cent) copulations. Unmarked or at least unidentifiable males were involved in the balance (20·5 per cent) of the matings observed during the 1965 period of study.

Discussion

Bird M3 accounted for 70·0 per cent of the observed copulations on the booming ground

in 1964, while bird M8 was involved in 74·4 per cent of the matings observed during the 1965 season. These two birds were the individuals controlling the largest territories during these respective years. During 1964, two males (M3 and an unmarked bird) accounted for 93·3 per cent of the observed copulations occurring on the booming ground, while in 1965 birds M8 and M20 were involved in at least 79·5 per cent of the matings. The second group of males using the booming ground in 1964 (including birds M6, M7 and M8) arrived on the booming ground after the mating season and therefore had no opportunity to mate; however, birds M7 and M8 were among the regular booming population during 1965.

It appears superficially that the size of the inner portion of the territory is directly correlated to mating success, at least more so than the peripheral portions (Table I). The total territory size is similarly associated with the incidence of copulations. Birds M3 and M8 occupied significantly larger primary portions of territories than other booming participants present on the booming ground during the 1964 and 1965 mating season, respectively, and were involved in 72·5 per cent of the observed matings during this study. The difference in sizes of the secondary and tertiary portions of the territories is less significant (Table I). The secondary and tertiary portions of the territory were also less vigorously defended than was the inner or primary portion,

It is apparent that the function of the greater prairie chicken booming population is only indirectly one of mating, since such a minority of the participating individuals are actively involved in mating with visiting females. The more direct function appears to be one of female attraction and concentration. The general activity of the ground is closely related to the presence of females. The booming sounds of an individual prairie chicken are lost in the myriad of prairie sounds, but the chorus of sounds emitted by an active booming ground can be heard for 1·6–6·4 km when wind velocities are low. Well established greater prairie chicken booming grounds are generally separated from each other by this hearing distance in northeastern Kansas.

Most of the functions generally ascribed to territories of avian species (Tinbergen, 1957) are inapplicable to booming territories, unless it be one commonly called 'survival of the fittest'. Certainly males which are more aggressive and control the largest territory on the booming ground are well fitted to perpetuate their kind by mating with a great majority of the females serviced at that breeding area. Such a behavioural trait might be of great value in the long-term maintenance of a bird population. The small number of male greater prairie chickens actually involved in copulations on booming grounds could plausibly limit the reproductive potential of the population, hence substantiating parts of the theory of Wynne-Edwards (1962). Further studies along this vein will be rewarding.

Summary

Greater prairie chickens are social birds in which males gather and perform a group display on a mating area during the spring. On these mating areas (booming grounds) male birds establish territories. This study was initiated to determine the relationship between booming territory size and mating success of individual birds.

Observations were made during fifty-three mornings at one booming ground in northeastern Kansas during 1964 and 1965. Exact locations for eight tagged males were determined at 15 min intervals during the period of study. From these location data, territories were determined for each marked male with concentric rings drawn to delineate areas of more intensive bird activity. Sixty-nine copulations were tabulated as a measure of mating success.

Two males, both of which controlled large booming ground territories, accounted for 72·5 per cent of the matings observed during this study period. The size of the inner portion of the territory and the total territory size appear to be directly related to mating success in booming populations of greater prairie chickens.

Acknowledgments

I am deeply indebted to Messrs Grover and John Simpson, owners of the property on which this study was conducted and to Mr Kenneth Miller, ranch foreman, for unlimited cooperation during the course of this study. This study was supported in part by a National Institutes of Health Grant No. GM 12301-01 and an equipment grant from the Penrose Fund of the American Philosophical Society.

REFERENCES

Brickenridge, W. J. (1929). The booming of the prairie chicken. *Auk*, **46**, 540–543.

Hamerstrom, F. N., Jr. (1941). A study of Wisconsin prairie grouse—breeding habits, winter food, endoparasites, and movements. Ph.D. Dissertation, University of Wisconsin.

Hamerstrom, F. N., Jr. & Hamerstrom, Frances (1955). Population density and behavior in Wisconsin prairie chickens (*Tympanuchus cupido pinnatus*). *Trans. int. ornithol. Congr.*, **11**, 459–466.

Herbel, C. H. & Anderson, K. L. (1959). Response of true prairie vegetation on major Flint Hills range sites to grazing treatments. *Ecol. Monogr.*, **29**, 171–186.

Robel, R. J. (1964). Quantitative indicies to activity and territoriality of booming *Tympanuchus cupido pinnatus* in Kansas. *Trans. Kans. Acad. Sci.*, **67**, 702–712.

Schwartz, C. W. (1945). The ecology of the prairie chicken in Missouri. *Univ. Mo. Stud.*, **20**, 1–99.

Tinbergen, N. (1957). The functions of territory. *Bird Study*, **4**, 14–27.

Wynne-Edwards, V. E. (1962). *Animal Dispersion in Relation to Social Behavior*, pp. 206–216. New York: Hafner.

(*Received 16 August 1965, revised 11 February 1966; Ms. number:* A364)

15

Reprinted from *J. Mammal.*, **34**, 425–435 (1953)

GEORGE A. BARTHOLOMEW and P. G. HOEL

REPRODUCTIVE BEHAVIOR OF THE MALE

No mammal behaves in a more spectacular manner during its breeding season than the bull Alaska fur seal. He goes without food or water for two months or more. He defends a territory by aggressive behavior so fierce that not infrequently mortal wounds result, and at the same time he maintains a harem which may contain as many as 100 cows. Nevertheless, no detailed study has ever been made of this remarkable performance. No specific data on such things as duration of harem maintenance, integrity of harems, and relation of harem size to reproductive performance are available in the literature. The present study attempts to supply such documentation.

Harem maintenance. —A census of the seals in each of the harems in the study area was taken daily between 8:00 and 11:00 A.M. The histories of four typical and adjacent harems are summarized in Figure 3.

Most of the males establish their territories before the females arrive (Pl. I). The females come ashore and join a harem by settling down wherever room is available. Consequently, males that have territories close to the water acquire harems earliest. Males near the water may obtain their full complement of females before the males farther inland acquire any females at all. There are two reasons for this. The most important factor is the gregariousness of the females. As long as space is available, females coming ashore join harems that are already established; the choice of a harem by the female appears to be made without reference to the male. Indeed, for the fur seal cow, mate selection as such does not exist. The second and less important factor is the extremely zealous harem maintenance by bulls. If a cow attempts to leave his harem, the bull will chase her and drive her back or seize her in his mouth and throw her back. But even so, a bull is never able to keep a cow in his harem for more than a few minutes if she is determined to leave.

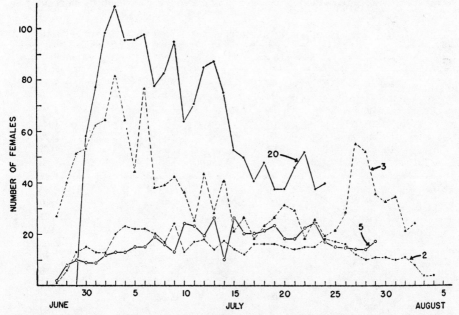

FIG. 3.—Histories of four typical and adjacent harems in Kitovi Amphitheater. The territories of male number 2 and male number 5 were largely bounded by ledges or crevices while the territories of male number 20 and male number 3 were not delimited by topographic features (see Fig. 4).

225

A territory-maintaining male that has no females will frequently dash to the edge of a harem adjacent to his territory and seize with his teeth a female by whatever part of her body happens to be nearest, carry her into his territory, and drop her. Within a few minutes, while the male is occupied with territory defense, or is lying down to rest, the female usually scurries back to the harem from which she was stolen. As a result, raiding is usually futile unless the male can accumulate four or five females. The females are so gregarious that a smaller group is unstable, at least in the early part of the breeding season. Consequently, a male must make many raids before he can establish even a small harem.

The size of a male's harem is more dependent on the location and topography of his territory than upon his diligence and vigor. Large harems develop first in territories which occupy relatively flat areas, which are topographically delimited in such a manner that the male need defend only a part of the perimeter, and which are conveniently accessible to the water but not immediately adjacent to it. If a territory abuts on the water, it is continuously disturbed by the attempts of nonterritorial males to hold females in the shallow water at the edge of the shore or to establish themselves on land.

In just a few days, early in the breeding season, a male occupying such a favorably located territory may acquire a hundred or more females, so many that they literally overflow his territory and spread into the areas of adjacent harem-less males which immediately preempt them. The new harems thus established then begin to grow at the expense of the big one. Thus, in some circumstances, harems grow rapidly and fluctuate widely in size from day to day (note harems 3 and 20 in Fig. 3).

On the other hand, if a male's territory is small and occupies an area delimited by rocks or cliffs in which there is room for only a limited number of females, the size of the harem may remain relatively constant throughout the breeding season (note harems 2 and 5, Fig. 3). In all cases, however, because of the arrivals and departures of pregnant and post-estrous females, the composition of a harem changes from day to day. When a pregnant female comes ashore, she may establish herself in the first harem she reaches or she may wander through several harems before settling down to bear her pup. If she happens to occupy a position that is near the boundary between two harems, she may be fertilized by the male of either territory. When she comes ashore after her periodic trips to sea she usually returns to the same section of the rookery in which she bore her pup,

TABLE 4.—*Daily total copulation rate of 20 harem bulls in Kitovi Amphitheater*

DATE	COPULATIONS PER HOUR	DATE	COPULATIONS PER HOUR
June 26	.00	July 16	1.71
27	.29	17	1.57
28	.13	18	1.00
29	.40	19	1.20
30	.66	20	1.29
July 1	.53	21	1.67
2	.33	22	1.43
3	1.00	23	1.42
4	2.00	24	.86
5	1.07	25	1.00
6	1.71	26	.71
7	1.50	27	.29
8	1.86	28	.57
9	2.00	29	.86
10	1.29	30	.86
11	2.00	31	.33
12	1.54	Aug. 1	.33
13	2.86	2	.50
14	3.71	3	.00
15	1.00	4	.00
		5	.00

PLATE I

UPPER. A view of part of the study in Kitovi Amphitheater on June 26, 1951. The bulls are on their territories and the cows have begun to aggregate in harems near the edge of the water.

LOWER. Same view, July 5, 1951. The number of cows has increased greatly and nearly all the bulls now have harems.

227

but she makes no effort to rejoin her previous harem. Instead, she settles down wherever she happens to find her pup.

Mean harem size. —Although it has long been known that polygyny is more strongly developed in the Alaska fur seal than in any other mammal, precise quantitative data have not previously been available. The ratio of breeding females to harem bulls in Kitovi Amphitheater during the breeding season of 1951 was 39:1. The estimated total number of cows which belonged to each of 19 harems is shown in Table 5. The mean number of females per harem was only 16, for during most of the summer fewer than half the breeding females were ashore at any one time.

Duration of harem maintenance. —Vigorous harem maintenance by bulls begins as soon as the first cows come ashore in June. During the 1951 breeding season in the study area this harem maintenance began to decline almost imperceptibly about July 21. By July 25, a general loosening of the social organization was conspicuous, and by August 2, no distinct harems remained. The period of harem maintenance in the 20 stable harems varied between 18 and 41 days with a mean of 31 days. There was no way of telling whether or not the time of a harem bull's departure from the rookery was correlated with the time of his arrival on the rookery, because most of the harem bulls had already established their territories when observations were begun. Although such imponderables as vigor and stamina are often of critical importance, other things being equal, a male with a territory in terrain that is easy to defend tends to hold his harem longer than a male with a territory that is difficult to defend. Most of the arriving and departing females moved through territory 1B, for it occupied the only stretch of shore which afforded convenient access to the harems in the study area (Fig. 4). For this reason also, most of the males that tried to move into the rookery from the sea attempted to establish themselves in territory 1B. As a result, throughout the breeding season this territory was a center of confusion and fighting. Territory 1B was occupied by a succession of males, none of which was able to withstand the attrition for more than 18 days and most of which could maintain themselves for only a few days. The resident male was continuously under stress. He was incessantly busy, either fighting off males that wanted to establish themselves, or attempting to incorporate departing or arriving females into his harem.

TABLE 5.—*The relation between number of copulations and size of harem. The methods for determining harem size and copulation number are described in the text. The males are ranked in order of decreasing harem size*

MALE NO.	NO. OF COPULATIONS	NO. OF FEMALES	RATIO: COPULATIONS TO FEMALES
20	142	153	.93
3	87	108	.80
1A	161	82	1.97
11	75	55	1.36
16	55	45	1.22
5	78	43	1.81
2	38	35	1.08
13	122	34	3.59
15	38	32	1.19
22	7	28	.25
11A	21	26	.81
6	38	22	1.72
10	0	19	.00
14	34	17	2.00
1	0	16	.00
1B	68	13	5.23
4	37	13	2.84
7	17	11	1.55
8	0	3	.00

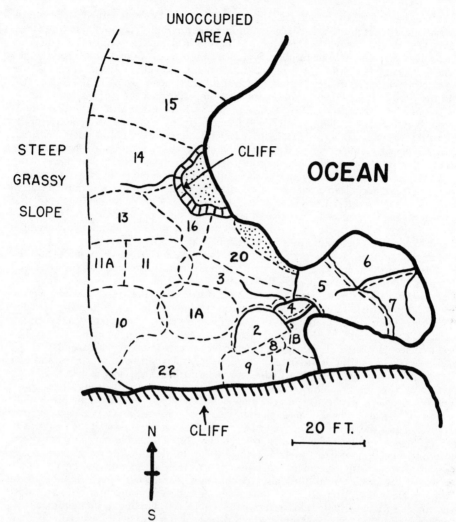

FIG. 4.—Distribution of territories in the study area in Kitovi Amphitheater, summer of 1951. Broken lines represent territory boundaries. Thin solid lines represent ledges or crevices. The nature of the terrain is shown in Plate I. Stippled areas, although occupied by seals, were not included in the study because they were not clearly visible from the blind.

Males number 6 and 7, both of which were still in residence, but not holding females or defending territories when the study ended, maintained their harems for at least 41 days. The reason for their long period of harem maintenance and residence seemed to be the isolated position and readily defended location of the areas which they held. The integrity of their harems was for the most part assured by the nature of the terrain. Although the two males maintained adjacent territories, there was little conflict between them because the boundary between their territories was determined by a rock ledge. The two territories were situated on the edge of a low cliff and were practically inaccessible from the sea. Ordinarily, no more than a vocal threat was needed to cause the withdrawal of males awkwardly attempting to scale the low cliff on the seaward perimeters of these territories. The two territories could be approached from the main rookery area only through the area occupied by male number 5, whose territory maintenance thus protected the landward sides of the two areas.

Although one might expect that a large harem would not be maintained for as long as a small one, such is not the case. For example, male number 3, an animal in his prime, held a very large territory and maintained a harem with a mean population of 38 cows for 38 days. In contrast male number 8, an old and decrepit animal with three canines missing, managed by bluffs and threats to hold a territory of about 60 square feet—little more than enough room in which to turn around. The mean size of his harem was two and the period of his harem maintenance was only 20 days.

Departure from the rookery. —During the first half of the breeding season the attachment of the bull fur seal to his territory is extremely strong. In most cases it is literally impossible for a man or a rival seal to drive him off his territory; even a potential threat to his territory by another bull will evoke a violent aggressive reaction. As the end of the breeding season approaches, the degree of aggressiveness and the strength of the territorial attachment decline markedly; bulls sometimes abandon their territories spontaneously, and often a very slight disturbance by another seal will cause a bull to leave the area he has defended so zealously and return to the sea. Sometimes a bull that stays in his territory after he has ceased any but the most perfunctory harem maintenance will watch with apparent unconcern while a subadult male wanders through the area which a week before he defended furiously against all intruders.

Harem maintenance and territory occupancy stop long before the females and the pups leave the rookery. Consequently, no organization is apparent on the rookeries during the late summer and early fall.

Sexual activity. —The elaborate pattern of activity involved in harem maintenance and territorial defense is significant only in so far as it allows a male to fertilize females. Therefore, from the evolutionary standpoint it is important to ascertain whether or not a male's reproductive success is correlated with his success in harem maintenance. The best available measure of a male's reproductive success is number of copulations. A tabular record was kept of the time of occurrence, and when possible the duration, of every copulation which took place in the study area during the period of observation. To furnish a basis for the interpretation of these numerical data, detailed descriptive records were kept of all the aspects of male behavior which were immediately sexual.

Most of the activities of harem maintenance are directed toward pregnant, pre-estrous, or post-estrous females, but male fur seals show direct sexual interest only in those females which are in heat. Although it is often possible for an observer to determine whether or not a female seal is in heat on the basis of her behavioral and postural mannerisms, the male seal appears to identify the estrous female almost exclusively by smell. One of the most frequent and characteristic performances of the harem male is his search for estrous females. A male will rouse himself, then move vigorously and rapidly about his harem, often completely circling it, sometimes tacking back and forth through it, sniffing energetically at the noses of his females. If one of the females smells as if she is in estrus or approaching estrus, the male then smells her vulva. When the female is in or near heat, she characteristically coöperates by arching her back and making her perineum more accessible. If the female is in estrus copulation ensues promptly, with the male playing the active role throughout. Quite frequently the male does not even bother to smell the female's vulva, but moves into copulatory position immediately after obtaining the appropriate olfactory clue from the female's nose.

By means of these periodic excursions through and around his harem, the male serves his females one after the other as they come into heat. Once the breeding season is well underway, the male operates methodically and relatively calmly, but early in the season as the first few of his females come into heat, he behaves in a very excited manner. He may copulate several times with the first of his estrous females, but with subsequent females a single copulation is the rule. Although there is a decline in a male's sexual interest as the breeding progresses, during most of the breeding period his reproductive capacity usually appears to be adequate even when several of his females come into heat on the same day. Even while copulating with one female, a male may show sexual interest in another female by sniffing energetically at her nose. Males frequently copulate with two females within a half hour, and once a male copulated with two females within 16 minutes. For a male to copulate four or five times in as many hours is not unusual, and in one instance a male which had been mating regularly for a number of days copulated with six different females in five hours and 17 minutes.

* * * * * * *

Efficiency of harem bulls. —In order to estimate by means of available data the reproductive efficiency of the harem bulls, it is necessary to assume that the copulation rate of males may be used as a measure of their fertilizing efficiency. A high level of sexual vigor does not necessarily indicate a high level of fertility; nor does low sexual vigor indicate low fertility, but other things being equal, it is reasonable to assume that those males which copulate with the most females fertilize the most females.

Table 4 gives the observed copulation rate per hour for the 41 days of observation. Since the day-to-day variation is extremely large, a smooth curve was fitted to these rates after they had been multiplied by 24. The resulting curve, which is shown in Figure 5, may be considered to be an estimate of the frequency distribution of copulations for the population under consideration. These estimates may be too large because the copulation rate for day time observations may not hold for all 24 hours of the day. However, no diurnal variation in the level of activity of the rookeries is apparent. Kitovi Amphitheater was visited at all times when there was enough light for observation (2:00 A.M., to 11:00 P.M.) and no significant differences in behavior associated with time of day were noted. If noise is a criterion, there is little diminution in male activity even during the brief hours of subarctic darkness.

Figure 5 also shows a curve that represents the estimated daily number of estrous females. This curve was obtained in the following manner. Differences between successive daily counts of pups were taken from the pup curve in Figure 1 to give estimates for the daily number of pup births. These estimates were graphed and smoothed to give a more refined estimate for the expected daily number of births. The resulting curve was translated two days to the left and treated as a refined estimate for the expected number of daily arrivals of new females. The validity of using the pup census for estimating the frequency distribution of female arrivals lies in the fact that each newly arrived female produces one pup and the fact that the mean time lag between arrival and parturition is two days. The curve for female arrivals obtained in this manner has been centered so that its maximum agrees with the maximum for the copulation curve. The justification for making the maxima agree is that the maximum number of copulations should occur when the maximum number of females are in heat and this latter date should occur a fixed number of days after the date when the maximum number of females arrived. If, for example, the copulation maximum occurred to the right of the translated maximum for female arrivals it would imply that at first the copulation rate decreased steadily, then increased for several days, and then decreased steadily again, which is unlikely.

With this centering it will be found that the difference in the time scales for the two curves is nine days. This is precisely the estimate obtained from the statistical model (1) for the time between arrival and estrus. The excellent agreement here with the statistical model based on daily censuses adds further validity to the conclusions previously made and also to the conclusions about to be made.

A comparison of the two curves of Figure 5, shows that the males are giving adequate service during the early and middle parts of the season; i.e., on the average slightly more than one copulation per estrous female, but that they are not doing so near the end of the season, i.e., not all the estrous females are being served. Since the copulation curve was based on the assumption that the copulation rate obtained from day time observations holds during the entire 24-hour period, and since the true rate may be somewhat lower, and since as pointed out earlier the female curve may be slightly higher than indicated in Figure 5, it follows that the deficiency of copulations near the end of the season may be more pronounced than Figure 5 indicates, and that this deficiency may extend considerably further back in time.

Figure 5 demonstrates that only during the first week of the breeding season was the average number of couplations per female as high as two. Thereafter the rate decreased rather rapidly, showing that for most of the females there was but one copulation. The copulation curve is quite reliable, provided the assumption that the copulation rate is independent of the hour of the day is valid. The average number of copulations per female for the entire breeding season based on this same assumption is 1.3.

If the copulation rates for the various males are treated in the same manner as the total copulation rates were treated in the preceding paragraphs, the estimates given in

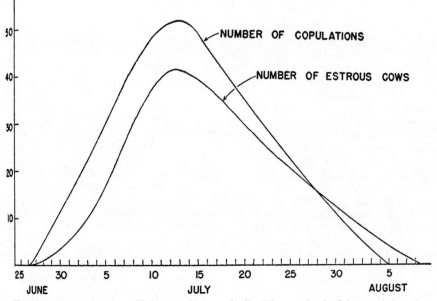

FIG.5.—Reproductive efficiency of harem bulls. The method of determining the two curves is described in the text.

Table 5 for the total number of copulations for each male will be obtained, based once more upon the assumption that the observed rate holds for the entire 24-hour period.

If it is assumed that the number of females belonging to a harem, and hence requiring service, is proportional to the sum of the daily observed number of females in the harem, then an estimate of the total number of females belonging to a given harem can be obtained. It is merely necessary to calculate the ratio of the daily census of all the harems, and then take this ratio of the known total number of breeding females. The total number of breeding females should be about the same as the total number of pups born, alive or dead. For Kitovi Amphitheater this total was estimated to be 780; hence the foregoing ratios were taken of 780. The resulting estimates for the number of females belonging to each harem are listed in Table 5.

An inspection of the fourth row of Table 5, which gives the ratio of copulations by the various males to the number of females in their harems, shows that the variability in estimated male activity is very large. If the 19 males are ordered with respect to the size of their harems and then are divided into three groups consisting of the first six, the second six, and the last seven, it will be found that the ratios for these three groups are 1.23, 1.49, and 1.70, thus showing a slight trend in the direction of increased efficiency of fertilization with decreased harem size. If the two largest harems are deleted, this trend will disappear; hence about all that can be said here is that males do not appear capable of servicing a harem of more than 100 females.

If the males are arranged in order of their Table 5 ratio values and the number of days of harem maintenance is treated as a correlated variable, it will be found that there is no relationship indicated for these two variables. A similar conclusion can be made for the variables harem size and onset of harem maintenance.

Although individual variations, both physiological and psychological, are undoubtedly important in determining a male's reproductive effectiveness, such factors do not lend themselves to analysis under field conditions. However, some of the apparent irregularities in Table 5 are related to a factor which is readily studied and which has proved to be important in many aspects of the social behavior of fur seals, namely territory location. For example, males number 1B and number 4, whose territories lay on the principal route of departure of the females from the study area, served many more females than would be expected on the basis of harem size. The high copulation rates of these two males are caused by the fact that they occasionally had coitus with females from more inland parts of the rookery when these females were on their way toward the water on the first part of their periodic trips to sea.

Efficiency of the fur seal social organization. —From the standpoint of reproduction, the social organization of the Alaska fur seal appears superficially to be inefficient. The sex ratio presumably is equal yet in a given year only one male in approximately 40 regularly serves the females. Although the copulation rate of this minority of bulls is adequate to take care of all the females coming into estrus during the early and middle parts of the breeding season, toward the end of the organized breeding period these males appear to be unable to serve all their females. Some of the females must, therefore, depend for fertilization on the subadult males that intercept them when they make their first trip to sea.

No matter how inefficient, wasteful, or imperfect the breeding organization may appear to be, however, if it enables the population to maintain itself, it must be considered adequate. That the present social setup is adequate is suggested by the steady growth of the fur seal population over the past 30 years. The population now appears to have become stable and any further increase does not appear imminent as long as the present system of management is maintained (Kenyon, Scheffer, and Chapman, ms.). It therefore becomes of interest to determine whether or not the maximum biotic potential of the species is currently being realized and this is merely to ask the question, "Is the maximum possible number of females being fertilized?"

The most satisfactory data for frequency of conception of female fur seals have been gathered by Wilke (Prelim. Study No. 67, Gen'l. Hdqtrs. Supreme Comm. Allied Powers, 1951) who found that of 191 females four years old or older which were collected in March, April, and May of 1950 off the coast of Japan, 83 per cent were pregnant. It is not known how many of the non-pregnant 17 per cent failed to conceive because of inadequate service by the males, and how many were sterile or had aborted. From this group of 191 females, however, of the 96 females six to ten years old—presumably at the height of their reproductive capacity—92 per cent were pregnant. This suggests that virtually all the females are served. Thus, it cannot be said that the social organization of the fur seal is reproductively inefficient despite the very large number of non-breeding males. Nor does it appear that the present rate of exploitation of the non-breeding males for commercial purposes is excessive. Apparently the present policies of management allow a sufficient escapement of three- and four-year-old males so that the fertilization of females, at least in the six to ten year age classes, is virtually maximal. Pearson and Enders (1951, Anat. Rec., 111:695-712) present figures for reproductive failure in another population of fur seals collected in 1949 which are in agreement with this thesis.

ACKNOWLEDGEMENTS

This work was carried out in collaboration with the United Sates Fish and Wildlife Service and was undertaken at the suggestion of Dr. Victor B. Scheffer, whose arrangements and help were of invaluable aid. Mr. Clarence L. Olson and Mr. Charles H. Anderson made available the facilities of St. Paul Island and gave every possible coöperation. Mr. Ford Wilke and Mr. Karl W. Kenyon were largely responsible for the difficult and often dangerous job of marking the seals. Their daily help greatly facilitated the study and, in addition, their extensive familiarity with fur seals gave a perspective to the investigations which would not otherwise have been possible.

V

Intraspecific Variation in Social Systems

Editor's Comments on Papers 16, 17, and 18

16   Anderson: Density, Social Structure, and Nonsocial Environment in House-Mouse Populations and the Implications for Regulation of Numbers
Trans. N.Y. Acad. Sci., **23**, 447–451 (1961)

17   Leuthold: Variations in Territorial Behavior of Uganda Kob *Adenota kob thomasi*
Behaviour, **27**, 255–256 (1966)

18   Paluck and Esser: Territorial Behavior as an Indicator of Changes in Clinical Behavioral Condition of Severely Retarded Boys
Amer. J. Mental Deficiency, **76**, 284–290 (1971)

We had long assumed that each species had its characteristic, largely genetically determined, social organization, some of which were territorial, others hierarchical. We now know that a single species may have several kinds of social organization. Many song birds typically hold a territory during the breeding season but at other times form flocks wherein a hierarchy prevails. Even within the breeding season a species may change its system depending upon the reproductive stage, density, or other environmental conditions. Thus, in the black-headed gull, breeding birds first establish a pair-formation territory, then a nesting territory at another location, and finally a group-feeding territory (Tinbergen and Moynihan, 1953).

John H. Crook (1964), after studying more than 60 species of weaver birds, concluded that the social system of a species was strongly influenced by the quantity and distribution of food and predator pressure.

Density also plays a part in the social organization. This may readily be seen in the confines of small aquaria where many fish can maintain a territory only a few inches in diameter. Greenberg (1947) found that several male sunfish may share a 10-gallon aquarium, each with its own territory; but with the further addition of males, the territorial system changed to a hierarchy. Perhaps because Greenberg was working with such highly confined fish, his work did not receive the credit it deserved. Only when Paul Anderson (Paper 16) published his more theoretical paper on house mice was it recognized that a social hierarchy and a territory represented two extremes on a continuum of social organization. In independent studies, Crowcroft and Rowe (1963) reached similar conclusions.

Paul K. Anderson is Professor of Zoology at the University of Calgary. His research is directed toward the ecological and evolutionary strategies developed by populations in interaction with their environments. He is concerned with the role of territorial behavior in regulating density, encouraging colonization of new and marginal habitats, and influencing the exchange of genetic information between populations.

More evidence for this intraspecific variation in social behavior appeared in the Uganda kob. Buechner (Paper 13) had discovered the lek system in kob. Later he and Leuthold observed that kob might also hold larger, single territories similar to classical breeding territories in birds (Paper 17). Leuthold postulates that the single large territory may be a proving ground from which the male may later move onto the lek and breed; hence there was a possible ontogenetic explanation for this shift in territorial type. He believes it more probable that kob are genetically dimorphic with respect to their social system. He also believes that the large single territory is the original type, still advantageous in small or low-density populations; at higher densities, the lek provides maximum efficiency of reproduction. More work needs to be done to separate the genetic from environmental determinants of social systems.

Leuthold's study of the Uganda kob was the basis for his doctoral thesis at the University of Zurich. Subsequently, he has continued to study the behavior and ecology of various hoofed animals of East Africa, where he is currently in charge of elephant research at Tsavo National Park, Kenya.

Paper 18 is the only one in this volume on human territorial behavior. Man is the most seriously overlooked species in this respect, *The Territorial Imperative* (Ardrey, 1966) not withstanding. Paluck and Esser's study is important because it shows that one can use the same methodology for humans as for other animals; that humans in the confines of a single large room may be territorial or hierarchical depending upon their social status; and that, following therapy, an individual may rise in rank and shift from territorial to hierarchical status.

Robert J. Paluck is in the Department of Psychology, Hunter College, where he divides his time among teaching, research, and clinical practice. He utilizes operant conditioning to alter the territorial structure of retarded children.

Esser is a psychiatrist and Director of the Central Bergen Community Mental Health Center in Paramus, New Jersey. He conducts research in social biology at Rockland State Hospital. He teaches man–environment relations in the College of Human Development, Pennsylvania State University, and at the College of Architecture at Virginia Polytechnic Institute. He has been instrumental in the founding of the Man and Environment Society.

References

Ardrey, R. 1966. *The Territorial Imperative*. Atheneum, New York. 390 pp.

Crook, J. H. 1964. *The Evolution of Social Organization and Visual Communication in the Weaver Birds (Ploceinae). Behaviour* (Suppl. 10). 178 pp.

Crowcroft, P., and F. P. Rowe. 1963. Social organization and territorial behaviour in the wild house mouse (*Mus musculus* L.). *Proc. Zool. Soc. London,* **140**: 517–531.

Greenberg, B. 1947. Some relations between territory, social hierarchy and leadership in the green sunfish (*Lepomis cyanellus*) *Physiol. Zool.,* **20**: 269–299.

Tinbergen, N., and M. Moynihan. 1953. Head flagging in the black-headed gull: its function and origin. *Brit. Birds,* **45**: 19–22.

16

Reprinted from *Trans. N. Y. Acad. Sci.*, Ser. 2, **23**, 447–451 (1961)
DENSITY, SOCIAL STRUCTURE, AND NONSOCIAL ENVIRONMENT IN HOUSE-MOUSE POPULATIONS AND THE IMPLICATIONS FOR REGULATION OF NUMBERS*

Paul K. Anderson

Department of Zoology, Columbia University, New York, N. Y.

The information summarized in FIGURE 1 of this paper has been arranged in graphic form to facilitate consideration of four questions: (1) Over what range of densities do *Mus musculus* populations occur? (2) What kind of social structure is assumed? (3) What behaviors lead to such social structures? (4) What is the significance of social structure in the regulation of population numbers?

Across the center of the figure density has been indicated on a logarithmic scale in terms of mice per unit area. Observed densities are shown below. Naturally occurring populations have been reported only in the lower 10 per cent of the density range; populations inhabiting stacked or standing grain (field commensals) achieving the highest densities, those occupying buildings (domestic commensals) possibly less dense on the average, and those populations not associated with man but derived from commensal ancestors (feral) the least dense. Although the information is probably available in the Soviet literature, I have found no records of density in the wild populations inhabiting the steppe grasslands of central Asia and, therefore, have considered the feral populations as approximating those in this presumably ancestral habitat. Overlap between experimental populations and those occurring naturally is slight and is difficult to evaluate since effective density, in contrast to the absolute density (mice/unit area), may have been higher where simplified (not complexedly structured) experimental environments permitted maximum social interaction.

Correlation between density and social structure is shown by observations summarized above the density scale. Three types of social organization are suggested.

Mice of laboratory stock[3,24,25] and wild-caught stock[12,27,28] have been observed to form social hierarchies in which a single male dominates one or more additional males. The subordinates appear to be of more or less equal rank. Dominance is established through violent combat among sexually mature males. Under similar conditions females do not exhibit aggressive behavior, and they become involved in combat only when they happen to be in the path of aroused or contesting males.

"Togetherness" is the least well-documented social system. A domestic commensal population[33] and a field commensal population[18] are the basis of this category. Neither was the subject of detailed behavioral study, but in both cases a lack of evidence of aggressive behavior was

*This paper, illustrated with slides, was the third of three papers presented at a meeting of the Section of Biological and Medical Sciences on February 13, 1961.

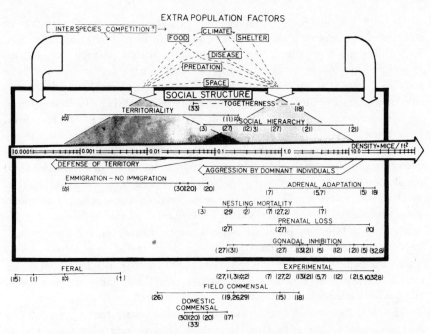

FIGURE 1. Relationships between habitat, density, social structure and density regulating factors in *Mus* populations. Density is shown on a logarithmic scale across the center. Numbers in parentheses refer to the bibliography. Density records for feral, domestic, and field commensal and experimental populations are given at the bottom of the figure. Within the heavy line indicating intrapopulation phenomena are shown records for social hierarchy and territoriality (*above density scale*) and for population responses tending to limit density (*below density scale*). Pathways for operation of extrapopulation factors are suggested at the top of the figure.

* Anderson, P. K. Unpublished data.

† Lidicker, W. Z. Unpublished data.

noted in association with high densities. This suggests that under certain circumstances aggressive behavior patterns may break down.

Territorial social organization in house mice conforms to both the broad definition of territoriality as occupancy of an area to the exclusion of other individuals of the same age and sex, and to the narrower definition of territory as a defended area. The literature contains one clear-cut observation[11] of territorial organization. In addition, I have found in a feral population (P. K. Anderson, unpublished data) that home ranges of mature males are almost wholly exclusive, and that ranges of mature females are largely so. Direct observation showed that most mature females aggressively defended their home ranges, but that tolerant associations of as many as five adult females were formed in especially favorable habitat. No instance of social contact between mature males was observed in this population, a fact that suggests that established males have little contact with each other at territory boundaries.

Davis[14] has suggested that the same pattern of individual behavior may lie at the root of both territorial and hierarchical social organization. This is a cogent argument, and information available permits suggestions regarding the course of events and the conditioning factors. It has been found in domestic mice that victory in combat is most likely for males that are in a familiar pen, are in the presence of familiar associates, have experience of prior victories, or are of larger size.[22,23,25] In a feral population young mice remain within the parental home range until puberty (P.K. Anderson, unpublished data), the time at which aggressive behavior has first been observed in mice of domestic stock.[24] In the feral population extensive wandering begins at this time, and social encounters with strange mice are probably frequent. Such encounters are most probable with larger and older, resident individuals. The latter enjoy the important advantages of previous experience of victory on familiar terrain, and successful establishment of the wanderers would appear likely only when unoccupied territory was reached. There the opportunity to establish familiarity with the terrain would make victory over later arrivals likely and territory could be established.

The development of social hierarchy seems likely to be the result of two changes in the situation described above: lack of opportunity to emigrate, and lack of escape cover. In the spatially complex environment of a feral population, and in spatially complex population cages, aggressive encounters only rarely involve physical contact (P. K. Anderson, unpublished data). The victim of an attack usually flees at the first rapid movement of the aggressor and is able to escape before the latter can close and complete his attack. Social dominance seems less likely to be achieved under these conditions. In respect to emigration, the instances of social hierarchy reported in the literature[3,12,21,27] have all involved confined populations. In such populations young mice would be vanquished in combat by established dominant males and become socially subordinate. Natural environments where social hierarchy should be looked for may be characterized as having abundant and highly localized food supplies, and sharp discontinuities between favorable and unfavorable habitat that might discourage emigration. The populations of stacks of grain (corn ricks) seem favorable for such social arrangements, although in such situations escape cover is abundant.

Where the opportunity for emigration exists an exodus seems to be the first response to crowding.[31] Therefore regulation of population density might occur through banishment of wanderers (P. K. Anderson, unpublished data) and social subordinates[4] to marginal environments where failure or delay in finding suitable breeding sites and mates would lower reproductive output and where vulnerability to predation, disease, or inclement weather would be high. The necessary corollary, inability of socially unestablished mice to penetrate established social units has been reported.[20] It thus appears that under commonly encountered environmental conditions social mechanisms alone may control population density and, as indicated in the diagram, these mechanisms tend to mediate the effects of extrapopulation factors. Calamitous environmental change (large arrows) can overcome the stabilizing effects of social structure and cause local or regional extinction. Conversely, extremely favorable conditions may cause the breakdown of the social mechanism and permit population explosions. In the extremely dense experimental populations commonly studied, other behavioral factors tending to regulate density (Cannibalism of young, nest destruction) and physiological maladjustments (abortion, resorption, gonadal inhibition, failure of lactation) have been reported. Such effects have been absent (for example adrenal adaptation in corn rick populations[29]) or observed rarely[29,30] under natural conditions. Gonadal inhibition, in particular, seems dependent on a degree of environmental and social stability not to be expected in the wild.[13,21] The conclusion suggested here, therefore, is that density regulation may involve such mechanisms only when the purely social factors prove inadequate. If such instances are as rare as the density records diagrammed here indicate, the gonadal-adrenal-pituitary relationship explored by Christian[5-9] would be involved only rarely, and its operation in population regulation may be abnormal or secondary to some other, undiscovered, primary function.

References

1. BAKER, R. H. 1946. A study of rodent populations on Guam, Marianas Islands. Ecol. Monog. 16: 393-408.

2. BROWN, R. Z. 1953. Social behavior, reproduction, and population changes in the house mouse (Mus musculus L.). Ecol. Monog. 23: 217-240.

3. CALHOUN, J. B. 1956. A comparative study of the social behavior of two inbred strains of house mice. Ecol. Monog. 26: 81-103.

4. CALHOUN, J. B. 1956. Behavior of house mice with reference to fixed points of orientation. Ecology. 37: 287-301.

5. CHRISTIAN, J. J. 1955. Effect of population size on the adrenal glands and reproductive organs of mice in populations of fixed size. Am. J. Physiol. 182: 292-300.

6. CHRISTIAN, J. J. 1956. Reserpine suppression of density dependent adrenal hypertrophy and reproductive hypo-endocrinism in populations of male mice. Am. J. Physiol. 187: 353-356.

7. CHRISTIAN, J. J. 1956. Adrenal and reproductive responses to population size in mice from freely growing populations. Ecology. 37: 258-273.

8. CHRISTIAN, J. J. & C. D. LEMUNYAN. 1958. Adverse effects of crowding on lactation and reproduction of mice and two generations of their offspring. Endocrinology. 63: 517-529.

9. CHRISTIAN, J. J. 1959. Control of population growth in rodents by interplay between population density and endocrine physiology. J. Wildlife Diseases. 2: 1-38.

10. CREW, F. A. E. & L. MIRSKAIA. 1931. The effects of density on an adult mouse population. Biol. Gen. 7: 239-250.

11. CROWCROFT, P. 1955. Territoriality in wild house mice. J. Mamm. 36: 299-301.

12. CROWCROFT, P. & F. P. ROWE. 1957. The growth of confined colonies of the wild house-mouse (*Mus musculus* L.). Proc. Zool. Soc. London. **129**: 359-370.

13. CROWCROFT, P. & F. P. ROWE. 1958. The growth of confined colonies of the wild house-mouse (*Mus musculus* L.): the effect of dispersal on female fecundity. Proc. Zool. Soc. London. **131**: 357-365.

14. DAVIS, D. E. 1958. The role of density in aggressive behavior of house mice. Animal Behavior. **6**: 207-210.

15. ELTON, C. 1942. Voles, Mice and Lemmings. : 1-496. Clarendon Press. Oxford, England.

16. EMLEN, J. T., JR., H. YOUNG & R. L. STRECKER. 1958. Demographic responses of two house mouse populations to moderate suppression measures with 1080 rodenticide. Ecology. **39**: 200-206.

17. EVANS, F. C. 1949. A population study of house mice (*Mus musculus*) following a period of local abundance. J. Mamm. **30**: 351-363.

18. HALL, E. R. 1927. An outbreak of house mice in Kern County, California. Univ. Calif. Publ. Zool. **30**: 189-203.

19. LAURIE, E. M. O. 1946. The reproduction of the house mouse (*Mus musculus*) living in different environments. Proc. Roy. Soc. London. **B133**: 248-381.

20. MACLEOD, C. F. 1959. The population dynamics of unconfined populations of the house mouse (*Mus musculus* L.) in Minnesota. Dissertation. Univ. Minn., Minneapolis, Minn. Unpublished.

21. PETRUSEWICZ, K. 1957. Investigation of experimentally induced population growth. Ekol. Polska A. **5**: 281-309.

22. PETRUSEWICZ, K. 1959. Further investigation of the influence exerted by the presence of their home cages and own populations on the results of fights between male mice. Bull. acad. polon sci. **7**: 319-322.

23. PETRUSEWICZ, K. & T. WILSKA. 1959. Investigation of the influence of interpopulation relations on the result of fights between male mice. Ekol. Polska. A. **7**: 357-390.

24. SCOTT, J. P. 1944. Social behavior, range and territoriality in domestic mice. Proc. Ind. Acad. Sci. **53**: 188-195.

25. SCOTT, J. P. & E. FREDERICSON. 1951. The causes of fighting in mice and rats. Physiol. Zool. **24**: 273-309.

26. SOUTHERN, H. N. & E. M. O. LAURIE. 1946. The house-mouse (*Mus musculus*) in corn ricks. J. Animal Ecol. **15**: 134-149.

27. SOUTHWICK, C. H. 1955. The population dynamics of confined house mice supplied with unlimited food. Ecology. **36**: 212-225.

28. SOUTHWICK, C. H. 1955. Regulatory mechanisms of house mouse populations: social behavior affecting litter survival. Ecology. **36**: 627-634.

29. SOUTHWICK, C. H. 1958. Population characteristics of house mice living in english corn ricks: Density relationships. Proc. Zool. Soc. London. **131**: 163-175.

30. STRECKER, R. L. 1954. Regulatory mechanisms in house mouse populations. The effect of limited food supply on an unconfined population. Ecology. **35**: 249-253.

31. STRECKER, R. L. & J. T. EMLEN, JR. 1953. Regulatory mechanisms in house mouse populations: the effect of limited food supply on a confined population. Ecology. **34**: 375-385.

32. WHITTEN, W. K. 1959. Occurrence of anoestrus in mice caged in groups. J. Endocrinol. **18**: 102-107.

33. YOUNG, H., R. L. STRECKER & J. T. EMLEN JR. 1950. Localization of activity in two indoor populations of house mice, *Mus musculus*. J. Mamm. **31**: 403-410.

Reprinted from *Behaviour*, **27**, 255–256 (1966)

Territorial Behavior of Uganda Kob

WALTER LEUTHOLD

SUMMARY

Territorial behavior of the Uganda kob (*Adenota kob thomasi*; Reduncini, Hippo-traginae) was studied in the Toro Game Reserve, western Uganda. Two types of territories were found: (a) small individual plots, 15-30 m in diameter, aggregated in tight clusters that are called leks, arenas, or territorial breeding grounds (TGs); (b) larger territories of 100-200 m diameter, distributed between the arenas, called single territories (STs).

The largely permanent TGs, to which most of the breeding is confined, provide the basis for a social organization of the kob population of *ca.* 15 000. A certain number of kob are, by tradition, attached to a particular TG, so that the total population is subdivided into units, each associated with one TG.

STs are spread out between the TGs; their size, number, and distribution vary with season and local conditions. There is an irregular gradient in size and density of territories from the center of a TG through the STs in its vicinity. STs may be aggregated in loose clusters used as temporary or seasonal TGs. Permanent TGs may arise from such clusters. Abandonment of existing and formation of new TGs are relatively rare.

The males on the STs are strongly attached to confined areas which they defend against intruding males. Competition for STs is not intense, but males are occasionally replaced. Males defeated from their STs join a male herd and may attempt later to reoccupy the same ST, often successfully. Whistling probably serves for marking the territory or for attracting females to it.

Herds of females often pass through or stay on STs, but the males do not possess harems. They court the female and attempt to copulate with them, but most females avoid their approaches. Few copulations occur on the STs; in several cases the females involved proved to be physiologically abnormal, and it is concluded that the males on STs do not contribute significantly to the reproduction of the population.

The daily activity of males on the STs is compared with that of males on TGs. The latter spend less time for feeding and have less food available on their territories; this, combined with the higher proportions of fighting and sexual behavior on TGs, is propably the main reason for the much higher rate of interchange of males on TGs compared with STs. Also, the degree of competition for territories is higher on TGs than STs.

The males of the kob population studied are, on the whole, divided into two categories: Those frequenting TGs, and those staying on STs. Both types join a male herd when they are not territorial. The age distribution among males on TGs and those on STs is largely equal. Some males occupied both territories on a TG and STs, but such cases are relatively rare. Two young-adult males first occupied a ST for some time, before they appeared on a TG, but this course of behavioral development does not seem to be the general rule.

Territorial behavior was found in several other kob populations; the relative number of STs and the development of TGs vary considerably between different areas.

Territoriality and lek behavior in other ungulates are briefly reviewed. The Uganda kob is the only antelope known, so far, to exhibit typical lek behavior. In addition, behavioral polymorphism such as the occurrence of different types of territories within the same population has not yet been found in any other species of antelopes.

The following conclusions pertaining to the Uganda kob are drawn: STs are the original form of territoriality, still prevalent in small or marginal populations. In large

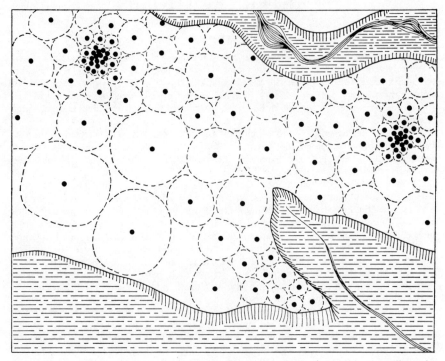

Fig. 9. Schematic distribution of territories, illustrating the — irregular — gradients in size and density from the centers of two TGs. Along the swamp (upper part) the STs are slightly concentrated; on a raised, level area (lower right) they form a cluster. The central parts of the territories (see Fig. 7) are marked with black dots; their exact boundaries are unknown (broken outline).

and dense populations the formation of TGs offers certain ecologic advantages, such as providing a social organization and a spacing mechanism to the population and ensuring maximum efficiency of reproduction. Despite these advantages of TGs over STs the latter have not disappeared. Either they provide some social advantage, as yet unknown, or their persistence ensures adaptive plasticity of local populations and the species as a whole, to meet emergencies brought about by changes in the environment.

REFERENCES

BOURLIÈRE, F. & VERSCHUREN, J. (1960). Introduction à l'écologie des ongulés du Parc National Albert. — Institut des Parcs Nationaux du Congo Belge, Bruxelles.

BROOKS, A. C. (1961). A study of the Thomson's gazelle (*Gazella thomsonii* Günther) in Tanganyika. — Col. Res. Publ. No. 25, London.

BUECHNER, H. K. (1961 a). Territorial behavior in Uganda kob. — Science 133, p. 698-699.

—— (1961 b). Territorial and reproductive behavior in Uganda kob. Unpubl. Report to National Science Foundation.

—— (1963). Territoriality as a behavioral adaptation to environment in Uganda kob. — Proc. XVI Intern. Congr. Zoology 3, p. 59-63.

—— (in press). Territoriality in an African antelope. A population study of behavior, reproduction, and ecology of the Uganda kob.

—— & SCHLOETH, R. (1965). Ceremonial mating behavior in Uganda kob (*Adenota kob thomasi* Neumann). — Z. Tierpsychol. 22, p. 209-225.

——, MORRISON, J. A. & LEUTHOLD, W. (1966). Reproduction in Uganda kob with special reference to behavior. — In: Comparative Biology of Reproduction in Mammals; ROWLANDS, I. W. (ed.). Sympos. Zool. Soc. London 15, p. 69-88.

BURCKHARDT, D. (1958). Observations sur la vie sociale du cerf au Parc National Suisse. — Mammalia 22, p. 226-244.

BURT, W. H. (1943). Territory and home range concepts as applied to mammals. — J. Mammal. 24, 346-352.

DARLING, F. F. (1937). A herd of red deer. — Oxford University Press.

DASMANN, R. F. & MOSSMAN, A. S. (1962). Population studies of impala in Southern Rhodesia. — J. Mammal. 43, p. 375-395.

DEKEYSER, P. L. (1956). — Le parc National du Nikolo-Koba. III. Mammifères. — Mémoires IFAN, No. 48, Dakar.

GRAF, W. (1956). Territorialism in deer. — J. Mammal. 37, p. 165-170.

HEDIGER, H. (1949). Säugetier-Territorien und ihre Markierung. — Bijdr. Dierkunde 28, p. 172-184.

—— (1951). Observations sur la psychologie animale dans les Parcs Nationaux du Congo Belge. — Institut des Parcs Nationaux du Congo Belge, Bruxelles.

HINDE, R. A. (1956). The biological significance of the territories of birds. — Ibis 98, p. 340-369.

KILEY-WORTHINGTON, M. (1965). The waterbuck in East Africa: Spatial distribution. — Mammalia 29, p. 177-204.

KOFORD, C. B. (1957). The vicuña and the puna. — Ecol. Monogr. 27, p. 153-219.

LEUTHOLD, W. (1966). Homing experiments with an African antelope. — Z. Säugetierkunde.

LINDSAY, D. R. (1965). The importance of olfactory stimuli in the mating behavior of the ram. — Animal Behaviour 13, p. 75-78.

LORENZ, K. (1937). Über die Bildung des Instinktbegriffs. — Naturwiss. 25, p. 289-300, 307-318, 324-331.

MAYR, E. (1963). Animal species and evolution. — Harvard University Press.

NICE, M. M. (1941). The role of territory in bird life. — Amer. Midl. Nat. 26, p. 441-487.

NOBLE, G. K. (1939). The role of dominance in the life of birds. — Auk 56, p. 263-273.

18

Reprinted from *Amer. J. Mental Deficiency,* **76**, 284–290 (1971)

Territorial Behavior as an Indicator of Changes in Clinical Behavioral Condition of Severely Retarded Boys

Robert J. Paluck

Letchworth Village, and Hunter College, City University of New York

AND

Aristide H. Esser

Letchworth Village, Thiells, New York

Individual territorial behavior of severely retarded boys in small groups was charted in an experimental dayroom during 2 trials separated by 20 months. All boys in Trial 2 staked out or fought for some specific territories which they possessed in Trial 1, thus suggesting rigidity of spatial behavior. Small changes in territorial behavior were correlated with changes in overall psychological functioning. Improvement in psychological health was related to: (a) the move from isolated or uncontested territories to popular or contested territories, and (b) the move from fixed territorial ownership to participation in the staff power structure as an expression of dominance.

The relationship of territorial behavior to clinical behavioral condition has only begun to be explored in humans. It has been documented that institutionalized adult schizophrenics and institutionalized mentally ill children show a heavy reliance on primitive modes of behavior, including territoriality (Esser, 1968; Esser, Chamberlain, Chapple, & Kline, 1965). It has also been demonstrated that territoriality is one of the strongest social ordering phenomena among institutionalized severely retarded boys (Paluck, 1971; Paluck & Esser, 1971). The phenomenon of territoriality in institutions has been shown to be systematically related to position in the dominance hierarchy (Esser, 1970), which in turn is systematically related to improvement in clinical behavioral condition (Esser, 1968). It seemed highly likely, therefore, that naturally occurring changes in territorial behavior among institutionalized severely retarded boys should reflect changes in clinical behavioral condition. The following experiment tested this hypothesis.

Five institutionalized severely retarded boys whose territorial behavior was charted in a previous study (Paluck, 1971; Paluck & Esser, 1971), were returned to the same experimental playroom after a period of 20 months. Their territorial behavior, under the same physical

and staff conditions as in the previous study, was charted again. In order to determine change in this behavior after 20 months, comparisons were made between each child's territorial behavior in the first trial and his territorial behavior in the second trial. Observations of the clinical behavioral condition of each boy (behavioral symptomatology and social adjustment) were made in both trials, and changes in these parameters were examined. Finally, changes in territorial behavior were correlated with changes in clinical behavioral condition to explore whether these two classes of behavior are systematically related.

Method

Subjects

The subjects for the study were physically healthy boys, ages 5 through 10 and with reported IQs of less than 50 at the time of the first study. They were selected from the population of a large institution for the mentally retarded. All boys were ambulatory, had a history of aggressive behavior, no gross psychotic symptomatology, and at least 2 years of continuous institutionalization prior to the original study.

In the original study, territorial behavior

was charted using three groups of seven boys each. At the time of the retesting, only five of the original boys were available. To make the retest group as comparable as possible to the original groups, two previously untested boys, who met the original criteria, were added to the retested five boys in order to make the group size seven. In addition, the two new boys were selected for specific degrees of aggressiveness (as judged by their regular staff), since the original three groups were set up with a comparable spectrum of aggressiveness among children in each group. As evidence of the success of the matching of groups, analysis of variance showed no significant difference between any of the four groups in the mean number of aggressive acts in the group per day during the study.

Experimental Environment

The experimental dayroom is described in detail elsewhere (Paluck, 1971; Paluck & Esser, 1971). In essence, large objects (slide, table, rocking boat, etc.) were arranged in a pattern of spatially distinct areas permitting territorial behavior (Fig. 1). An observer's desk was isolated in the corner of the playroom. Square grids of .9m were laid out with tape on the floor to facilitate location observations described below.

Procedure

During the original study, 21 boys in three separate groups of 7 boys each were observed during a 30-hour control condition spanning 1 week. During this period all three groups were treated with the same paradigm: Two experienced attendants were in charge of the children, one in the morning, one in the afternoon, alternating the morning-afternoon schedule on a daily basis. Aggressive behavior was reprimanded by the staff, and positive social reinforcement was offered for constructive and cooperative behavior. In addition, staff activity involved care and attempts to initiate play and other constructive behavior. This is the treatment to which the boys are accustomed in the institution. The five retested boys were from two of the original groups, and all had territories during

the original control period (referred to as Trial 1). These boys together with the two new boys were given identical treatment by the same staff in the same experimental dayroom for a 30-hour period during a week 20 months later (referred to as Trial 2).

Data Collection and Analysis

During both trials, data on the location of each individual were taken every 10 minutes in terms of the grid square he was on (Fig. 1). These data were analyzed to yield information as to which boys used specific areas of one playroom significantly more than the other boys during each trial, (elaborated in Paluck, 1971). In essence, the location data in each group were analyzed square by square to see which squares were used significantly more by one child than by the other children. Adjacent squares which showed a high occupancy by the same child were combined and called one territory. For example, in the slide area (Fig. 1), one child may control just the top of the slide, i.e., square No. 122; this would be one territory. Another child may control the front of the slide (square No. 112) and the mat in front of the slide (square No. 102); this latter child would be designated as having one territory whose area was two spaces. A territory could therefore be any size, as long as all of it was controlled by one specific subject. A child was designated as controlling a particular territory when the number of times he was observed in that area during the week was greater than two standard deviations beyond the mean of the number of observations of the other six children during the week.

During both trials, observations were also made of the clinical behavioral symptomatology and social adjustment of each boy. These observations were made by the senior author, the attendants, and the observer. The behaviors observed were: stereotypies (repetitive idiosyncratic movements), escape behavior (desperate attempts to flee from the room whenever the door was opened), communicative behavior (vocalization and gesture), amount of social interaction and amount of aggressiveness to staff and children. Boys were designated as improved or deteriorated in

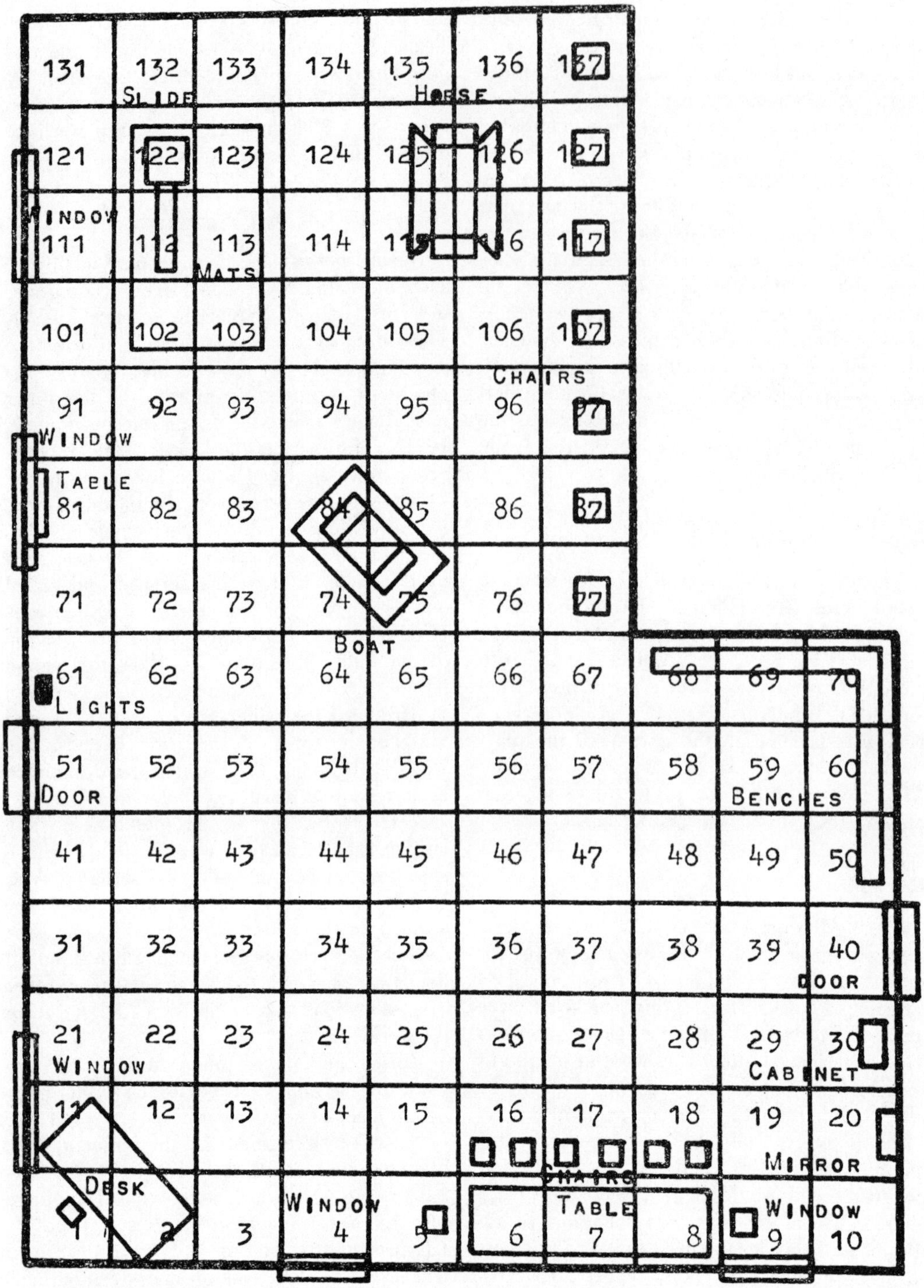

Fig. 1. Experimental dayroom.

these categories only when the change was so obvious that there was no disagreement in independent ratings by the project staff. Changes in territorial behavior across time in the five retested boys were compared to these changes in symptomatology and social adjustment.

Results

In both trials, individual boys showed clear preferences for specific areas and appeared to remember their relationship to these areas. For instance, immediately upon entering the room for the second trial, three of the retested boys went directly to spots they had either staked out or fought for during the first trial. These selections did not go uncontested. Many aggressive territorial incidents, though fewer than during Trial 1, occurred during Trial 2.

There was remarkable consistency across trials, both in the choosing of specific territories and in the amount of time boys spent in them. In Trial 1, Subject ML was on the bike or with the bike on bench No. 50 during 62 percent of the observations, while no other child had either the bike or the bike and the bench more than 7 percent of the observations. In Trial 2, ML was in this territory during 32 percent of the observations, while no other boy was in it more than 6 percent of the observations. Subject ML again took over the bike and the bench despite the fact that he was in a different group of boys and had not been in that setting for 20 months. Likewise, all the boys except one (RG) staked out some territories which were identical or morphologically similar to their past territories (Table 1). Although Subject RG did not stake out the Trial 2 territory he possessed in Trial 1, he did stake out the territory (table top) in Trial 2 which he had bitterly fought for but lost in Trial 1. Thus all boys showed territorial behavior in Trial 2 remarkably similar to that shown 20 months before.

The consistency over time was not perfect, as Table 1 shows, but the variations show a systematic pattern when related to changes in clinical behavioral condition. For example, one of Subject ML's territories in Trial 1 was a little-used, empty space outside but near the mat and slide area. At that time, it was

unclear why he spent time there, since he did not participate in activities. In the second trial, he abandoned this empty square and took over the highly contested top of the slide. Concomitant with this change from isolated to dominant status with respect to the slide, was his considerable improvement in the ability to communicate needs to staff and children with gesture, he being nonverbal (Table 1). In addition, he showed considerably less of the desperate escape reaction he showed so prominently during Trial 1 whenever the staff opened the door. His decreased occupancy of the door territory (from 62 percent in Trial 1 to 32 percent in Trial 2) with a concomitant increase in his occupation of the slide area, reflects this change in escape behavior. He seemed less fearful and more adjusted to group life. Thus, ML's improvement in clinical condition was reflected in the taking of a much more desirable, socially integrated, and contested territory.

Another boy, Subject AR, established a Trial 1 territory around the observer's desk. A considerable portion of his time then was spent trying to rip up data sheets and generally being annoying. Twenty months later he spent a considerable amount of time helping the attendant and exercising the power which status by association gave him over other children. He was also observed to be less physically aggressive toward children and more sociable to everyone. Thus, for AR behavioral improvement was related to finding a more general way of asserting dominance than establishing a fixed territory; that is, he joined the staff power structure. While staff might be considered AR's "territory" in Trial 2, in the sense that he habitually used the space around her, other attributes of his association with staff differed from typical territorial behavior. For example, staff was not simply a space to defend or retreat to. Staff had a social role of considerable power, and association with her and her functions led to more power for AR. Moreover, AR did not defend the space around staff as others did the space around fixed objects. He had enough power to permit others to be near staff, since he could and did take space back again without a fight. For these reasons we say that AR's improved behavioral condition was re-

TABLE 1

COMPARISON OF TERRITORIAL BEHAVIOR IN TWO ONE-WEEK TRIALS SEPARATED BY 20 MONTHS

Name	Age at first trial	IQ at first trial	First trial group[a]	Territories in first trial	Territories in second trial	Change in behavioral condition over 20 months
AR	10	37	I	Space around observer's desk	None	Less aggressive to children; less annoying to staff; followed staff around as helper
CH	7	33	I	Horse	1. Horse 2. Mat	None
TH	7	31	II	1. Large chairs 2. Highly contested large table top	1. Large chair 2. Isolated small table	Less interaction; more stereotypy
RG	10	35	II	1. Large chairs 2. Horse	1. Highly contested large table top 2. Area around door	More interaction; less stereotypy
ML	9	37	II	1. Bench near door 2. Bike 3. Empty space two squares from slide	1. Bench near door 2. Bike 3. Top of slide	Communicates more effectively; less escape behavior

[a] Reported in Paluck & Esser, 1971.

flected in a move from fixed territory ownership to a position of dominance through participation in the staff power structure.

Two other boys, Subject RG and Subject TH, had shown mild autistic symptomatology in Trial 1. RG showed the stereotyped behavior of rocking and humming while holding his ears and also burst into fits of aggression on occasions which seemed to have no environmental stimulus. Both of these autistic behaviors were far less frequent 20 months later, and he also interacted more and with greater appropriateness. In Trial 1, TH had shown on occasion the stereotypy of biting one hand while punching into the air with the other. Also, he occasionally walked around randomly, with dazed or tearful expressions. Both of these autistic behaviors were more apparent 20 months later.

RG's improvement and TH's deterioration were reflected in their territorial behavior. They were in the same group in the first trial, and battled each other for the large table territory which TH won. Twenty months later, RG controlled the large table, and TH retreated to the possession of a morphologically similar but smaller and less desirable table-cabinet. Thus, for TH, deterioration was re-

flected in loss of ability to control desirable territory. For RG, clinical improvement was reflected in increased territorial ability. Moreover, RG's other change in territorial behavior, from chair holder to controller of the door and light switch area, was a move to a more highly used and therefore more desirable area.

Discussion

Territorial behavior is probably an early evolutionary means of regulating social contacts. If we hypothesize that territorial behavior is a more primitive social behavior pattern than general dominance behavior, then it would follow that we increase our systematic attention to the need for territory where progressively impaired functional levels and decreasing intelligence is diagnosed. In other words, the less there is of cortical function, the more institutionalized man will have to adapt to his environment with subcortical mechanisms, which include territoriality.

The degree of constancy of territorial behavior in form and quantity across such a long period of time adds to the earlier evidence (Paluck, 1971; Paluck & Esser, 1971) of the extraordinary degree in which these severely retarded boys behave in the territorial mode in a permitting environment. The fact that all boys fought for some territories which were identical or morphologically similar to territories fought for 20 months before suggests that rigidity in behavior pertaining to the use of space is another crucial aspect of the behavioral rigidity of the institutionalized severely retarded (Zigler, 1963).

In this group of boys, the following appear to be territorial signs of improved clinical condition: (a) the move from an isolated or uninteresting and uncontested territory to an interesting and populated one, like Subject ML's move from an isolated spot near the slide to the slide itself, and (b) the move from fixed territorial ownership to participation in the staff power structure as an expression of dominance.

The fact that changes in clinical behavioral symptomatology appear to have their correlates in changes in territorial behavior does not mean that changes in specific behavior exclusively *cause* changes in territorial behavior or that changes in the capacity to establish territories *cause* changes in specific behaviors. Improved capacity for communication of intentions, for example, makes territorial ownership easier; the security afforded by territorial ownership reduces anxiety (Paluck, 1971; Paluck & Esser, 1971), and makes communication easier. Thus, territorial behavior interrelates with other parameters of overall psychological functioning in the institutionalized individual. Changes in overall functioning, including territorial behavior, may be caused by treatment procedures in the institution dealing with specific behaviors or by other factors such as differential rates of growth in intellectual capabilities, size, and strength of individuals. In the latter case, for example, a boy who grows faster than his peers has increased capacity to control his peers, and thus his general functioning, including territorial behavior, may improve. (In the present study, none of the boys appeared to have grown at a rate significantly different from the other boys between Trial 1 and Trial 2). But regardless of the cause of the change in overall psychological health, results of this study showed that territorial behavior appears to be an aspect of the change, and therefore, an indicator of that change.

Since territorial behavior is a simple social behavior, in the sense that manipulation of spatial relationships forms the background of more complex social functions, it is possible that changes in territorial behavior may become apparent before changes in the more complex functions. For example, one possible interpretation of Subject ML's initial establishment of a territory in an empty spot outside the slide mat area was that it was an autistic act, since his overt behavior in it bore little relationship either to the slide or to the space itself. However, after Trial 2, it was obvious that the staking out of that spot was a first tentative step in an effort to use and control the nearby slide. This effort represents the emergence of some capacity to deal effectively with his dominance strivings which are a crucial aspect of improvement in general psychological health (Esser, 1968). Thus, when we see a patient change territorial behavior, we may be in the position of understanding whether his behavior is deteriorating or improving.

The possibility of prediction leads to the possibility of therapeutic intervention. Knowledge of ML's territorial behavior puts us in a position to help reinforce him in getting his share of the slide activity and thus promote social normalization. Knowledge of the meaning of changes in territorial behavior might also be used as an indicator of when to use other aspects of the therapeutic armamentarium, e.g., drugs. As a corollary, records of changes in territorial behavior might be used to test whether therapeutic procedures are actually working. Our work suggests, therefore, that a considerable amount of descriptive and predictive power can be obtained with the systematic recording of the sequential spatial locations of individuals.

R. J. P.
Department of Psychology
Hunter College of the City
  University of New York
695 Park Avenue
New York, New York 10021

References

Esser, A. H. Dominance hierarchy and clinical course of psychiatrically hospitalized boys. *Child Development*, 1968, 39, 147–158.

Esser, A. H. Interactional hierarchy and power structure on a psychiatric ward—ethological studies of dominance behavior in a total institution. In C. Hutt and J. Hutt (Eds.), *Behaviour Studies in Psychiatry*. Oxford: Pergamon, 1970.

Esser, A. H., Chamberlain, A. S., Chapple, E. D., & Kline, N. S. Territoriality of patients on a research ward. *Recent Advances in Biological Psychiatry*, 1965, 7, 37–44.

Paluck, R. J. Territoriality and operant conditioning of institutionalized severely retarded boys. Unpublished doctoral dissertation, Columbia University, 1971.

Paluck, R. J., & Esser, A. H. Controlled experimental modification of aggressive behavior in territories of severely retarded boys. *American Journal of Mental Deficiency*, 1971, 76, 23–29.

Zigler, E. Rigidity and social reinforcement effects on the performance of institutionalized and non-institutionalized normal and retarded children. *Journal of Personality*, 1963, 31, 258–269.

VI

Food and Energy Relationships
in Territorial Systems

Editor's Comments on Papers 19, 20, and 21

19 **Wolf and Stiles:** Evolution of Pair Cooperation in a Tropical Hummingbird
   Evolution, **24**, 759–773 (1970)

20 **Verner:** Evolution of Polygamy in the Long-Billed Marsh Wren
   Evolution, **18**, 252–261 (1964)

21 **Orians:** The Ecology of Blackbird (*Agelaius*) Social Systems
   Ecol. Monogr., **31**, 285–286, 290–295, 297–299, 301–306, 311–312, (1961)

The importance of the territory in ensuring adequate food (for at least some species) has been stated by many persons, including Altum and Howard. Hinde (Paper 5, p. 361) concludes that "while territorial behaviour is primarily concerned with food in a few species (e.g., hummingbirds, Pitelka, 1942), in most cases the food value is not significant." But the problem refuses to lie dormant. Brown (see Paper 23) suggests that a more constructive approach to the study of the role of food is to identify the conditions under which the defense of a territory is economically feasible. Whether an animal will find it profitable to establish and defend a territory will depend upon the balance between the amount of resources to be obtained and the energy expended in securing and defending the territory against competitors. The added risk of predation resulting from the male's conspicuousness while defending the territory is an additional factor. An individual will defend a territory only if the advantages more than offset the disadvantages and increase its overall fitness (i.e., increase the chances of maximizing its progeny). Where the adult must bring food to a central point such as nest or den, the rate at which the adult can bring food to the young will be a function of the richness of the food resource and the degree of competition. The more sparse the food supply, the farther the individual must travel to secure adequate food. There will come a point in some habitats where the animal must forage so far that breeding becomes impossible.

Additional evidence for the importance of food as a function of territory comes from the correlation between body weight and territory size in birds (Schoener, 1968) and mammals (McNab, 1963).

Orians (1971) has expanded on these relationships in a recent review. He summarizes his discussion of food as follows: "We should expect, however, that average territory sizes should be correlated with average food availability in different habitat types. An important part of future territory studies will involve quantification of this correlation and its behavioral basis."

The necessary measurements referred to by Orians are exceedingly difficult to obtain, especially among omnivorous feeders. A trio of ornithologists are collaborating in an attack on the problem in hummingbirds and sunbirds. These birds rely largely upon nectar for food and defend food territories. Larry L. Wolf and F. Gary Stiles (Paper 19) have recently been joined by Frank B. Gill, Curator of Ornithology, Philadelphia Academy of Natural Sciences, in attacking different aspects of the problem. They are able to measure the daily production of the minute quantities of nectar in individual blossoms and thereby estimate the total energy available on a given day to a bird within its territory. They have then been able to relate nectar production and flower abundance with the duration and size of territory held and have seen how these factors affected the mating system.

The evolution of a mating system is intimately linked with the nature and abundance of the food supply. Where food is abundant the male, in birds at least, may be freed from the necessity of parental care and thus be able to secure more than one mate. Verner's study of the long-billed marsh wren (Paper 20) is important because he shows that the success of the male in acquiring one or more mates depends upon such characteristics of the territory as size and amount of emergent vegetation. Verner did not measure the amount of food within the territory, but he felt that the most successful territories did provide the most food. Zimmerman (1971) found very similar relationships in the dickcissel.

Jared Verner has worked closely with Gordon Orians (see Paper 21) in studies of the evolution of mating systems. He has continued his work on evolutionary ecology and mating systems at Central Washington State College and at Illinois State University, where he is now Professor of Biological Sciences.

What factors limit the size of colonial nesting has been a perplexing problem. Since food is not obtained within such territories and nesting sites are rarely limiting, what prevents unlimited growth of the colony? Gordon Orians (Paper 21) has one explanation based upon food abundance and energy expenditure in tricolored blackbirds, which may nest in huge colonies of several thousand birds. When a newcomer joins a breeding colony, it makes many trips back and forth between the colony and the food source. Orians postulates that the bird is gaining a measure of the energy it must expend by flying back and forth in obtaining food against the degree of competition for this food as measured by the total display of the territory-holding males already established in the colony. There are always far more males in the colony at the beginning of the nesting season than later, suggesting that, in fact, many males may have found it unprofitable to attempt to breed.

Orians's boyhood interest in birds developed into professional involvement with the evolution of vertebrate social systems. He has concentrated on blackbirds because they are easy to study and have such a great variety of social systems. Of special interest to him have been problems of habitat selection, mating systems, spacing patterns, and foraging behavior. His students Jared Verner and Mary Willson are continuing in the same line. Orians's writings have had a strong influence in the field of evolution of mating systems (Orians, 1969, 1971; Orians and Willson, 1964; Orians and Christman, 1968).

References

McNab, B. 1963. Bioenergetics and the determination of home range size. *Amer. Naturalist,* **97**: 133–140.

Orians, G. H. 1969. On the evolution of mating systems in birds and mammals. *Amer. Naturalist,* **103**: 589–603.

Orians, G. H. 1971. Ecological aspects of behavior. *In* D. S. Farner and J. R. King (eds.) *Avian Biology,* Vol. I, 513–546. Academic Press, New York. 586 pp.

Orians, G. H., and G. M. Christman. 1968. *A Comparative Study of the Behavior of Red-winged, Tricolored, and Yellow-headed Blackbirds.* Univ. Calif. Publ. Zool. 84. 1–85.

Orians, G. H., and M. F. Willson. 1964. Interspecific territories of birds. *Ecology,* **45**: 736–745.

Schoener, T. W. 1968. Sizes of feeding territories among birds. *Ecology,* **49**: 123–141.

Zimmerman, J. L. 1971. The territory and its density dependent effect in *Spiza americana. Auk,* **88**: 591–612.

19

Reprinted from *Evolution*, **24**, 759–773 (1970)

EVOLUTION OF PAIR COOPERATION IN A TROPICAL HUMMINGBIRD

Larry L. Wolf

Department of Zoology, Syracuse University, Syracuse, New York 13210

F. Gary Stiles[1]

Department of Zoology, University of California, Los Angeles, California 90024

Received March 9, 1970

In contrast with the many species of birds that form pairs during the breeding season, hummingbirds, in general, have a promiscuous mating system. The male mates with the female, and the female carries out the remainder of the reproductive effort, including nest-building, incubating the eggs and caring for the young, unaided by the male. The only reports to the contrary are for *Colibri coruscans*, the Sparkling Violet-ear of the Andes of South America, in which the male may aid in incubating the eggs and caring for the young (Moore, 1947; Schafer, 1954). However, Ruschi (1965) reported that only females attended the six nests of this species he studied, one in the wild and five in captivity. A pair bond in most hummingbirds, then, is limited to the short time during which a male and a female must cooperate to insure successful copulation. During the breeding season, the typical situation in hummingbirds is for males to hold mating and feeding territories at flowering plants, from which they attempt to exclude all other hummingbirds. In many cases, the female is treated aggressively by the male, both before and after mating, in a manner similar to the treatment accorded other conspecific intruders into his territory.

The present report discusses pair relations of *Panterpe insignis*, the Fiery-throated Hummingbird, during the breeding season. In this species a longer-term pair bond is apparently formed. The male still does not participate in the actual nesting effort, but indirectly does aid the female in

obtaining food for herself and probably for the young.

Study Area and Methods

The observations were made from 21 to 29 August, 1969 at two sites along the Pan American highway, in the vicinity of Villa Mills, Costa Rica (09°34′N; 83°41′W) at an elevation of approximately 3100 meters. The area is primarily second growth with some remnants of oak forest still standing.

The actual study sites were in second growth areas, one a steep northern-facing roadcut on the highway itself and the other a small clearing about 100 feet from the highway. Our observations were restricted to three easily observed territories—two in the roadcut (Areas 1 and 2) and one in the clearing (Area 3) (see Figs. 1, 2). There were at least three additional *Panterpe* holding territories adjoining that of the male we watched in the center of the clearing. These adjacent territories included areas of such irregular topography that it would have been nearly impossible to follow the movements of individuals holding them. Observations of known individuals were made at distances of 100 feet or less with the aid of 10x and 7x binoculars, and totalled more than 45 man-hours. Observation periods usually lasted at least 2 hours. Generally one of us observed the birds and timed their activities with a stopwatch, while the other recorded the data. Prior to the timed observations, most individuals we watched were caught in mist nets and marked with one or two spots of airplane dope on the midback—a unique color combination for each individual. We will henceforth refer to a marked bird by this color

---

[1] Present address: Department of Ornithology, American Museum of Natural History, New York, New York 10024.

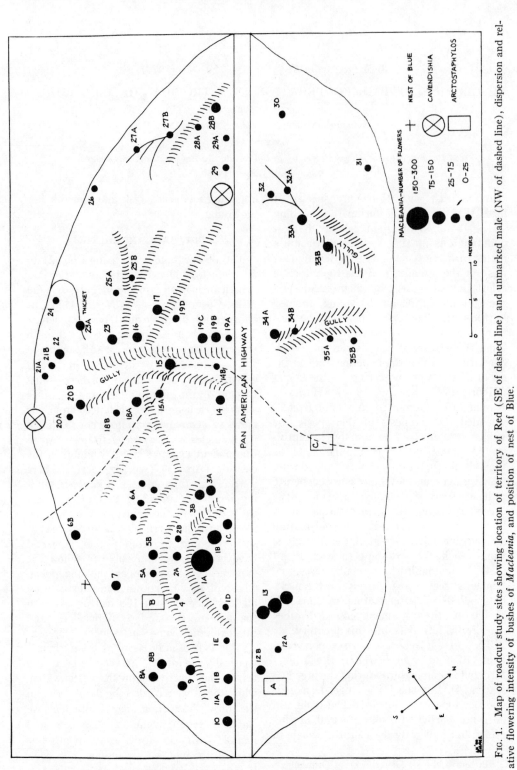

Fig. 1. Map of roadcut study sites showing location of territory of Red (SE of dashed line) and unmarked male (NW of dashed line), dispersion and relative flowering intensity of bushes of *Macleania*, and position of nest of Blue.

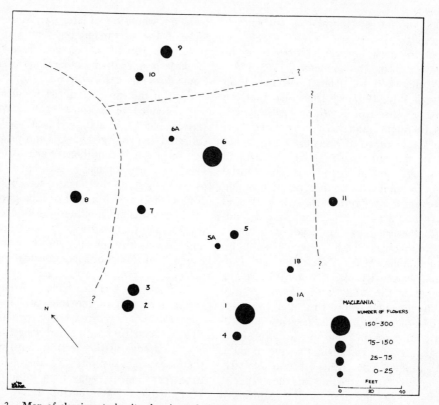

FIG. 2. Map of clearing study site showing relation of territory of WB (inside incomplete rectangle of dashed lines) to adjacent males and dispersion and flowering intensity of *Macleania* in territory of WB.

combination; for example, "White-Blue" means the individual with a white and a blue spot on the back. We noted no differences in behavior between marked and unmarked birds. All of the main participants were marked at two of the three study areas. Although neither bird was marked at one site in the roadcut, they both had such restricted patterns of movement that it was not difficult to follow them. Most *Panterpe* that are territorial or regularly using a restricted area are very consistent in choice of perches, routes of flying, and other behavioral patterns associated with exploiting the area (Wolf, MS). This consistency further facilitates recognizing individuals that are not marked.

Panterpe is sexually dimorphic in size (Wolf, 1969), so initial sex identification of the marked birds was based on wing and tail measurements (Table 1). At the end of the observations most of the important individuals were collected to verify sex identification.

GENERAL BIOLOGY OF *PANTERPE*

The Fiery-throated Hummingbird is limited to the mountains of Costa Rica and western Panama (Slud, 1964), usually above 2100 meters elevation. It is most common in second growth habitats and cleared areas where flowering shrubs and epiphytes are abundant. *Panterpe* is a year-round resident at Villa Mills; during at least part of the year it is sympatric with the Green Violet-ear (*Colibri thalassinus*), the Rivoli Hummingbird (*Eugene fulgens*) and the Volcano Hummingbird (*Selasphorus flammula*). These hummingbirds along with the Slaty Flower-piercer (*Diglossa plumbea*) are the only birds at Villa Mills to regularly exploit nectar of flowers as an energy source.

On the basis of our observations of *Panterpe* during the summers of 1965, 1966, 1968 and 1969 and from December, 1966 through July, 1967, we concluded that the breeding season of *Panterpe* probably begins in August and is over by December. The actual range of breeding dates is not known and only one nest has been found (Wolf & Stiles, in prep.). Circumstantial evidence points to a relation between timing of breeding of *Panterpe* and flowering of *Macleania glabra* (Ericaceae) and, secondarily, flowering of *Gaiadendron poasense* (Loranthaceae), both common epiphytes around Villa Mills. *Macleania* and *Gaiadendron* usually bloom from July to December in the study area. *Panterpe* also begins breeding before the other two large species of hummingbirds, *Eugenes* and *Colibri*. *Eugenes* and *Colibri* are both altitudinal migrants and do not arrive in numbers around Villa Mills until after early September and breeding of these species probably does not begin until a majority of the birds arrive.

RESULTS

Flower and nectar availability change seasonally around Villa Mills. Flowering activity in the summer of 1969 was somewhat different from the four previous summers. Normally in August several species of plants visited by hummingbirds, particularly *Macleania glabra, Miconia and Gaiadendron poasense,* are coming into bloom nearly simultaneously. However, in 1969 only *Macleania* was blooming abundantly. The onset of blooming of *Gaiadendron* at Villa Mills was much less synchronous than usual and *Miconia* was nearly finished blooming. In 1969, of the species at which *Panterpe* males usually are territorial at this time of year, only *Macleania* was abundant enough in the areas we investigated to support territorial males of *Panterpe*. We found no *Gaiadendron* clumps large enough to support a territorial *Panterpe,* although some of the clumps just starting

to bloom would do so later on. More than 95% of the nectar-foraging bouts of the territorial birds we observed were at *Macleania,* and all the other bouts were at *Cavendishia* or *Arctostaphylos.*

Each of the three territories we observed was defended by a single male *Panterpe* (see Table 1). Each territorial male we collected had enlarged gonads and probably was in breeding condition. Each male was also in the early stages of a complete molt, as were most of the females we collected. Blue (marked with a spot of blue paint on the back) was not molting, possibly related to the fact that she was already nesting, and she was also the only *Panterpe* we collected that had no subcutaneous fat deposits.

Red was the resident male in Area 1, in the eastern half of the roadcut. Two female *Panterpe* occurred fairly regularly in this first territory. Blue had a nest at the southeast edge of this territory and visited the flowers in the territory three or four times per hour for one to 4 minutes per visit (Table 2). We concluded that she did most of her foraging at flowers being defended by Red. Usually we saw her as she came down the hill from the nest to forage and were able to watch her until she settled onto the nest to resume incubating. White, on the other hand, was less regular in her visits, although she usually stayed as long per visit (Table 2). Neither White nor Blue stayed in the territory very long after foraging. A third female, Yellow, was a fairly regular visitor at the onset of observations, but she was consistently chased away by the resident male, Red. After 25 August she rarely appeared in the territory.

In Area 1, 25 flowering clumps of *Macleania* were visited by the hummingbirds (Table 3). Fourteen of these clumps (56%) with 50% (585) of the total flowers were visited regularly and nearly exclusively by the male; six (24%) of the bushes with 32% (375) of the flowers were visited by Blue; four (16%) of the bushes with 16% (183) of the flowers were visited by White; and one bush with 2% of the flowers was visited

TABLE 1. *Characteristics of* Panterpe insignis *observed in Areas 1, 2, and 3.*

Color code	Sex	Gonads (ova or left testis)	Wing (mm)	Tail (mm)	Weight (gm)	Status	Fat	Other
Red	M	3.5 × 3.0 mm	64	42	5.3	resident area #1	light	starting molt
Blue	F	ova, 0.5 mm	58	38	5.1	nesting female area #1	none	brood patch (?)
White-Blue	M	3.3 × 2.8 mm	64	43.5	5.7	resident area #3	light-mod.	starting molt
Green-Yellow	F	ova, 0.1–1.5 mm	57	36	4.9	female area #3	light	starting molt
——	M	2.5 × 2.2 mm	65.5	45	6.1	resident area #2	mod.	starting molt
——	F	ova, 0.1–1.3 mm	57.5	40.5	5.4	female area #2	mod.	starting molt
——	M	3.2 × 3 mm	64.0	42.0	5.8	replaced male area #1	very light	starting molt
White	F		58	39	5.3	female area #1		
Yellow	F		58	39	5.1	female area #1		

TABLE 2. *Frequency and duration of visits to territory 1 by* Panterpe *females Blue and White.*

Bird	Hours of observation	No. visits per hour	Average interval between visits (minutes)	Average duration of visit (minutes)
Blue	4.25	4.0	10.3 (6–16)[1]	2.7 (1–4)
White	3.00	3.3	12.9 (6–29)	3.6 (1.5–10)[2]

[1] Range of values in parentheses.
[2] White probably was not nesting and spent some time sitting in territory 1.

by Yellow during the first days of our observations. There was little overlap in utilization of the bushes by the several regular foragers (Table 3). White sometimes attempted to forage at several bushes used by the male, but generally she was quickly chased from these bushes. Blue, the nesting female, was using the choicest feeding site, a bush with more than 200 flowers, and a few smaller nearby bushes. The male had to visit more bushes and probably had to expend more energy than the female to collect the same total amount of energy. The other two females that utilized this area visited fewer bushes and fewer total flowers. Both of these females probably foraged in other areas between visits to Red's territory.

The resident male, Red, initiated aggressive encounters with the three marked females visiting his area (Table 4). He successfully chased Yellow away more than 90% of the time and Yellow soon stopped visiting the area. White was able to feed at bushes 7, 8, and 9, in spite of attempts by Red to chase her away. When White attempted to forage at 6a, she was regularly

TABLE 3. *Flower availability and visitations in Area 1—Eastern roadcut.*

Bush	No. flowers[1]	Visits by: Red[2]	Blue[3]	White[4]	Total visits
1 A	240	4 (5.2)	72 (94.8)	0 (0)	76
1 B	31	2 (8.7)	21 (91.3)	0 (0)	23
1 C	81	29 (100)	0 (0)	0 (0)	29
1 D	18	0 (0)	16 (100)	0 (0)	16
1 E	10	0 (0)	4 (100)	0 (0)	4
2 A, B	60	34 (100)	0 (0)	0 (0)	34
3 A, B	116	115 (100)	0 (0)	0 (0)	115
4	23	1 (3.3)	29 (96.7)	0 (0)	30
5 A, B	33	38 (100)	0 (0)	0 (0)	38
6 A, B	63	36 (87.8)	0 (0)	5 (12.2)	41
7	50	3 (23.1)	0 (0)	10 (76.9)	13
8 A, B	93	1 (5.6)	0 (0)	17 (94.4)	18
9	40	0 (0)	0 (0)	4 (100)	4
10	40	27 (100)	0 (0)	0 (0)	27
11 A	48	41 (100)	0 (0)	0 (0)	41
11 B	54	0 (0)	20 (100)	0 (0)	20
12 A, B	43	15 (100)	0 (0)	0 (0)	15
13	73	40 (100)	0 (0)	0 (0)	40
14	52	38 (100)	0 (0)	0 (0)	38
Totals	1168	424	162	36	

[1] 25 August 1969.
[2] 9½ hours 22, 24, 25, 27 August; numbers in parentheses are percent of total visits.
[3] 8⅔ hours 22, 24, 25 August.
[4] 5 hours 24, 27 August.

TABLE 4. *Aggressive interactions by Red.*[1]

	Species encountered														Total	Encounters with intruders per 100 flowers available
	Diglossa			Female *Panterpe* Blue			White			Unidentified *Panterpe*			*Selasphorus*	*Eugenes*		
At flowers utilized by:	D°	D+	D-[2]	D°	D+	D-	D°	D+	D-	D°	D+	D-	D+	D+		
Red	15	12	71	0	0	0	0	6	0	1	8	2	27	6	148	25.3
Blue	4	4	27	7	3	23	0	0	0	0	1	0	1	7	77	11.7
White	0	1	0	0	0	4	6	5	27	0	11	0	3	7	64	14.1
Totals	19	17	98	7	3	27	6	11	27	1	20	2	31	20	289	

[1] Based on 22½ man-hours of observation.
[2] D° = approach intruder, but no overt encounter.
 D+ = chases intruder away.
 D- = attacks, but fails to displace intruder.

chased away by Red, and often retreated to one of the bushes at which she could forage relatively undisturbed. Similarly, Blue was able to remain at certain bushes for foraging and even chased the male a few times at bushes she regularly visited. We saw no attempts by Blue to forage at bushes regularly used by the male.

The western half of the roadcut, area 2, was occupied by an unmarked male and female. Apparently no other *Panterpe* foraged in the territory with any regularity, although the male occasionally had encounters with other *Panterpe* at the edges of his territory at bushes he regularly visited. The two resident birds perched within the territory defended by the male. The female normally perched in or at the edge of a small tree and shrub thicket near several of her foraging sites. The male regularly used a small number of perches, often sitting near the foraging female. We saw two copulations and several apparent copulations by this pair; these took place in the thicket where the female regularly perched. A peculiar buzzy chittering, in our experience given only by the female (?) during copulation, was heard on several occasions when both birds were out of sight in the thicket.

There was little overlap of bushes at which the two birds in area 2 foraged (Table 5). The male visited about 60% of the 26

bushes in the area. These bushes contained about 58% of the total flowers. The remainder of the bushes were visited by the female, but rarely or not at all by the male.

The pattern of flower utilization was somewhat different in the clearing, area 3. The territorial male and female shared several large flower clumps, but the male used one of the clumps much more frequently than the female (Table 6). There was an indication that the male and female might use different portions of the large clumps they shared. However, we had too few observations to make definite statements about division of the available sources. As in the other two areas, the male initiated encounters with the female foraging at some of the flowers. The female was able to maintain her foraging position and sometimes to displace the male.

In summary, in each territory we found a single resident male and one or more females that were allowed to forage in the area. In two cases the flowers visited by the females were distinct from those visited by the male; in the third case there was marked overlap. In each area the female(s) had access to the richest foraging sites. In the clearing these sites were shared with the male, but at the roadcut they were visited only by the female.

TABLE 5. *Flower availability and visitations in Area 2—Western roadcut.*[1]

| Bush No. | No. flowers | Visits by: | | Total visits |
		Male[2]	Female	
14 B	15	0 (0)	4 (100)	4
15	35	0 (0)	5 (100)	5
15 A	15	2 (100)	0 (0)	2
16	37	0 (0)	7 (100)	7
17	30	0 (0)	8 (100)	8
18 A	42	12 (100)	0 (0)	12
18 B	23	6 (100)	0 (0)	6
19 A)				
19 B)	119	0 (0)	24 (100)	24
19 C)				
19 D)				
20 A	8	1? (0)	3 (100)	3 + 1?
20 B	55	6 (75)	2 (25)	8
21	20	6 (67)	3 (33)	9
22	38	4 (100)	1? (0)	4 + 1?
23 A	47	0 (0)	11 (100)	11
23 B	30	0 (0)	?	?
24	10	0	0	
25	4	0	0	
26	20	2 (100)	0 (0)	2
27	38	3 (100)	0 (0)	3
28	52	2 (100)	0 (0)	2
29	7	1 (100)	0 (0)	1
30	1	0	0	
31	11	1 (100)	0 (0)	1
32	18	1 (100)	0 (0)	1
33	77	2 (100)	0 (0)	2
34	62	? (100)	0 (0)	?
35	11	3 (100)	0 (0)	3
Total	825	51	67	

[1] Data obtained 0824–1030, 26 August 1969.
[2] Percent of total visits to each *Macleania* by each sex indicated in parentheses.

DEFENSE OF FLOWERS BY TERRITORIAL MALES

As noted earlier each territorial male occasionally made aggressive flights toward the females that regularly foraged in his territory. We saw a male actually chase a female from one of "her" bushes only once. Often the female responded aggressively to an encounter initiated by the male and the male stopped the encounter. However, each male was able to defend the flowers in his territory from other hummingbirds. This suggests that a more or less co-dominant relationship existed between the male and the female(s) sharing the resources of a territory.

The male at the clearing defended or at-

TABLE 6. *Flower availability and visitations in Area 3—Clearing.*[1]

Bush No.	No. flowers	No. visits male	No. visits female	Total visits
1	260	14	3	17
1 A	24	0	0	0
1 B	19	2	0	2
2	75	4	0	4
3	90	0	0	0
4	60	0	0	0
5	40	0	1	1
5 A	16	0	0	0
6	285	11	10	21
6 A	15	2	1	3
7	43	5	1	6
8	77	9	1	10
Total	1004	47	17	64

[1] Data recorded 0750–0950, 28 August 1969.

TABLE 7. *Aggressive interactions by WB in clearing.*[1]

Bush No.	Diglossa D°	D+	D-	Panterpe GY Female D°	D+	D-	Panterpe RB Male D+	Unidentified Panterpe D+	Lampornis D+	Eugenes D+	Selasphorus D+	Total
1	0	4	8	0	0	1	1	10	1	0	2	27
1 a												
1 b												
2			1									1
3												
4												
5						1				1		2
5 a								1				1
6	0	0	12	1	0	7	0	5	0	1	0	26
6 a	0	1	0	0	0	1				2	0	4
7	0	0	1					2				3
8	0	1	2	0	1	0						4
9												
10												
11												
Total	0	6	24	1	1	10	1	18	1	4	2	68

[1] Data recorded 0750–0950, 28 August 1969.

tempted to defend most and probably all of the flowers at which he was foraging (Table 7). However, adjacent territorial *Panterpe* shared a few clumps of *Macleania* with this male along territorial boundaries. Since his primary foraging sites overlapped with those being visited by the female, he, in effect, was defending the flower sources used by the female.

At the roadcut sites the males defended the energy sources of the females as well as their own, even though they did not use the energy available in the flowers visited by the females. Our most conclusive data on this point were from area 1 where we made our most extensive observations. As can be seen in Table 4, Red defended or attempted to defend flowers which the females were using as well as those he was using. Relative to the number of flowers available, Red defended the bushes used by Blue and White less often than those at which he foraged. However, since the latter were more peripherally located and therefore more likely to attract intruders, one cannot necessarily conclude that levels of defense were different.

Red consistently tried to chase *Diglossa* from *Macleania,* but was rarely able to dis-place a foraging male and had only slightly better success with females (Table 4). At other times of the year, when alternate food sources were readily available, territorial *Panterpe* were able to displace *Diglossa* from a nectar source in the territory with a high degree of success (Wolf, MS).

We have many fewer records of defense of flowers by the unmarked male in the western half of the roadcut. However, he did chase one or more female *Eugenes* from the flowers visited by the resident female. He also attempted to chase both sexes of *Diglossa* from the female's flowers with about the same degree of success as Red.

FACTORS INFLUENCING EXCLUSIVENESS OF FORAGING SITES WITHIN A TERRITORY

There appear to be two possible ways for a male and female *Panterpe* to share the flowers in the male's territory; either (1) certain flower clumps will be used exclusively by one bird or the other, or (2) the two may share some or all clumps. The pattern of flower utilization that occurs is probably related to the energetic efficiency of foraging. Other things being equal, the pattern of exclusive use of different shrubs by male and female, as seen in areas 1 and

L. L. WOLF AND F. G. STILES

TABLE 8. *Territory size, flower abundance, and flower utilization in three territories of* Panterpe.

Territory No.	Area (Sq Ft)	Total bushes	No. of bushes used male	female	Total flowers	No. of flowers used male	female	Flowers per 1000 ft²	Flowers per bush
1	5,650	25	15	6–Blue 4–White	1168	609	376–Blue 183–White	207	46.7
2	11,300	29	20	12	825	407	341	73.0	22.3
3	11,500	12	7+	6+	1004	774±	720±	87	83.7

2, is more efficient. If only one bird uses a given set of flowers, it can consistently visit the flowers with the most available nectar simply by avoiding those it used on the previous visit. In this manner, less time and energy will be spent at flowers recently depleted of nectar by another bird. However, when a large proportion of the flowers in a territory concentrated in a very few clumps, it may not be energetically feasible for one member of the pair to be excluded from these richer sites. In this case sharing will become advantageous, but there may be a tendency for the two birds to use different parts of the clumps, as we observed in the clearing. Flower exclusiveness at a shared clump tends to minimize the overlap of actual flowers used, and increases foraging efficiency.

Table 8 presents data on area and flower abundance for the three territories. Territories 1 and 2 are similar in the dispersion of the food supply, having a large number of bushes with relatively few flowers per bush. Territory 3 has the food supply concentrated in relatively fewer clumps, but about the same area and flower density as territory 2. Since the division of resources is similar in territories 1 and 2, and different in territory 3, it appears that dispersion of the food supply is the most important factor determining foraging patterns.

In the two roadcut areas, several of the largest clumps of flowers would be required to reach a total of more than 50% of the energy resources in the territory. However, in the clearing the two largest clumps contain approximately 54% of the flowers there (see Tables 3, 5, and 6). If the male in the clearing allowed the female exclusive use of these two clumps, less than half of the resources of the territory would be available to him. Although this might still leave him with about the same number of flowers as the roadcut males, the flowers would be much harder to defend due to their scattered distribution and the presence of adjacent territorial males. Conversely, the combined total of flowers in the largest clumps in the clearing is about 150% that available to the roadcut females, and is presumably more than enough to support the female in the clearing, especially as she did not limit her foraging to those clumps. With such a localized and superabundant resource, there is probably little disadvantage to sharing it, especially if the actual flowers used by the two birds are different.

Competition probably also affects the pattern of flower utilization by members of a pair. Potential competitors include other *Panterpe* (other females, intruding males, or males trying to establish or to hold adjoining territories), other hummingbird species, and the coerebid *Diglossa*. In the clearing, at least three other *Panterpe* males held territories adjacent to that of White-Blue, and one or more of these shared foraging sites with him along territorial boundaries. The boundary between the two roadcut males was much more definite, and there was very little overlap of foraging sites. As might be expected, there appeared to be fewer aggressive encounters between adjacent territorial males at the roadcut than at the clearing. Where territories overlap and disputes between neighboring males are frequent, the energy requirements of the territorial male will probably be higher and the efficiency of energy accumulation by the male will be more critical. A male requiring less energy because of fewer aggressive encounters could better afford to exploit a less rich or less localized food source and in this situation a male might be more likely to allow a female exclusive use of a rich flower clump. The amount of flowers present in a territory may itself affect the frequency of aggressive encounters, as a rich territory will probably attract more intruders than one with less available nectar. At the same time, a large territory will require more movement and hence energy to defend than will a small one, all other things being equal (Gibb, 1956).

The resultant time and energy expenditures of a territorial male due to these various factors will affect his energetic balance, which in turn will determine whether the male can allow a female exclusive foraging rights to one or more clumps of flowers. The energetic balance of the male will depend upon the relationship between the amount of energy available and his efficiency at exploiting it, versus the energetic costs of territorial defense. The most important factors in determining the balance are the number and dispersion pattern of the flowers, and the intensity of competition. Exclusive use of rich clumps by a female will occur when competition is relatively low and/or the food source is relatively evenly distributed, provided the total amount of food present is sufficiently large and not too widely scattered.

Exclusiveness of foraging site is probably more important for nesting females than for males. Females forage quickly and return to the nest for another period of incubation or brooding and have little or no time available for defending a nectar supply. A nesting female must be able to obtain sufficient energy in a short time to allow her to incubate or brood for extended periods (6 to 16 or more minutes for Blue, Table 2) before foraging again. Similarly, a female with young must be able to collect enough food to feed herself and one or two young in a sufficiently short time that the young do not suffer from exposure. Foraging effi-

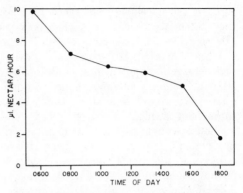

FIG. 3. Rates of nectar production of individual flowers of *Macleania glabra*. Rates are plotted at the end of the sample interval. Data obtained 28 and 29 August, 1969.

ciency for an incubating female becomes especially critical on cloudy, windy, and rainy days when midday temperatures at this time of year at Villa Mills reach as low as 9C to 10C. We noticed that Blue delayed her appearances at the flowers when it was raining, but would finally leave the nest anyway if the rain continued long enough.

A further energetic consideration for both sexes is the diurnal decline in rate of nectar production by *Macleania*. Figure 3 shows measurements of nectar production rates in fresh flowers at the roadcut on 28 and 29 August, 1969. These data are from different portions of each day, but indicate relative rates of production. Probably little nectar accumulates during the day since the birds forage over the major bushes thoroughly one or more times per hour. Near the end of the daylight foraging period, when both sexes are attempting to accumulate reserves of energy for the night, energy production per hour is low and still declining. A nesting female with eggs or small young is doubly pressed as she probably does not enter torpor at night, as at least some individuals of *Panterpe* are capable of doing (unpublished experiments). Howell and Dawson (1954) showed that an incubating female Anna Hummingbird (*Calypte anna*) did not enter torpor at night, probably as a device to maintain a high temperature for the developing embryos. Loss of this potential device for saving energy (Pearson, 1950; Lasiewski, 1963; Hainsworth and Wolf, 1970) increases the overnight energetic requirements. Additionally, a female feeding young must collect enough energy reserves to sustain the young throughout the night. Some energy reserves might be accumulated as fat when nectar production is high and energy demands are reduced at the midday peaks of environmental temperatures. We did not measure diurnal changes in body fat in *Panterpe*, but Blue, the nesting female, had no subcutaneous fat deposits when collected at 1:30 PM; all other *Panterpe* collected at about the same time had some fat deposits (Table 1).

DISCUSSION

Evolutionarily, it is highly unlikely for a male to expend time and energy to defend flowers visited by a female(s), unless he also derives some ultimate benefit by allowing her to forage on the area. The most obvious benefit is that the female(s) have mated with, or will mate with, the male, and that the offspring will be genetically related to the male. By allowing the female(s) to forage on the territory he is, in effect, helping to insure that the female, which undertakes the nesting effort alone, can acquire sufficient energy to successfully rear the two young. This indirect aid helps to maximize his genetic contribution to the next generation. The female, Blue, with the nest on the edge of territory 1 probably had mated with Red, but this could not be proved; neither was the relation of White to Red known. In area 2 we observed two copulations and several attempted copulations involving the resident pair. Although mating data are available only from this single pair it seems likely that nesting females allowed to forage in the territory of a male have in fact mated with that male.

There is probably an advantage, in terms of time and energy requirements, for a female to nest close to the territory where she forages. However, given the flying ability of hummingbirds, a female could also nest some distance away at a point more or less between the territories of two males, with each of which she has mated. Such a female could visit either territory but would do so less frequently than a female using one or the other territory exclusively. White may have been a female in this situation. Such an arrangement would be advantageous for a female unable to find a territory in which the male allowed access to enough flowers to meet her total energetic requirements. However, nesting at too great a distance from a good food source would increase the dangers of exposure of the eggs or young during bad weather, while the female was away foraging. It might be ultimately advantageous for a male to permit a female to do part of her foraging on his territory if he fathers the offspring of some percentage of such females, provided the energetic cost of defending the flowers she uses does not reduce his overall potential genetic contribution to the next generation.

It is hard to explain the evolution of pair cooperation if the resident male aided the reproductive output of another male, i.e., if the resident male had not mated with the female(s) he allowed to forage in his territory. Cooperation between unmated organisms may be at a selective advantage if the organisms are related closely enough genetically (Williams and Williams, 1957; Hamilton, 1964). In the case of *Panterpe* with but two young per nest, it seems unlikely that the three individuals, as in Area 1, are sufficiently related genetically for a similar explanation to hold.

One might ask why it is advantageous for a male to limit his matings to only those females that his territory will support rather than attempting to mate with every possible female and not aiding, even indirectly, in the nesting effort. The simplest and most reasonable explanation depends on the abundance and dispersion of the food sources during the nesting season. Around Villa Mills *Macleania* and *Gaiadendron*, the primary food sources, are extremely abundant, but are often highly localized. Males defend food sources that are sufficiently rich to make such defense feasible in terms of time and energy (Brown, 1964). Undefended sites may be scattered, and will probably be visited by other competitors. Presumably females forced to forage at undefended areas will have a lower average reproductive success. The inefficiency of exploiting undefended sites puts females under strong selective pressure to mate with those males which defend rich foraging areas. The rapid replacement of Red by another male after the former was collected strongly supports the idea of limited, rich foraging sites from which some members of the population are being excluded. A male probably can increase his genetic contribution by aiding one or more females in successfully raising their young rather than by not aiding a larger number of less successful females.

A further question, then, is why the male does not directly aid the nesting effort of one or two females. This pair relationship would reduce the time and energy a male had available for territorial defense and for other matings and presumably would decrease the number of females with which he could mate. This reduction potentially would decrease the number of young genetically related to the male that could be produced during that breeding season and would put the male at a potential selective disadvantage.

However, it would be advantageous for a female to mate with a male that directly aided in the nesting effort if, by doing so, she could increase her reproductive output (see Orians, 1969 and included references). If it were possible for a pair of hummingbirds working together to raise more than two young, then presumably at some point in the evolutionary history of the family direct male aid in nesting would have evolved. That this has apparently not occurred, in spite of the great ecological and taxonomic diversity of hummingbirds, suggests that such cooperation would not increase reproductive output.

The mating system of *Panterpe* is diffi-cult to classify. The difficulty revolves around whether there is actually a pair bond formed or whether nesting females, through their aggressiveness, can retain for their own use a share of the resources in the territory of a male. The latter possibility seems unlikely as there is an active avoidance of certain foraging sites by the male rather than an exclusion of the male by the female. If, then, there is a pair bond formed, but the male may mate with more than one female, the mating system probably should be called polygynous, the only known example among hummingbirds. It is a peculiar form of polygyny in that the male apparently aids only indirectly in the nesting effort of his mate(s).

If the pair cooperation evolved to enhance the nesting success of a female with which a male mated, it follows that selection would be strong for aggressiveness in males to facilitate defending as large an energy supply as possible. The more energy that is available, the more females that could efficiently exploit the resources in the course of nesting. Furthermore, a male should defend the source against all potential competitors, including females with which he has not mated. This strict territorial defense should lead to increased behavioral adaptations to insure that, eventually, a female is recognized as a potential mate and not simply as a competitor for energy.

In the initial stages of pairing a female probably searches for the best territory and attempts to forage in the area. The resident male responds aggressively to an intruding female and chases her from the territory. If the area is sufficiently rich in nectar the female will return and will gradually establish a foraging area. The male begins to recognize, and probably become habituated to, the female, eventually mates with her and establishes a more or less co-dominant relationship. Once the resources defended by the male are divided between the number of birds that the territory will support, additional females are always driven away and not allowed to forage. In our study,

Yellow apparently was attempting to acquire foraging rights to bushes 12 and 12a in Red's territory, but was continually chased and eventually ceased visiting the area.

This proposed sequence of activities leading to a co-dominant sharing of the resources on a territory and also leading to mating suggests the further conclusion that female *Panterpe* are not directly choosing male *Panterpe* with which to mate, but rather that they are choosing energetically adequate territories in which to forage. The dispersion of the major food sources and the population pressure by *Panterpe* insures that all good sites will have one or more resident, territorial males. A similar phenomenon was important in leading to polygamy in Long-billed Marsh Wrens (*Telmatodytes palustris*; Verner, 1964). Probably females choose ecological parameters rather than male characters in many bird species in which female choice plays an important role in mating. This might at first glance seem to contradict the importance of sexual selection in many birds that are strongly dimorphic in characteristics associated with reproductive behavior. However, as Selander (1965) and Orians (1969) pointed out, sexual selection actually includes all characteristics that are important influences on mate choice. Territories, in their view, are externalized secondary sexual characteristics much as are the constructed bowers of bower birds (Gilliard, 1963).

If females choose ecological parameters rather than male characters, many sexually dimorphic characters usually considered to have evolved in response to mate choice by females, may be directly involved in inter-male dominance relationships. Matings would not be determined by females, but would be based on the results of these inter-male encounters. By lumping these two alternative techniques of establishing mating rights under the heading of sexual selection, one loses the distinction between characters that evolved because of their importance in mate attraction per se, and those that are important in maintenance of mating priority. Obviously, these two are not mutually exclusive—many characters could function in both contexts. But the selective basis of certain characters could be clarified by subdividing what Darwin (1871) and Selander (1965) called sexual selection into sexual selection, acting on characters associated with mate attracting, and aggressive selection, acting on characters related to the maintenance of social position, especially with respect to mating. Pitelka (1942) and Wolf (1969) suggested that many of the bright colors of hummingbirds, normally present only in males, have evolved for their importance in territorial defense rather than courtship. And aggressive selection would seem to be the primary evolutionary pressure on the tremendous size dimorphisms of some pinnipeds. In several species of seals, females are held in harems and mate with males that have established their dominance in male-male encounters, in which size probably plays an importance role. Sexual dimorphism in many grouse species may also be produced through aggressive selection. This assumes that most of the displays in which the dimorphic characters are important are male-male displays to establish and maintain dominance positions on the mating leks. The females probably have little choice but to mate with the dominant males. If true, it shows how full understanding of the grouse displays, currently considered as sexual displays, has been hindered by thinking oriented toward sexual selection and toward the idea that females choose males on the basis of secondary sexual characters. A hinderance of this sort in the evolutionary analysis of behavior and morphology would argue for more precise definitions, in this case separating sexual and aggressive selection.

SUMMARY

During the breeding season individuals of both sexes of *Panterpe insignis*, a montane hummingbird occurring in Costa Rica and Panama, use the food resources in a territory being defended by a male. The male and female(s) normally forage at different flowers, usually on different bushes. Exclusiveness of utilization probably increases efficiency of the acquisition of energy. Efficiency of utilization by both sexes and the ability of the male to defend the flowers are presumably related to the abundance and dispersion of the flowers throughout the site. When the food source is abundant, but highly clumped and the population of hummingbirds is high most of the flowers that can be efficiently exploited are defended by territorial males. A territorial male can indirectly insure the greatest possible reproductive success of the female(s) with which he mated by defending the feeding sources used by the female(s). This defense can be effective only if the female is allowed to forage in the territory regularly defended by the male.

One of the females regularly visiting flowers in the territory of a male had a nest with eggs at the edge of the territory. In another territory the resident male and female copulated several times. We propose that the male in whose territory a female(s) is feeding has mated with, or will usually mate with that female. We found no evidence that the male directly aids in the nesting effort. However, female hummingbirds lay only two eggs and most can raise two young unaided by a male. Direct male aid apparently would not select for a larger clutch, so males are at a selective advantage if they mate with several females.

Females probably select the best territory and not the most "attractive" male. So the proximate selective force on sexually dimorphic characters that are important in aggressive encounters is probably the ability to hold a territory, not the ability to attract a mate. This extreme case of sexual selection is termed aggressive selection to facilitate the understanding of evolutionary pressures on sexual dimorphism.

ACKNOWLEDGMENTS

This study was supported by the National Science Foundation (GB-7611) and the Chapman Memorial Fund of the American Museum of Natural History. Arthur and Nancy Weston provided some field assistance. The Organization for Tropical Studies furnished logistical and administrative aid for our field work in Costa Rica. Linda Ziemer prepared the maps.

LITERATURE CITED

BROWN, J. L. 1964. The evolution of diversity in avian territorial systems. Wilson Bull. 76: 160–169.

DARWIN, C. 1871. The descent of man and selection in relation to sex. Appleton, New York. 2 vol.

GIBB, J. 1956. Food, feeding habits and territory of the rock pipit *Anthus spinoletta*. Ibis 98:506–530.

GILLIARD, E. T. 1963. The evolution of bowerbirds. Sci. Amer. 209:38–46.

HAINSWORTH, F. R., AND L. L. WOLF. 1970. Regulation of oxygen consumption and body temperature during torpor in a hummingbird, *Eulampis jugularis*. Science 168:368–369.

HAMILTON, W. D. 1964. The genetical evolution of social behavior. I, II. J. Theor. Biol. 7: 1–16, 17–52.

HOWELL, T. R., AND W. R. DAWSON. 1954. Nest temperatures and attentiveness in the Anna Hummingbird. Condor 56:93–97.

LASIEWSKI, R. C. 1963. Oxygen consumption of torpid, resting, active, and flying hummingbirds. Physiol. Zool. 36:122–140.

MOORE, R. T. 1947. Habits of male hummingbirds near their nests. Wilson Bull. 59:21–25.

ORIANS, G. H. 1969. On the evolution of mating systems in birds and mammals. Amer. Natur. 103:589–603.

PEARSON, O. P. 1950. The metabolism of hummingbirds. Condor 52:145–152.

PITELKA, F. A. 1942. Territoriality and related problems in North American hummingbirds. Condor 44:182–210.

RUSCHI, R. 1965. Observacoes sobre a nidificacao, incubacao e cuidados com a prole em *Colibri coruscans* (Gould), realizado unicamente femea. Bol. Mus. Biol. "Mello-Leitao"; 45:1–9.

SCHAFER, E. 1954. Sobre la biologia de *Colibri coruscans*. Bol. Soc. Venez. Cienc. Nat. 15: 153–162.

SELANDER, R. K. 1965. On mating systems and sexual selection. Amer. Natur. 99:129–141.

SLUD, P. 1964. The birds of Costa Rica. Bull. Amer. Mus. Natur. Hist. 128:1–430.

VERNER, J. 1964. Evolution of polygamy in the long-billed marsh wren. Evolution 18:252–261.

WILLIAMS, G. C., AND D. C. WILLIAMS. 1957. Natural selection of individually harmful social adaptations among sibs with special reference to social insects. Evolution 11:32–39.

WOLF, L. L. 1969. Female territoriality in a tropical hummingbird. Auk 86:490–504.

20

Reprinted from *Evolution*, **18**, 252–261 (1964)

EVOLUTION OF POLYGAMY IN THE LONG-BILLED MARSH WREN

Jared Verner[1]

Department of Zoology, University of Washington, Seattle, Washington 98105

Accepted November 15, 1963

Evolution of polygyny in a species whose young require parental care, to which the male can contribute directly, presents certain interesting problems. With each additional mate, the male may have more young to care for simultaneously; less time being devoted to each may influence their survival. Even if polygamy increases the male's total contribution to future generations, as compared to monogamy, the reverse would probably be true for each of the females mated to him. Since it is generally accepted that in birds females select mates from among available males, it seems unlikely that selection would favor those females that bred with males already mated. This has probably led to the general belief that polygamy in avian mating systems is a by-product of an unbalanced sex ratio (cf. Chapman, 1928; Kendeigh, 1941a; Mayr, 1941; Ryves and Ryves, 1934; Skutch, 1935; Williams, 1952), an assumption which is not necessarily true but which has probably led to a stagnation of thought concerning evolution of polygamy in birds.

Fisher (1958) and Kolman (1960) indicate that sex ratio at fertilization would be adjusted by selection so that average parental expenditure to rear males equals that to rear females. Willson and Pianka (1963) point out that if sexual selection results in sexual dimorphism expressed during the period of parental care, there may be a change in average expenditure per sex with a consequent change in sex ratio. Thus, mate preference, which in my opinion should have a profound influence on the mating system, may also have an influence on the sex ratio. This does not preclude the possibility that sex ratios may

be altered by other means, in which case polygamy would be a preferable alternative to not breeding at all. Obviously each situation requires separate analysis.

The present paper is an outgrowth of an ecological study of two populations of the Long-billed Marsh Wren (*Telmatodytes palustris*) in Washington State. Field work was carried out at Seattle during the breeding seasons of 1961 and 1962, while biweekly observations were made during the 1962 breeding season on a population at Turnbull National Wildlife Refuge, 6 miles south of Cheney, Spokane Co., 225 miles east of the Seattle population and isolated therefrom by the Cascade Mountains. Seattle's population (*T. p. paludicola*) is resident while Turnbull's (*T. p. plesius*) is migratory.

Two marshes—the "Red" and the "Blue"—were utilized at Seattle in 1961. These areas were largely eliminated by dredging by April, 1962 so a new study area—the "Yellow Marsh"—was utilized in 1962. At Turnbull two sites were worked, one (Blackhorse Lake) extensively, the other (Beaver Pond) only to determine breeding density, nest-building rates, and nesting cycles. The Red Marsh occupied 6.3 acres, the Blue Marsh 3.3 acres, the Yellow 3.2 acres, the Blackhorse only 0.5 acre, and the Beaver Marsh 1.0 acre. Density of the wren population at Turnbull was nearly eight times that in the Red Marsh and more than three times that in the Blue and Yellow marshes at Seattle.

The Seattle marshes were predominantly of cattail (*Typha latifolia*) with small patches of bulrush (*Scirpus acutus*) in the Blue and Yellow marshes. All had a few small willows (*Salix* sp.) scattered singly, and all were extensive marsh patches, in contrast to the narrow strips and lenses of cattail and bulrush characteristic of the

[1] Present address: Department of Zoology, University of California, Berkeley, Calif. 94720.

pond margins at Turnbull. Large areas of both the Blue and Yellow marshes grew from a dense mat of roots and decaying plant materials. These sections floated, hence they never became covered with water when the lake level rose, an important factor in the suitability of these areas for breeding by wrens.

Both populations are partially polygamous, and all males exhibit behavior conducive to multiple mate acquisition (Verner, 1963a). Briefly, the courtship activity of the male includes advertising, by means of song, from an area referred to as a courting center, in which a number of courting nests have been constructed. When a female first enters a territory, the male leads her to his various nests and displays nearby as she examines them. If the female accepts the male, she utilizes one of his courting nests or initiates a new nest for breeding. The male begins construction of a new courting center in another portion of his territory, usually before the female commences laying. The same general situation prevailed in both populations but the roles of males in parental care differed. At Seattle males normally fed nestlings and fledglings throughout the season, whereas the males at Turnbull did so only at the end of the season. Asso-

ciated with this difference are differences in the mating systems.

POLYGAMY AND THE POPULATION SEX RATIO

Data on population sex ratios are meager and difficult to obtain because non-breeding individuals are not readily sampled (Stewart and Aldrich, 1951; Hensley and Cope, 1951; Orians, 1961). Armstrong (1955) and Orians (1961) have questioned the notion that polygamy is a by-product of a skewed sex ratio. The Black Rosy Finch (*Leucosticte atrata*) is monogamous but all data suggest a high preponderance of males in adult populations (French, 1959). Among polygamous species the Redwing (*Agelaius phoeniceus*) had a 1 : 1 sex ratio in 94 successfully fledged young (Williams, 1940); and breeding European Wrens (*Troglodytes troglodytes*) apparently have an even sex ratio (Kluijver et al., 1940; Armstrong, 1955). The same is true of the Long-billed Marsh Wren in the present investigation.

In the polygamous population with a 1 : 1 sex ratio some individuals of the polygamous sex must not breed. Table 1 presents data from Seattle wrens relating to the known adult population at different

TABLE 1. *Sex ratios of adult marsh wrens at Seattle; 1961 season above, 1962 season below*

| Date | Territorial males | | | | Adult females | ♂ : ♀ ratio |
	Monogamous	Bigamous	Bachelor	Total		
May 6	9	1	4	14	11	1.27 : 1.00
May 11	8	2	4	14	12	1.17 : 1.00
May 25	6	3	5	14	12	1.17 : 1.00
June 13	9	1	3	13	11	1.18 : 1.00
June 19	5	2	6	13	9	1.45 : 1.00
June 25	4	2	6	12	8	1.50 : 1.00
July 8	4	1	4	9	6	1.50 : 1.00
Average	6.4	1.7	4.6	12.7	9.9	1.28 : 1.00
April 18	6	1	1	8	8	1.00 : 1.00
May 22	4	2	0	6	8	1.00 : 1.33
May 31	3	2	1	6	7	1.00 : 1.17
June 2	3	2	1	6	7	1.00 : 1.17
June 20	4	2	2	8	8	1.00 : 1.00
July 23	3	1	3	7	5	1.40 : 1.00
Average	3.8	1.7	1.3	6.7	7.2	1.00 : 1.06

times during the breeding season. Days listed are those on which the second female of a bigamous male began laying. At no time did the sex ratio deviate significantly from unity. Furthermore, in every case during 1961 there were unmated, *territorial* males in the population when females paired with males already mated, and in two cases there were four bachelors available. Moreover, one of the bachelors was always either an immediate neighbor or only one territory away from the mated male selected. In 1962 a territorial bachelor was available on all but one occasion; however, the surplus of males was never so great as in 1961. On that one occasion, at least one and perhaps two were available less than a quarter mile away.

At Turnbull, also, bachelors were occasionally present when bigamous matings were initiated. Only a small segment of the Turnbull marshes was studied, however, and this was all highly productive. One territory there was examined at a different locality on July 5 and found to contain a number of courting nests appropriate for that time in the breeding season, but none had been used for breeding. If selection favors females preferring mated males to bachelors, there must be some advantage accruing to those females. Possible selectice forces operating on polygamous birds are therefore reviewed below.

OPERATION OF SELECTION IN POLYGAMOUS CASES

Disadvantages to males.—A polygamous male spends more time advertising for and courting females than a monogamous male, and demands of territorial defense may also be greater. This reduces time available for each female and care of her young, with a resultant lowered reproductive success per female. These activities would also make the male conspicuous longer so possibly increase the chance of his mortality. Furthermore, the more females a male has, the greater the risk of stolen copulations, with the result that he would help rear another male's offspring.

Advantages to males.—Males may leave more offspring with each additional female, even though production per nest is not as high. According to Orians (1961), if polygamy results in greater offspring production per male, "selection will always occur in favor of polygamy. . . . The system is self-accelerating once it has started, so that once some polygamy has been established, fewer and fewer offspring are required per female mate . . . to keep the selection going." However, if this results in lowered production per female in relation to that had they mated monogamously, then sexual selection should decelerate evolution of polygamy, and a balance between these selective forces would result.

Disadvantages to females.—Any mate of a polygamous male receives less time from him than she would from a monogamous mate. This "lost" time could be used to protect the female or nest in warning of danger, assisting in mobbing potential predators, and to obtain and transport food to the female or her young, thus increasing the probability of offspring survival. In general it seems to be true that territories of polygamous males are not increased in size in proportion to the number of females possessed (but see Linford, 1935); each female may therefore have a reduced foraging area.

Advantages to females.—Imagine a polygamous species in an area containing several territorial males, each with a single mate, and two territorial bachelors. A female entering the population could select from all males and all territories. If a similar population were monogamous, however, the new female must select one of the bachelor males (or none at all), even if their territories provide a low probability of nesting success. Therefore, there may be a distinct advantage of polygamy to females in increasing the probability of their breeding in a more suitable area with a superior male. Such an advantage must outweigh the disadvantages of polygamy before the system can evolve. Obviously the population must be large enough that some males are forced to occupy such poor territories that females breeding there would have lower reproductive success, even with the assistance of the male, than they would on a better site without significant assistance from the male. This may help to explain mating system variations within the same species.

Although in most well-known populations of the marsh wren from 30 to 50% of the males have been polygamous, Kale (1961) reported only 3.2% polygamous at Sapelo Island, Georgia. Kale (in litt.) suggests that the shape of territories may influence the mating system of these wrens. At Sapelo Island, the wrens nest in the narrow strips of *Spartina* bordering tidal ditches. Only the ends of their long, linear territories must be defended, so the male must continually fly back and forth between ends of his territory. This obviously limits the total area that can be defended, and, as Kale suggests, may prevent defense of a sufficient area to permit more than one female to nest therein. A second factor would seem to bear consideration in this case, however. There appears to be a very low fledging success of Sapelo Island wrens (42% in 1958, 20% in 1959); the population may not reach sufficient numbers during the breeding season for some males to be forced to occupy very marginal habitats.

Most populations of the European Wren are also polygamous, but some are not (Armstrong, 1955). Those that are not occupy islands in the northern part of the British Isles, where Armstrong suggests the habitat is so bleak that both parents are required to provide enough food for the young. From this it follows equally well that all populations of the European Wren should be *monogamous* but that clutch size should vary in different environments, if clutch size is ultimately determined by food supply according to Lack's (1947) hypothesis. It seems more likely that mortality among monogamous European Wrens is high enough before the breeding season that males are rarely or never forced to occupy very inferior habitats.

Circumventing disadvantages of polygamy.—Clearly what may be adaptive for one sex may not be so for the other; compromises between conflicting selective pressures must occur. For instance, if all females in a male's harem placed their nests close together, the male could fulfill some duties, such as warning or mobbing, for all females at once. Both sexes would benefit. The male would benefit further because it would facilitate prevention of stolen copulations. On the other hand, this arrangement would overload foraging areas immediately around the nests; the females might have to forage farther away. This problem may be overcome in part by foraging outside of territories, which seems to be common among polygamous birds.

The courting period of the male would be reduced if all his females commenced laying synchronously, but this would increase the risk of stolen copulations. Moreover, all females would have young simultaneously, reducing the care per young by the male. Synchrony would also increase demands on available food supply at one time. There may be an advantage to staggering the cycles of the various females, depending upon the nature of production of food sources, with the added advantage of permitting the male to devote more time to each brood. Staggered cycles, however, would limit the number of females any male could obtain.

At Seattle, a statistically significant number of females of bigamous males adjusted their cycles to have less than 2 days of overlap in their nestling periods, a fact believed to be correlated with Seattle males' habit of feeding young (Verner, 1963b). Thus, the first brood could get full care from the male during the nestling period and divided care during the fledgling period, the reverse being true of the second brood. Since males normally do not carry food to nestlings until the oldest is at least 2 days old, this much overlap can be tolerated. At Turnbull, where males did not feed young until the end of the season, cycles of females in polygamous associations were not significantly staggered to avoid overlap of nestlings. The fact that these males did feed young at the end of the season is likely related to the fact that the probability of a male attracting another female and successfully rearing a brood by her approaches zero as the season wanes. Selection should thus favor increasing male parental care then.

A male cannot have more than two females without incurring extensive overlap of young; the timing of cycles by Seattle wrens should therefore limit the number of females any male could have concurrently to two. On the other hand, Turnbull males should not be so restricted. Of ten polygamous males at Seattle, all were bigamists, while of five at Turnbull, four were bigamists and one a trigamist. Moreover, Welter (1935) reported regular bigamy with a possible case of trigamy among New York and Minnesota marsh wrens, in which males did not feed nestlings, while Kale (1961) reports 3% bigamy and no trigamy among Georgia marsh wrens in which male parental care is apparently common. Similarly, cases of trigamy have been reported in those populations of the European Wren in which males contribute little to the young (Kluijver et al., 1940; Armstrong, 1955), but Kendeigh (1941b) reports regular male parental care, no trigamy, and a low rate of bigamy in House Wrens (*Troglodytes aedon*) he studied.

From other families of birds as well there is evidence that staggering of cycles has accompanied the evolution of polygamy (see Tinbergen, 1939, for references). Only after more species have been studied in greater detail can we hope to suggest what general conditions result in the evolution of staggered, random, or synchronized cycles of females in polygamous associations.

BASES OF MATE PREFERENCE

Selection of territories.—In species where pairing takes place in the territory, females may select a mate on the basis of the occupying male, his territory or both. It seems improbable, however, that selection would not favor some judgment of territory. In the few species for which banding data are adequate, there is evidence that females, as well as males, breed the following year more consistently in the same area than with the same mate (e.g., Kendeigh, 1941b; Nice, 1943; Lack, 1946; Mayfield, 1960). Territorial parameters most effective in determining where females breed should be correlated with the probability of breeding success. By using a male's pairing success as a measure of the suitability of his territory, it is possible to suggest what some of these parameters are. Two criteria to estimate pairing success have been used: (1) the number of females obtained concurrently, (2) earliness of first pairing. In the following section, various features of marsh wren territories will be compared in relation to the pairing success of males defending them. Data are summarized in table 2.

The average sizes of territories of bachelor, monogamous, and bigamous males, measured about the time bigamous matings were formed, were successively larger. Since bigamous associations were established several times during a season, the same male's territory may enter into the average each time; but the territory boundaries had usually changed between each measurement. At Turnbull only the four territories on Blackhorse Lake were considered, since this was the only area of intensive observations. The fact that the bigamous territory averaged smaller than the monogamous one on Blackhorse may be partly the result of qualitative differences in the territories as suggested below. I was aware of no significant increase in the size of a male's territory as a result of his acquiring a mate; territory sizes were established before pairing, and so could have been a basis for mate preference.

Available data indicate that marsh wrens have an almost exclusively animal diet, especially aquatic forms and terrestrial

TABLE 2. *Territorial parameters (sq ft) in relation to pairing success; sample sizes in parentheses*

Marsh	Bachelors	Monogamous	Bigamous	Trigamous
		(a. Territory size)		
Red	12,802 (12)	19,655 (11)	34,835 (3)	
range	2,567–28,000	5,306–34,230	30,396–38,748	
Blue	9,787 (7)	14,215 (9)	17,394 (4)	
range	7,376–15,567	10,887–20,426	12,624–23,688	
Yellow	9,297 (4)	15,636 (12)	19,247 (5)	
range	4,508–12,136	11,525–25,797	16,169–23,458	
Blackhorse	2,071 (3)	5,080 (15)	3,673 (4)	2.589 (1)
range	1,812–2,589	2,589–6,957	2,589–4,757	
		(b. Emergent area)		
Blue	5,263 (7)	11,430 (9)	16,596 (4)	
% total	50.00	83.44	94.13	
Yellow	7,839 (4)	8,839 (12)	11,274 (5)	
% total	82.94	57.13	57.23	
		(c. Brush canopy)		
Red	1,269 (12)	2,133 (11)	3,641 (3)	
% total	10.36	11.66	10.35	
Blue	714 (7)	2,257 (9)	110 (4)	
% total	6.33	16.05	0.82	
Yellow	2,248 (4)	1,187 (12)	1,465 (5)	
% total	22.53	8.55	7.51	
		(d. Bulrush)		
Blackhorse	906 (3)	1,369 (15)	1,699 (4)	1.424 (1)
range	647–1,424	680–2,556	1,424–1,974	
% total	43.7	22.5	46.2	55.0

forms with aquatic life stages. Thus the extent of emergent vegetation in a male's territory should have a bearing on his pairing success. Since all of the Red and Blackhorse marshes were 100% emergent, only data from the Blue and Yellow marshes are presented in table 2. There is a trend toward increasing pairing success with increasing emergent area.

Much food for Seattle young was obtained from willow thickets. Where these bordered on territories, a zone usually about 10 feet wide adjacent to the marsh was extensively utilized and this has been included in the territory. In spite of the apparent importance of such willow stands to the wrens, the correlation between the extent of the territory with a shrub canopy and success of occupying males is very weak. Male 7, one of the most successful males, had almost no shrub canopy in his territory but his was one of the largest territories and had the greatest amount of emergent vegetation. Male 11, also very successful, had little shrubbery available but had an extensive bulrush stand in which he foraged heavily. Moreover, he was noted foraging in willows of other males during one period when he was bigamous and had nestlings in one nest. The importance of bulrushes for foraging may result from their being more thoroughly broken down during the winter than is cattail, thus allowing more light to penetrate to the water surface and promoting higher production there. Bulrush seemed to be of considerable importance to wrens at Turnbull where it constituted a more predominant segment of the total habitat. Observations showed that by far the greater proportion of foraging took place in it, and the only trigamous male observed in this study possessed one of the smallest territories observed but one consisting almost exclusively of bulrush.

Other observers have indicated that territorial features may be related to pairing success. Welter (1935), Kendeigh (1941b), and Tompa (1962) all suggest that size may be important (but see Nice, 1937), and Linsdale (1938) and Kale (1961) implicate quality.

Selection of males.—It is often implicit in discussions of song rates that the more songs a male delivers, the more likely he is to attract females. Marler (in Thorpe, 1961) found evidence that in Chaffinches (*Fringilla coelebs*) females responded preferentially to higher rates of song delivery. However, success in attracting females is not equivalent to success in forming pair bonds. To the contrary, there must be an upper limit to song rates beyond which any additional time channeled into song reduces that available for activities more essential to reproductive success. Moreover, the motivation for song is complex and poorly understood (see Andrew, 1961), and it is possible that too much song is indicative of motivation not conducive to breeding success.

In 1962, records of song rates prior to the first pairing of six males show that the male with the highest song rate was among the latest to have a breeding nest under way, and the male with the next to slowest rate was the first to pair (table 3). Order of pairing, assuming this to be indicative of pairing success, was more closely correlated with features of territories. Male

25 did not fit this picture. His song rate was very low, but his territory ranked second in size and third in emergent area. It is possible that the absence of adequate nesting cover in the area of his courting center was a major factor in his late pairing. When he finally paired, the breeding nest was placed so precariously that I tied it firmly in place to prevent its dislodging.

The number of nests completed before pairing could not be correlated with pairing success; some males acquired mates with less than one completed courting nest, while others had completed many before pairing. If anything, there may be an inverse correlation between earliness of pairing and the number of nests completed, since most males began building about the same time and continued to do so up to and after the time of pair formation. That the possession of nests is, however, an important link in the chain leading to pair formation is suggested by two cases. In 1961, courting nests of Male 3 were systematically removed before or soon after they were completed. He continued to build for approximately 2 weeks then disappeared (color-banded). In 2 weeks he returned, reclaimed his old territory, and resumed nest building. These were also removed, and the male disappeared permantly 10 weeks later. Although females were twice observed in his territory, he never secured a mate. Male 27 apparently had a low tendency to build courting nests. He began 11 days later than the average date of first construction by other males and only brought one nest to 20% comple-

TABLE 3. *Characteristics of males and features of their territories (sq ft) up to the date of first laying*

Male	First egg	Avg songs per min		Nests completed	Date courting nests begun	Territory size	Emergent area	Shrub canopy
		A.M.	P.M.					
23	Apr. 3	3.16	1.85	3.5	March 24	21,017	21,017	3,159
30	Apr. 9			1.1	March 26	14,169	12,610	2,434
32	Apr. 11	4.80	3.44	3.0	March 23	7,288	7,288	1,139
29	Apr. 14			3.7	March 24	11,797	6,780	2,102
24	Apr. 19	6.61	3.92	3.8	March 23	11,186	5,491	20
25	Apr. 19	2.33	2.12	5.0	March 26	16,746	9,254	34
26	Apr. 20	4.70	1.84	3.7	March 25	13,356	4,339	1,119
27	Never	3.73	2.65	—	April 4	10,915	3,763	0

tion. At least one female remained on his territory for a day or two but did not form a permanent bond.

The process of pair formation has been shown in a number of species to involve what Tinbergen (1951: 47) discusses as reaction chains. A special set of sign stimuli by one sex releases a particular response in the other, which serves to release the next response in the first sex, etc. Any break in this chain prevents completion of the act. A break can occur, however, not only when one sex fails to respond but also when it responds inadequately or too strongly. For example, Marler (1956) observed that if female Chaffinches reacted overaggressively toward their mates, they were deserted. Hinde (1959) further demonstrated that a Chaffinch "which shows strong aggressive responses to a male model may also behave aggressively to a female model." It seems reasonable, therefore, to assume that a male may have such a strong aggressive tendency as to preclude effective pair formation under normal circumstances. Circumstantial evidence from one male in this study suggests that this may have occurred.

Male 16 first entered the Red Marsh on April 2, 1961, after other males were already established. Initially he established himself in an unoccupied area, but within a day he began challenging Male 2, who had a mate nearly ready to lay. Eventually Male 2 was forced into a small segment of his territory, and Male 16 claimed the greater share, including that portion in which the female was lining a nest. This was the only occasion on which a newcomer evicted an established male from his area of major activity, suggesting that Male 16 was more aggressive than normal. The female remained with Male 16, perhaps because she was so near to laying and already had the nest lined, although there is no certainty that she was not inseminated by Male 2. After completing the nesting cycle, she deserted Male 16 to pair with Male 13, who had remained unpaired

up until that time, suggesting that he or his territory was inferior. This was one of only 3 cases out of 16 in which known females changed mates while their original mate still lived. Perhaps Male 16's overaggressive behavior caused his abnormally acquired mate to desert him, indicating that he might never have acquired a mate had he established himself in the marsh at the beginning of the season when other males were not paired.

Although I believe that territorial parameters are probably more important in determining where a female breeds than is the character of the occupying male, it is impossible to separate the two. The best male may, by definition, be the one able to defend the best territory. Nevertheless, there are many species (e.g., most Anseriformes; Brewer's Blackbird, *Euphagus cyanocephalus*, Williams, 1952) in which pair formation takes place in winter flocks. In those cases, females must select on the basis of male characters alone. Whatever these may be, they must be closely correlated with breeding success. Finally, the chance involved in pair formation (see especially Nice, 1937; Lack, 1946) complicates the task of determining those features most important in the selection of mates.

SUMMARY

Polygamy in avian mating systems has generally been interpreted as a by-product of an unbalanced sex ratio, an assumption which is not necessarily true but which has apparently led to a stagnation of thought concerning evolution of polygamy in birds. Monogamy in populations with highly skewed sex ratios is known, and polygamy occurs in populations with even sex ratios. The Long-billed Marsh Wren falls into the latter group.

Field studies in two separate populations of the marsh wren in Washington State revealed that bigamous associations were formed even when territorial bachelors were present in the vicinity, the bachelors being judged to occupy inferior territories. A female selecting a mate under these conditions might rear more young by pairing with a mated male on a superior territory than with a bachelor on an inferior one, notwithstanding the fact that she would obtain less help from her mate. A necessary correlate to this idea is a high density of mature males, such that some are forced to occupy the marginal sites. Monogamous populations of normally polygamous species (e.g., the marsh wrens of Sapelo Island, Georgia; European Wrens on various islands in the northern British Isles) may not occur in sufficient numbers during the breeding season to force some males into marginal habitats.

An attempt is made to suggest some characteristics of males and their territories that are related to pair formation. There is evidence that the size of a male's territory and the total amount of emergent vegetation in it are correlated with his success in acquiring mates. Males' song rates tend to show an inverse correlation to pairing success, and males that failed to construct courting nests never paired. Whether features of territories or of males are important in pair formation, those factors leading to a high probability of pairing success must somehow be correlated with a high probability of breeding success.

ACKNOWLEDGMENT

This study was carried out under the supervision of Dr. Frank Richardson; it is a pleasure to acknowledge his numerous helpful suggestions and criticisms. Dr. Gordon H. Orians, who served in an advisory capacity for one year during the absence of Dr. Richardson, has also made many valuable criticisms throughout the course of the study. Dr. R. C. Snyder has read the manuscript and offered useful suggestions. Many colleagues contributed significantly, both with direct assistance in the field and discussions of data. I especially wish to mention Dr. T. H. Frazzetta, M. E. Kriebel, K. P. Mauzey, E. R. Pianka, C. C. Smith, L. W. Spring, and M. F. Willson. My wife, Marlene, in addition to her endless encouragement, processed many of the raw field data presented here. Facilities were provided by the Department of Zoology, University of Washington, and financial assistance was provided by National Science Foundation predoctoral fellowships (1960–1963).

LITERATURE CITED

ANDREW, R. J. 1961. The displays given by passerines in courtship and reproductive fighting: a review. Ibis, **103a**: 315–348, 549–579.

ARMSTRONG, E. A. 1955. The wren. London, Collins.

CHAPMAN, F. M. 1928. The nesting habits of Wagler's Oropendola (*Zarhynchus wagleri*) on Barro Colorado Island. Bull. Am. Mus. Nat. Hist., **58**: 123–166.

FISHER, R. A. 1958. The genetical theory of natural selection. 2nd ed. New York, Dover.

FRENCH, N. R. 1959. Life history of the Black Rosy Finch. Auk, **76**: 159–180.

HENSLEY, M. M., AND J. B. COPE. 1951. Further data on removal and repopulation of the breeding birds in a spruce-fir forest community. Auk, **68**: 483–493.

HINDE, R. A. 1959. Some factors influencing sexual and aggressive behaviour in male Chaffinches. Bird Study, **6**: 112–122.

KALE, H. W. II. 1961. A study of breeding populations of Worthington's Marsh Wren (*Telmatodytes palustris griseus*) in the salt marshes of Sapelo Island, Georgia. M.S. Thesis. Univ. of Georgia Library, Athens.

KENDEIGH, S. C. 1941a. Birds of a prairie community. Condor, **43**: 165–174.

——. 1941b. Territorial and mating behavior of the House Wren. Illinois Biol. Monog., **18**: 1–120.

KLUIJVER, H. N., J. LIGTVOET, C. VAN DEN OUWELANT, AND F. ZEGWAARD. 1940. De levenswijze van den winterkoning *Troglodytes tr. troglodytes* (L). Limosa, **13**: 1–51.

KOLMAN, W. A. 1960. The mechanism of natural selection for the sex ratio. Amer. Nat., **94**: 373–377.

LACK, D. 1946. The life of the Robin. London, Witherby.

——. 1947. The significance of clutch-size. Ibis, **89**: 302–353; **90**: 25–45.

LINFORD, J. H. 1935. The life history of the Thick-billed Redwinged Blackbird, *Agelaius phoeniceus fortis* Ridgway in Utah. M.S. Thesis. Univ. of Utah Library, Logan.

LINSDALE, J. M. 1938. Environmental responses of vertebrates in the Great Basin. Amer. Mid. Nat., **19**: 1–206.

MARLER, P. 1956. Behaviour of the Chaffinch *Fringilla coelebs*. Behaviour Supplement 5. Leiden, Brill.

MAYFIELD, H. 1960. The Kirtland's Warbler. Bull. Cranbrook Inst. of Sci., **40**: 1–242.

MAYR, E. 1941. Red-wing observations of 1940. Proc. Linn. Soc. New York, **52–53**: 75–83.

NICE, M. M. 1937. Studies in the life history of the Song Sparrow. I. Trans. Linn. Soc. New York, **4**: 1–247.

——. 1943. Studies in the life history of the Song Sparrow. II. Trans. Linn. Soc. New York, **6**: 1–328.

ORIANS, G. H. 1961. The ecology of blackbird (*Agelaius*) social systems. Ecol. Monog., **31**: 285–312.

RYVES, B. H., AND H. H. RYVES. 1934. The breeding habits of the Corn-bunting as observed in North Cornwall: with special reference to its polygamous habit. Brit. Birds, **28**: 2–26, 154–164.

SKUTCH, A. F. 1935. Helpers at the nest. Auk, **52**: 257–273.

STEWART, R. E., AND J. W. ALDRICH. 1951. Removal and repopulation of breeding birds in a spruce-fir forest community. Auk, **68**: 471–482.

THORPE, W. H. 1961. Bird-song. Cambridge Monog. in Exp. Biol., **12**: 1–143.

TINBERGEN, N. 1939. The behavior of the Snow Bunting in spring. Trans. Linn. Soc. New York, **5**: 1–94.

——. 1951. The study of instinct. Oxford, Clarendon.

TOMPA, F. S. 1962. Territorial behavior: the main controlling factor of a local Song Sparrow population. Auk, **79**: 687–697.

VERNER, J. 1963a. Aspects of the ecology and evolution of the Long-billed Marsh Wren. Ph.D. Dissertation. Univ. of Washington Library, Seattle.

——. 1963b. Song rates and polygamy in the Long-billed Marsh Wren. Proc. XIII Int. Orn. Cong.: 299–307.

WELTER, W. A. 1935. The natural history of the Long-billed Marsh Wren. Wilson Bull., **47**: 3–34.

WILLIAMS, J. F. 1940. The sex ratio in nestling eastern Red-wings. Wilson Bull., **52**: 267–277.

WILLIAMS, L. 1952. Breeding behavior of the Brewer Blackbird. Condor, **54**: 3–47.

WILLSON, M. F., AND E. R. PIANKA. 1963. Sexual selection, sex ratio and mating system. Am. Nat., **97**: 405–407.

21

Reprinted from *Ecol. Monogr.*, **31**, 285–286, 290–295, 297–299, 301–306, 311–312 (1961)

THE ECOLOGY OF BLACKBIRD (*AGELAIUS*) SOCIAL SYSTEMS

Gordon H. Orians

Museum of Vertebrate Zoology and Department of Zoology, University of California, Berkeley, California*

INTRODUCTION

The conspicuousness of adaptive radiation in morphology tends to conceal the fact that often the slight differences between closely related species give no clues to their widely differing ecologies, because many of the important differences between species are the result of behavioral and not morphological adaptations. This study analyses the role of social organization of the Red-winged Blackbird (*Agelaius phoeniceus*) and the Tricolored Blackbird (*A. tricolor*) in the different ways in which these two species exploit their environment.

Knowledge of avian social systems began with natural history studies, but certain phases, such as territoriality, early attracted special consideration. In the 1930s, social systems began to be studied from the viewpoint of the comparative ethologist, who is primarily interested in the motivational and evolutionary aspects of behavior patterns, but whose publications contain a wealth of information about many ecological features of avian social systems. The mathematical approach to population parameters has provided a basis for considering the consequences of changes in social system characteristics upon basic population parameters, but biologists have in general been suspicious of this approach, which seemed to rest upon assumptions of doubtful biological validity. The result is rather widespread failure to realize the significance of certain features of social systems in quantitative terms, and failure to record and publish relevant information. Finally, the study of social systems from the modern ecological viewpoint has lagged behind other approaches because few observers have made use of the background of a century of Darwinian thinking in evaluating their observations.

In this study I have considered all features of social systems to be the products of natural selection just as are any physiological or morphological adaptations. To the question whether or not differences between social systems are adaptive, three types of answers are possible. Firstly, it may be assumed that the particular features of a social system are surely adaptive. Secondly, it may be assumed that the traits are purely fortuitous, without selective significance. Thirdly, it may be assumed that the particular traits are not adaptive but that they are associated with other, as yet unrecognized, differences which are adaptive (Maynard Smith 1958). In this paper I shall attempt to interpret as far as possible the characteristics of social systems in the light of the first of these three assumptions. The second is rejected because it is sterile as a basis for research and because the widespread and consistent differences to be discussed cannot be without selective significance. The third can never be easily accepted, for unless this statement of faith is followed by attempts to discover the traits of adaptive significance and their connection with the supposedly unadaptive trait, nothing is really explained. Furthermore, no such case involving polygenic traits has been shown to be true, and separation of desirable from undesirable traits will almost certainly occur with time.

Because the closely related and morphologically similar Red-winged and Tricolored blackbirds differ strikingly in their social organization, they are excellently suited to studies developed from the premises just given. Furthermore, these and other species of the family Icteridae are common, easily observed, and well-known. Moreover, their social systems range from routine territoriality to extreme coloniality and from monogamy to promiscuity and parasitism. My studies of these two blackbird species were carried out in north-central California from 1957 through the spring of 1960. The behavioral aspects of this investigation are being treated separately, and a report on autumnal breeding in the Tricolored Blackbird has already been published (Orians 1960).

* * * * * * *

* Present address: Dept. of Zoology, Univ. of Washington, Seattle 5, Washington.

THE SOCIAL SYSTEMS

1. THE RED-WINGED BLACKBIRD

A striking feature of Redwing social organization during the fall and winter is the segregation of the sexes which, though by no means complete, is very pronounced. As already mentioned very few females roosted at Jewel Lake, and though both sexes roosted abundantly at Colusa, most flocks seen feeding during the day were wholly or largely composed of one sex.

The enormous numbers of Redwings in the Sacramento Valley in the autumn obtain most of their food from agricultural land. I did not examine any stomachs, but by observing feeding birds it is easy to determine what they are taking. In late summer the birds concentrate upon seeds of the water grass (*Echinochloa crusgalli*), which is abundant around the edges of all rice fields and ditches in irrigated country, and rice, which is then coming into the milk stage. The greatest damage to the rice occurs at this time though the birds continue to utilize it heavily until it is harvested. Mechanical methods of harvesting rice leave large amounts of grain scattered on the ground among the stubble which the blackbirds continue to use until the fields are plowed. At this time of year, newly sprouting alfalfa fields are also used as sources of insects. As the autumn progresses, more and more fields are harvested and plowed, and feeding conditions become progressively worse. This is partly offset by the flooding of many fields to attract ducks for hunting, because these fields are not plowed and also produce many insects. Nonetheless, by early January the populations of Redwings and Tricolors in the Sacramento Valley are greatly reduced from their mid-autumn level. One can drive great distances in late winter and see few blackbirds where earlier there had been millions. It is not certain where they go at this time, but they probably move to the San Joaquin Valley and other areas to the south where agricultural practices are different.

* * * * * * *

Nesting Habitat. Redwings nest in a wide variety of habitats (Allen 1914, Sherman 1932, Todd 1940, Nero 1956a), but most nests are located in emergent vegetation, particularly cattails. In California they commonly nest in vegetation bordering irrigation ditches, roadside and fencerow vegetation, riparian situations, weed and brush patches, cropland such as alfalfa and cereal grains, and even upland areas of mixed chaparral and grass. The chief requirement is apparently vegetation strong enough to support the nest surrounded by suitable feeding grounds. Burned cattail areas are used before the new growth sprouts if enough charred stumps remain. In fact, at the East Park Reservoir in 1959, burned areas were chosen instead of dense, unburned patches when both were available on one marsh area. On Brooks Island, the Redwing is a common breeding bird throughout the island, nests being located in bushes of poison oak (*Rhus diversiloba*) and coyote brush (*Baccharis pilularis*) even on the tops of the main ridges. I have not found this situation duplicated

elsewhere, but I know of no mainland area with such varied, ungrazed vegetation as is found on the island.

Time of Breeding. The most complete studies of breeding chronologies were made at Jewel Lake in 1958 and East Park Reservoir in 1959. The major features of the breeding season for both areas are summarized in Figs. 6 and 7. Less complete observations at Jewel Lake in 1957 and 1959 show that, with minor modifications, the same pattern held for those years as well. For example, the arrival of females, the beginning of chasing of the females by the males and the start of precopulatory displays were within one week of their 1958 time in 1959. Egg-laying in 1957 began two days earlier than in 1958. Because of burning, events were delayed in 1960 at the East Park Reservoir, but in the unburned areas, nesting began four days earlier than in 1959.

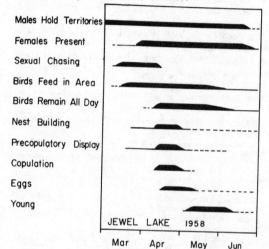
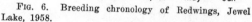
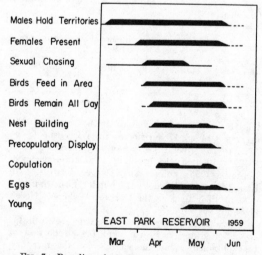

FIG. 6. Breeding chronology of Redwings, Jewel Lake, 1958.

FIG. 7. Breeding chronology of Redwings, East Park Reservoir, 1959.

Territory. The territories of the males are defended by means of song, displays and chasing, but little is known about the variations in territory size and the factors influencing them. Linford (1935) found that territories of polygamous males were twice the size of those of monogamous males, but Nero (1956b) found no such relationship. I also failed to find any correlation between number of females and territory size.

The East Park Reservoir afforded the opportunity to study territory size in marsh areas of contrasting characteristics. Progressively more food is obtained on the territories in the small clumps of isolated cattails, the peripheral strips of cattails along the main marsh, and the main marsh itself, in that order. Territories were substantially the smallest in the isolated cattail clumps, larger in the peripheral strips, and largest in the main marsh (Table 7). Territory size was also determined for a portion of the Haskell Ranch marsh for both 1959 and 1960. In 1959, territories averaged larger than at the East Park Reservoir; but in 1960, they were comparable to territories at the periphery of the main marsh at the reservoir (Table 8). There are no other data from the valley floor with which to compare the results obtained at

TABLE 7. Size of Red-winged Blackbird territories, East Park Reservoir, 1959.

Situation	Number of territories	Average size (sq. ft.)
Isolated cattail clumps surrounded by grassland.......	21	2,512
Strip of marsh at the edge of the reservoir..............	17	8,477
Main marsh area, including both central and peripheral territories................	22	10,653

TABLE 8. Size of Red-winged Blackbird territories, Haskell Ranch.

Year	Number of territories	Average size (sq. ft.)
1959.....................	10	13,720
1960.....................	16	8,575

the Haskell Ranch. Nero (1956b) reported the average size of 17 territories in Wisconsin to be 3,550 sq ft. Average size, however, increased from 1947 to 1953 as the breeding population declined. Linford (1935) found much larger territories in Utah (average: 31,603 sq ft) but his birds gathered most of their food on the territories whereas Nero's birds did not.

There is thus a general correlation between the size of Redwing territories and the proportion of food obtained within the confines of the territory. However, it is doubtful whether food *per se* is the *proximate* factor by which territory size is regulated.

The available evidence suggests that many bird species use features of vegetative physiognomy as their major cues in evaluating environmental suitability (Lack 1940), though the mechanisms by which this is accomplished are unknown. That this is also the case with the Redwing is suggested by change in territory size in response to stage of vegetative succession (Martin 1960) and by the response to burning of the marshes. The exceptionally complete burn at the East Park Reservoir marsh in 1960 left large areas devoid of emergent vegetation until the new growth appeared. In these areas the Redwing territories were initially several times larger than in 1959, but as the vegetation grew, additional birds inserted themselves, and territories became smaller (Table 9) though never as small as in the previous year.

TABLE 9. Size of Red-winged Blackbird territories on burned marshes, East Park Reservoir, 1960.

Roadside study area	Number of territories	Average size (sq. ft.)
April 29.............	3	26,500
May 15..............	4	19,875
North study area		
April 29.............	2	32,300
May 15..............	4	16,150

Food for the young may be gathered either on the territory or adjacent to it. The cattail areas at the south end of the main marsh at the East Park Reservoir were surrounded with sedge meadows from which much of the food was gathered, but often the birds flew across the road to an alfalfa field. Much food was gathered within the territory among centrally located territories. Where oak parkland adjoined the marsh, the birds frequently foraged among the grass and trees.

Several types of evidence suggest that territorial behavior is limiting breeding density on the study areas. Firstly, territorial challenges by newly arriving males are common much of the breeding season, and they may be vigorous and prolonged. For example, on April 19, 1959, I watched an intruding male, easily identified by his more orange epaulets, attempting to take over a territory for more than an hour. When first discovered at 0730, the intruder was submissive to the resident male, but by 0745 he began to give full song spreads on perches and in flight over the territory and began diving at the resident male, each time evoking a chase. By 0800 he was at times flying over the territory unchallenged by the resident, and his attacks were intensified so that he hit the resident while diving. By 0824 he was displaying to females flying over the territory, and had apparently succeeded in taking over, but at 0836 the resident male became more vigorous in his defense of the territory and the intruder left. He returned again at 0842 but was immediately chased by the resident and left again. By 0900 I had seen no further sign of him nor did he reappear later.

On April 30 at one of the isolated cattail patches I observed another unsuccessful territorial challenge which lasted intermittently from 1330 to 1445. The challenger held a nearby territory without cattails or other emergent vegetation which could support a nest. These are extreme cases, but the frequency of occurrence of territorial challenging by both sexes suggests that more birds would settle if they could. Nero (1956b) has reported at length on this aspect of territoriality in Wisconsin Redwings.

A second line of evidence is provided by the behavior of birds which I had trapped and banded. Twice, males which I had trapped fought to regain their territories from new males even though the duration of their confinement could not have been longer than a few hours. Nero (1956b) reported this also.

To test the matter further, the males from an area at the East Park Reservoir, containing 7 territories, were shot on May 8, leaving only one color-banded male whose vocalizations were being studied. The following morning, this male and a bird from across the stream had expanded their territories to include most of the vacated area, and though this was late in the season, there were five replacements by May 17 (Fig. 11). Since this was later in the season than any new areas were occupied in this region, it is likely

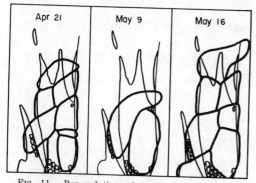

FIG. 11. Repopulation of a small marsh by male Redwings following shooting of the established territorial males. East Park Reservoir, 1959.

that the invading individuals were birds which had been prevented from breeding by the prior occupation of all territories by other males.

On the basis of these preliminary results, the experiments were continued in 1960. A section of the Haskell Ranch marsh and another area at the East Park Reservoir were selected as removal sites. Males were removed from the Haskell Ranch marsh eight times and from the East Park Reservoir five times (Table 10). At both sites first-year males, which do not normally hold territories, did so as removals continued. All birds had functional testes. How long they would have been able to defend their territories successfully is not known. One of the replacement adult males, on the other hand, had nonfunctional testes.

Observations following shooting demonstrated that replacement was often quite rapid. Dr. Leopold observed the Haskell Ranch area on the morning of April 12, the day following the first shooting, and

TABLE 10.   Red-winged Blackbird removal experiments, 1960.

HASKELL RANCH		EAST PARK RESERVOIR	
Date	Number of ♂♂ shot*	Date	Number of ♂♂ shot
April 11	5	April 11	2
April 20	5	April 28	3
April 23	3	May 7	3 (1)
April 28	2 (1)	May 15	3 (1)
May 8	3	June 8	2
May 15	3 (1)		
May 30	3 (1)		
June 16	2		

* Figures in parentheses indicate first-year males.

found all territories reoccupied. Several times I observed a replacement to occur within an hour and once within fifteen minutes. Particularly during April, when activity is at its peak, it would probably be possible to get daily replacement, so that the removals actually made give no idea of the number of birds which could be taken from a marsh during a season, nor what the seasonal pattern of time required for replacement might be.

Ever since the publication of Howard's (1920) book, territoriality has attracted considerable attention, but progress has not been commensurate with the effort expended (Hinde 1956). Data from blackbirds suggest some new avenues of approach. The role of territorial behavior in limiting the density of breeding birds, strongly indicated for the Redwing, should be tested for more species. Howard believed that density was limited by territorial behavior, but his view has been challenged by Lack (1954). Stewart & Aldrich (1951) and Hensley & Cope (1951) observed repopulation following shooting in coniferous forest insectivorous birds, but their experiments were performed during a spruce budworm outbreak and the results may not be generalizable.

The role of different factors in influencing territory size may profitably be explored by studying variability in territory size in different habitats. Some species, such as the Redwing, change their spacing system with habitat, providing clues to its significance. The value of comparative studies of closely related species has been largely ignored, but often such species differ strikingly in their territorial behavior. This aspect of blackbird spacing will be discussed following the presentation of data on the Tricolored Blackbird.

Mating System. It is well known that the Redwing is polygynous, the females maintaining territories within the larger territories of the males. Females regularly breed when they are one year old though it is not known if they always do so. Males do not normally breed until two years of age, though they have been observed holding territories (Beer & Tibbits 1950) and, rarely, breeding (Wright & Wright 1944, Nero 1956b) when one year old.

First-year males, some of them reproductively mature, were common around the marshes and attempted to occupy territories. Some held small areas for short periods of time, but I had no evidence that they ever succeeded in fertilizing any females, nor is it known whether any of the first-year males which held territories after removals copulated with females.

Determining the number of males on a given marsh is a relatively simple task, but females are much more difficult to count. Counting all the nests in the area only gives a rough estimate of the number of females because of the many repeat nests following failure. Consequently, I was able to determine the actual sex ratios in only a small portion of the Redwings I studied. Precise figures are also rare in the literature. On my study areas the number of females per male has ranged from one to six. My data and those from the literature are given in Table 11.

TABLE 11. Red-winged Blackbird sex ratios.

Source	Number of ♂♂	Number of ♀♀	Average number ♀♀ :♂
Smith (1943).....	23	37	1.61
"	40	110	2.75
"	42-46	115-117	2.50-2.78
Nero (1956)......	25	49	1.96
This study:			
E. Pk. Res....	29	108	3.72
Haskell R.....	13	37	2.84

They suggest the possibility of geographical variation in sex ratio, but in the absence of data on temporal variation nothing definite can be said.

* * * * * * *

2. THE TRICOLORED BLACKBIRD

Non-breeding Period. Outside the breeding season, Tricolored Blackbirds feed in the same situations as Redwings and mixed flocks are common. Roosts are located in the same types of habitat and are often shared between the two species. Tricolors roost later in the evening than Redwings, and, in my experience, the sexes do not segregate at any time. As in the Redwing, there is a mass exodus from the Sacramento Valley in the late winter, birds being absent from large areas for several months. Since the Tricolor is not known to occur in large numbers outside the Great Valley of California, it is likely that the bulk of the population moves to the San Joaquin Valley where personnel of the U. S. Public Health Service, working on encephalitis control, observe them in enormous numbers.

* * * * * * *

Time of Breeding. Extreme synchrony, as found at the East Park Reservoir, is characteristic of most colonies of Tricolored Blackbirds (Tables 14 and 15). Even in colonies as large as 50,000 to 100,000 nests, all eggs may be laid within one week. The number of nests started daily in a large colony (Haskell Ranch) and a small colony (Lake Isabella) are shown

in Figs. 16 and 17. On the other hand, some colonies, such as the one at the Capitol Outing Club in 1959 and 1960, grow through the addition of new birds on their peripheries so that, while any given area is uniform, different parts of the colony vary. For example, at the Capitol Outing Club on June 5, 1959, young were being fed in nests in the northeast part of the colony, farther southwest all females were incubating, and at the extreme corner of the colony nests were still being constructed. This type of colony organization has been noted before (Tyler 1907, Dawson 1923, Lack & Emlen 1939), and Dr. Leopold has observed it in previous years at the Haskell Ranch.

* * * * * * *

Territory. Territory sizes in dense Tricolor colonies are difficult to measure accurately, but by estimating distances between neighboring males I have determined that territories are usually 35 sq ft or less in dense vegetation although they may be larger in less suitable cover. The area is defended without aerial displays from a low platform of bent cattails. The tops of the vegetation form neutral ground over which prospecting males and females move without being attacked. It is only when an intruding male actually moves lower down into the vegetation that he is chased. The male defends his territory only for that week when the females are nest building and actively soliciting copulation.

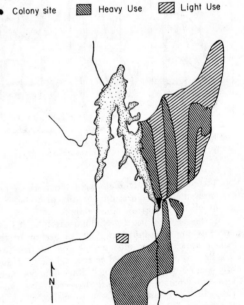

FIG. 18. Feeding grounds of Tricolored Blackbirds during nest building and incubation periods at the East Park Reservoir, 1959.

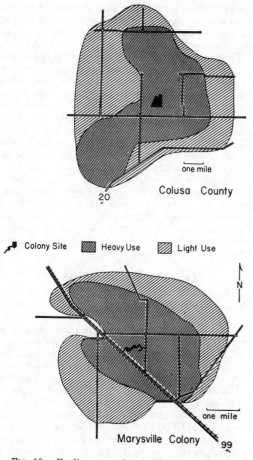

20″

Colusa County

Colony Site   Heavy Use   Light Use

N

one mile

Marysville Colony   99

FIG. 19. Feeding grounds of Tricolored Blackbirds at the Colusa and Marysville colonies, 1959.

Once the clutches are complete he leaves and may not visit the territory again until the young hatch.

To support the food needs of thousands of young birds, a large area must be exploited by the adults and this area forms the ecologically significant territory. Figs. 18 and 19 show the pattern of utilization around the East Park Reservoir and the Marysville and Colusa colonies in 1959. At these colonies, birds travelled up to 4 miles from the colony site and more than 30 sq mi of land were exploited for food. At the Marysville colony, conditions were excellent for observing changes in feeding pattern during the nesting period.

Details of the temporal pattern of environmental utilization are given in the thesis manuscript deposited in the library of the University of California. The general picture which emerges from these observations is that Tricolors react quickly to any changes in the surrounding environment which make food supplies more readily available. As soon as pastures were flooded or a crop cut or raked, thou-

sands of birds descended upon the newly exposed insect supply. The source of food is apparently communicated to others by the direction from which incoming birds approach the colony. I observed no special behavior which might have assisted with this, but communication was nonetheless efficient.

* * * * * *

Colony-size Limitation. Evidence has been presented for the Redwing indicating that territorial behavior limits the density of the breeding population. In the Tricolored Blackbird, territory size varies little from colony to colony unless the vegetation prevents the nests from being as close together as they normally are in undisturbed cattails. Furthermore, except in rare instances, the total nesting space is only partly utilized by the colony, so that territorial behavior would be ineffective in preventing additional birds from settling to breed. Nevertheless, when enormous numbers of individuals must be fed from a fixed spot, the relationship between colony size and food supply is critical because a colony too large for the surrounding environmental resources might be a complete failure. Certain evidence strongly indicates that there is some mechanism of colony-size adjustment, although its details are yet to be determined.

Firstly, colony size is correlated with the suitability of the surrounding environment. In the grazing lands of the foothills I have never found colonies larger than a few thousand nests. The agricultural country of the valley supports larger colonies, the largest being in the rice-growing areas where rich insect supplies are produced in the shallow water (Table 18).

TABLE 18. Tricolored Blackbird colony sizes.

Habitat type	NUMBER OF COLONIES WITH		
	<1,000 nests	1,000-10,000 nests	>10,000 nests
Foothills.............	7	2	1
Valley cropland (no rice).	3	2	3
Rice country...........	0	3	7

Secondly, territorial challenges are frequent during the colony-establishment period. Territories are often taken over by a new male while the resident is absent for a few minutes feeding, but such intruders are quickly expelled by the owners when they return. At any time during the colony-establishment period there are many unsettled birds which continually move back and forth over the colony looking for unoccupied territories. In fact, from a distance it appears that most birds are wandering aimlessly through the colony area. Actually, the established males are all singing and displaying low in the vegetation, and the movements are almost entirely composed of unestablished birds. This is not correlated with availability of territory sites because it is equally true whether there is a shortage of nest sites or whether only a small portion of the marsh is oc-

cupied. Apparently intruders attempt only to substitute themselves for already established birds
rather than to increase the colony size.

Thirdly, in all colonies observed at the time of
territory establishment, the number of birds present
was always greatly in excess of the number which
actually remained to breed. At the East Park
Reservoir in 1959, about twice as many birds were
present the first few days as bred. At the Haskell
Ranch in 1959, about three or four times as many
birds as nested were present the first few days. Some
of this overflow may have moved to the Marysville
Colony about eight miles to the northeast, but even
here the colony at its maximum extended nearly one-
fourth mile farther along the drainage channel than
the limit of actual nests. In none of these cases was
there a shortage of nesting sites.

The evidence suggests that during the first few
days of colony establishment an assessment is made
of the food supply available in the surrounding environment by means of mass feeding flights. During
this period the birds make what appears to be an
excessive number of feeding trips to the surrounding
country, and by watching from a blind it can be
determined that unestablished birds make far more
trips than established ones. These mass feeding
flights form the most conspicuous activity around
colonies at this time. At the East Park Reservoir in
1959, I observed 17 mass feeding flights, involving
most of the birds in the colony, in 6.5 hrs on April
20, the day the colony started. On April 21 I observed 14 such flights in 4.75 hrs and on April 22,
5 in 2.25 hrs. This yields an average of 2.7 flights
per hour. It seems unlikely that such a rate of feeding is necessary for the adults merely to gather the
amount of food they need.

This is simply a special case of the general
phenomenon of environmental evaluation among
birds. Many species are known to adjust their clutch
sizes and/or territory sizes to food supply of the
environment, and it is well known that colony size
in many colonial species is in some way adjusted to
the capacities of the environment to support breeding (see references in Lack 1954). Such an adjustment could be made in the Tricolored Blackbird
through the mass feeding flights.

Data for the Redwing and Tricolored Blackbird
relative to spacing can be summarized as follows.
In the Redwing, territorial behavior strongly limits
density, forcing part of the population into less suitable areas and probably totally preventing some individuals from breeding. Fighting over territories
begins early and is most severe in areas where territory size is ultimately the smallest. Variability is
related to habitat in two ways: (1) The nature of
the nesting vegetation may influence territory size,
as was shown following burning. Under undisturbed
conditions, however, this is likely to be of minor
importance. (2) More important is the nature of
the surrounding feeding grounds. Territories are
largest where most of the food is obtained on them
and smallest where the least food is obtained on them.
Territory size is unrelated to the number of females
building nests within it, nor is it related to the action
of known predators, though this point is less certain.

In the Tricolored Blackbird, territories are uniformly small unless the vegetation is not dense enough
to permit such a high concentration of nests. Territorial behavior does not limit density. Instead, the
important variable, colony size, changes with environmental conditions, being smallest in the grazing
and dry farming areas and largest in the rice growing
areas. There is suggestive evidence of a mechanism
of colony size limitation.

Therefore, whereas in neither species is there a
"food territory" in the classical sense, the spacing
within the systems is intimately related to the exploitation of the environment, and the known patterns of variability in territory size can be attributed
primarily to it. More data will be needed to clarify
the roles of other factors.

TIME AND ENERGY BUDGETS

The amount of time and energy which a bird de-
votes to different activities must inevitably influence
its survival and reproductive rates. It follows that
there exists for a species in a given environment an
optimal time and energy budget. It is of particular
theoretical interest to investigate the conditions in-
fluential in determining the relative significance of
different patterns of time and energy budgeting
(Hutchinson 1957, Fisher 1958:47). The general
evolutionary trend has been to reduce both the number of gametes produced and the amount of energy
devoted to their production. At the same time
there have been increases in the energy content per
female gamete, and the time and energy devoted to
the care of those few offspring produced. It is not
surprising that these trends are correlated since
giving extended care to offspring is incompatible with
producing enormous numbers of them, and production
of large gametes is incompatible with production of
large numbers of them. Beyond these obvious trends,
however, there are many unstudied variations in the
time and energy budgets of species producing similar
numbers of gametes of approximately equal energy
contents.

There are three major ways in which a species
can modify its expenditure of time and energy.
Firstly, the total energy expenditure may remain
approximately the same but its distribution among
different activities varied. Secondly, the total energy
budget may be increased, and thirdly, it may be decreased. The amount of time spent on reproductive
activities may vary in like manner. It is the purpose
of this section to present quantitative estimates of
time and energy expenditures in the Redwing and
Tricolored Blackbird.

These estimates are of necessity rather crude.

Firstly, the lack of adequate physiological data forces me to make assumptions about the energy demands of certain activities which may not be highly accurate. Secondly, the field data are based upon only a few individuals, whereas observations in other areas have shown that the pattern varies geographically. No previous attempt has been made to establish the budgeting of time and energy in natural populations, but Pearson (1954) made an estimate of the daily energy requirement of an Anna Hummingbird (*Calypte anna*). Hence, in spite of the various difficulties involved, crude attempts will nevertheless be ventured because the differences between the two species of blackbirds are so striking and because of their theoretical importance.

Ideally, one should present estimates of the entire annual time and energy budgets but the data do not justify such extended treatment and non-breeding differences appear to be minor. Instead, I have limited the comparison to those features in which the two species differ most strikingly, namely territorial defense and feeding of the young. The slight differences in the time and energy devoted to nest building and egg laying are ignored. These restrictions serve to concentrate attention upon the major differences, in terms of energetics, between the two social systems, preparing the way for a discussion of the evolution of these differences. Since no attempt is made to quantify the entire time and energy budget, it is impossible to express any time and energy expenditures as fractions of the whole. I have therefore expressed them as percentages of energy increase above the resting metabolic level, or as additional hours of time expenditure, as the case may be.

My most complete information on the Redwing is based upon several males studied intensively at Jewel Lake in 1958. During February the males spent about fifteen minutes on the marsh in the morning, defending their territories, after the departure of the main roosting flock. They then left for the day, returning in the evening shortly before the main roosting flock at which time they also engaged in territorial behavior. In March, the time spent on the marsh gradually increased to about 3.5 hrs in the morning, but the evening arrival time did not appreciably change. On the average, about two extra hours were spent on the territories during this month. By the end of the first week of April the birds remained all day and nesting was soon underway. This pattern continued for about two months until nesting was completed, after which the birds again left the area.

Once the males remained all day, about ¾ of their time was spent on the territory; the rest on nearby feeding grounds. From my notes I have determined that about ¼ of the time spent on the territory was occupied with actual defense of the territory, either by means of vocalizations and displays or through actual chasing and combat (see later).

As females are much more difficult to watch, I do not have comparable quantitative data, but they spend much less time in territory defense than males, and the period of time during which they do so is shorter. Territory defense at Jewel Lake lasted from mid-March until the hatching of the eggs, but during the incubation period the frequency of contacts between females was low as incubating birds stirred from their eggs only when new females arrived on the area. During the period of active territorial defense, I estimated that about ⅛ of the female's time is so spent.

Once the young hatch, the pattern of activity suddenly changes for the females, but not for the males, which continue much as before. On the average, a female visits the nest at least once every fifteen minutes with food. Most of her time is spent among

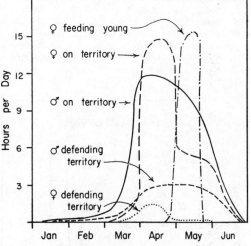

FIG. 20. Time and expenditure during the breeding season by a typical pair of Redwings.

the grass searching for food, only about 1.5% of it being required for flying to and from the nest. The remainder of time is devoted to feeding herself, preening, and resting. An estimate of the time expenditure of a typical male and female Redwing is summarized in Fig. 20.

In the Tricolored Blackbird, the pattern is strikingly different. Prior to the start of nesting almost no time is devoted to activities concerned with breeding, but activity is intense as soon as the colony forms. Since continued observations of individuals in these colonies is so difficult, estimates are based upon group behavior, supplemented by observations of individuals from a blind. Males devote about ½ of their time to territory establishment and defense during that one week period when nests are built and eggs laid. Thereafter, such activity ceases for the remainder of the breeding period. Since nests are started the first day, females spend almost no time in aggressive behavior. Once the eggs are laid, all territorial behavior on the part of all birds stops.

During the colony-establishment period considerable energy is devoted to the conspicuous mass feeding flights. About 26 minutes of every hour were devoted to feeding flights, a portion of which apparently forms a part of the environmental assessment of the breeding birds.

During the nestling period, both sexes actively bring food to the young, but in contrast to the Redwing, the major expenditure is in flying from the nest to the feeding area and back again. Since areas up to four miles from the nest are utilized when feeding the young, virtually half of the adults' time must be spent in flight, leaving much less time for gathering food than is available to the Redwing. If as much time were spent on foot by Tricolors, the rate at which food could be delivered would be greatly reduced, and the reproductive rate lower, although this is partially offset by male participation in feeding of the young. Since the clutch size of the Tricolor is only slightly less than that of the Redwing, it is apparent that searching time has been reduced substantially. An estimate of the time expenditure of a typical male and female Tricolor is given in Fig. 21.

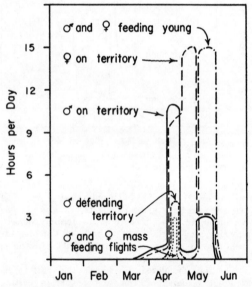

Fig. 21. Time expenditure during the breeding season by a typical pair of Tricolored Blackbirds.

Knowing the frequency of occurrence of different behavioral patterns during the nesting cycle, it is possible to calculate the energetic drain incurred by individuals of the two species. I am assuming that the physiology of avian protoplasm is similar to mammalian, an assumption supported by recent work of James R. King (pers. comm.). Estimates are based upon data given in Brody's book (1945) and Pearson's (1950) work on hummingbirds. Behavior

concerned with territorial defense may be divided into three categories: vocalizations, displays, and chasing and combat. The first two are energetically much more efficient means of accomplishing the objective and are consequently prominent in avian territorial behavior. For the purposes of calculation I assume that the energy required to produce song and other vocalizations raises the metabolic level of the bird 10 per cent above its resting level. This is equivalent to the additional energy required for standing as opposed to lying in man and several domestic animals (Brody 1945). Displays are assumed to double the metabolic rate much as walking does for man. Flight is assumed to require five times as much energy as resting, as found by Pearson for hummingbirds. Even if the flight of other birds is found to require an increase in energy less than that incurred by hummingbirds, the general picture obtained here will not be seriously altered.

Approximately six vocalizations per minute were given by male Redwings on their territories at Jewel Lake. Of these, five, mostly songs, were directly concerned with territory. Since the average duration of a song is slightly less than 1.5 seconds, about seven seconds per minute were devoted to this activity. Three displays concerned with territory were given per minute, averaging two seconds each, for a total of six seconds per minute. Flights and fights occupied about 1.5 seconds per minute. Thus, as mentioned earlier, ¼ of the bird's time is devoted to activities of territorial maintenance. Calorie-wise, the vocalizations require an increase of 0.7% in energy expenditure, the displays an increase of 5% and the flights and fights another increase of 5%, for a total increase of 10.7%. Furthermore, this 10.7% additional energy must be obtained in ¾ the time otherwise available for this purpose, and time available for other activities is correspondingly reduced. Since the male takes no part in feeding the young, his reproductive energy expense is restricted to this category until the young leave the nest.

In females the duration of territorial defense is only about one-half that of the males, and all forms of territorial behavior are indulged in less frequently, especially chasing and fighting. I have used an energy increase of 5% as an approximation of female territorial energy expenditure. However, once the young hatch, female time and energy expenditure changes radically. Assuming that walking on foot searching for food doubles the metabolic rate of the bird, the energy increase of females is about 157.5% above the resting level, about 150% of this coming via the search on foot and the remainder in flight between the nest and the nearby feeding grounds (Fig. 22).

In the Tricolor, the energy devoted to territorial defense and maintenance is greatly reduced in both sexes. All such activity takes place within the period of one week, and no energy is devoted to it during the incubation and nestling periods by either sex. Using mass behavior observations I have estimated that for one week, male Tricolors are at least twice as active

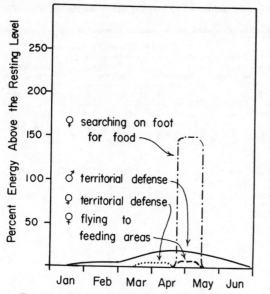

FIG. 22. Energy expenditure during the breeding season by a typical pair of Redwings.

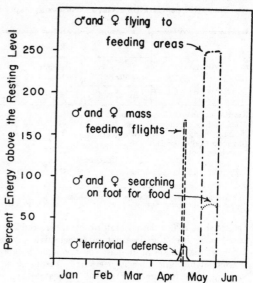

FIG. 23. Energy expenditure during the breeding season by a typical pair of Tricolored Blackbirds.

in territorial defense as male Redwings, but that females are much less so than female Redwings. To this estimate must be added the energy expense of mass feeding flights, one half of which will be assumed to be in excess of that merely needed to sustain the adults.

The major energy expenditure accompanies feeding the young because of the great distances flown. Using the calculations of time spent flying and walking given in Fig. 21, it can be concluded that the energy increase totals about 317%, 250% of which is expended in flying and 67% in walking, just the reverse of that found in the Redwing. The other contrast is that both sexes are involved. Energy expenditure is estimated in Fig. 23 for a typical male and female.

In determining the total energy requirements of a social system it is important to consider not only the energy demands of the activity, but also the duration of that demand. In comparing the two systems I have therefore expressed the energy expenditure, firstly, in terms of the period during which it occurs and, secondly, in terms of the total energy increase for the year (Table 19).

Clearly the colonial system of the Tricolor is more demanding of energy but less demanding of time than the territorial system of the Redwing. This is due to the fact that most time-consuming events are energetically less demanding than events compacted into short periods of time. Since the colonial system of the Tricolor is energetically more expensive, the species would stabilize at a lower population level, other things being equal (Slobodkin 1953). However, it is probable that the system evolved because it made other things unequal, and that the population level

was increased by the adoption of the nomadic colonial system under the particular conditions to which it has been exposed during its history (see later).

Because of the high rate of food gathering, the Tricolor colonial system demands more favorable environmental conditions in which to operate than the Redwing territorial system. These more exacting requirements may help to explain the peculiarly spotty distribution of the species during the breeding season. A detailed study of food supplies available in different feeding areas, in relationship to their distance from the nests and the frequency of their utilization, would be most rewarding.

One of the major differences between the species is the early occupation of territories by male Redwings. Since testis maintenance is probably energetically cheap, only a very slight advantage for the males to be in reproductive condition early is needed to offset the energy loss incurred through the long-term maintenance of functional gonads. The advantages of early testis maturation are (a) early occupation of territories with the attendant advantages of prior residency, (b) the advantage gained by being able to mate with the first females to come into breeding condition (Fisher 1958), and (c) the ability to inseminate females whenever the opportunity arises. The relative importance of these factors varies with the particular mating system employed by the species and the ecology of the area. In many regions early occupancy of territories by the males is prevented by ecological unsuitability of the nesting area prior to the time of nesting.

In contrast, since the maturation of ovaries and the production of eggs is energetically very expensive, selection can be assumed to favor such metabolic exertion when and only when the chances for success-

TABLE 19. Comparative social system energy expenditure (expressed as % increase above the resting metabolic level).

Activity	REDWING						TRICOLORED BLACKBIRD					
	Duration		Energy increase %		Total energy increase %/yr.		Duration		Energy increase %		Total energy increase %/yr.	
	M	F	M	F	M	F	M	F	M	F	M	F
Territory defense......	10 wks	5 wks	10.7	5.0	2.2	0.5	1 wk	1 wk	13.0	1.0	.25	.02
Mass feeding flights....	—	—	—	—	—	—	4 days	4 days	165	165	1.8	1.8
Feeding young........	2 wks	2 wks	T	157.5	T	7.8	2 wks	2 wks	317	317	12.6	12.6
(flight to feeding area).	—	—	—	(7.5)	—	(0.3)	—	—	(250)	(250)	(10)	(10)
(search on foot).......	—	—	—	(150)	—	(7.5)	—	—	(67)	(67)	(2.6)	(2.6)
Total............			10.7	162.5	2.2	8.3			330	318	14.65	14.42

Note: Duration of territory defense in the male Redwing may last up to 20 weeks but for part of this time the territory is occupied morning and evening only. The figure of 10 weeks represents an amount of time roughly equivalent to the total hours of full-time occupation.

ful breeding follow with a high probability. Furthermore, early breeding does not carry selective advantage for the female as it does for the male because the modal breeding time is necessarily the most advantageous if the breeding period is to stabilize, as it does. Hence, females are at a selective advantage if they come into breeding condition only upon arrival at a suitable breeding area where environmental conditions are favorable and a male is present. Thus, we should expect the female gonadal cycle to lag behind that of the male and the stimulatory effects of the male upon ovarian maturation to be strong, and this is the case.

Viewed in this light, the early occupation of territories by male Redwings in California, where winters are mild and the species is non-migratory, is reasonable, but male Tricolors fail to occupy territories prior to the time of breeding under the same environmental conditions. One of the requirements for adaptation to nomadism, the need for rapid response to suitable environmental conditions whenever and wherever they are encountered, leads to close group synchrony. Since the time and place of suitable breeding sites are unpredictable for nomads, no advantage can be gained through attempted occupation of sites in advance of the main group of birds. Instead, close flock organization at all times is most advantageous and the male's chances of leaving offspring are greatest if he remains with the group.

*   *   *   *   *   *   *

of the Tricolored Blackbird and Mallard. Condor **24**: 31.

Emlen, J. T. 1941. An experimental analysis of the breeding cycle of the Tricolored Red-wing. Condor **43**: 209-219.

Fischer, R. B. 1953. Winter feeding of the Red-wing (*Algelaius phoeniceus*). Auk **70**: 496-497.

Fisher, R. A. 1958. The genetical theory of natural selection. 2nd ed. New York: Dover.

Formosov, A. N. 1937. Materials on the ecology of aquatic birds according to observations made on the lakes of the State Naurzum reservation territory (Northern part of the Kasach SSR). Menzbir Memorial Volume (Sbornik): 551-593. (English summary 593-595.)

Friedmann, H. 1949. The breeding habits of the weaver-birds. A study in the biology of behavior patterns. Smithsonian Rept. for 1949: 293-316.

Grinnell, J. & A. H. Miller. 1944. The distribution of the birds of California. Pac. Coast Avifauna No. 27.

Gullion, G. W. 1953. Territorial behavior of the American coot. Condor **55**: 169-186.

Heermann, A. L. 1853. Notes on the birds of California, observed during a residence of three years in that country. J. Acad. Nat. Sci. Phila. V, 2nd ser., 2: 259-272.

Hensley, M. & J. B. Cope. 1951. Further data on removal and repopulation of the breeding birds in a spruce-fir forest community. Auk **68**: 483-495.

Hinde, R. A. 1956. The biological significance of the territories of birds. Ibis **98**: 340-369.

Howard, H. E. 1920. Territory and bird life. London: Murray.

Hutchinson, G. E. 1957. Concluding remarks. Cold Spring Harbor Symposia on Quantative Biology, V 22. Population Studies: Animal Ecology and Demography: 415-427.

Keast, A. 1959. Australian birds: Their zoogeography and adaptations to an arid continent. In: Biogeography and Ecology in Australia: 89-114.

Kennedy, C. H. 1950. The relation of American dragon fly-eating birds to their prey. Ecol. Monogr. **20**: 103-142.

Kolman, W. A. 1960. The mechanism of natural selection for the sex ratio. Amer. Nat. **94**: 373-377.

Lack, D. 1940. Habitat selection and speciation in birds. Brit. Birds **34**: 80-84.

————. 1954. The natural regulation of animal numbers. Oxford: University Press.

Lack, D. & J. T. Emlen. 1939. Observations on breeding behavior in Tricolored Redwings. Condor **41**: 225-230.

Linford, J. H. 1935. The life history of the Thick-billed Redwinged Blackbird, *Agelaius phoeniceus fortis* Ridgway in Utah. (M. S. Thesis, Utah Library.)

Lorenz, K. Z. 1949. Über die Beziehungen zwischen Kopfform und Zirkelbewegung bei Sturniden und Ikteriden. In: Mayr, E. ed. Ornithologie als biologische Wissenschaft. Heidelberg: Carl Winter.

MacArthur, R. H. 1960. On Dr. Birch's article on population ecology. Amer. Nat. **94**: 313.

Macklin, P. R. 1958. Spittle insects as food of the Red-winged Blackbird. Auk **75**: 225.

LITERATURE CITED

Allen, A. A. 1914. The Red-winged Blackbird: A study in the ecology of a cat-tail marsh. Abstr. Proc. Linn. Soc. New York, Nos. 24-25 (1911-1913): 43-128.

Axelrod, D. I. 1958. Evolution of the Madro-Tertiary Geoflora. Bot. Rev. **24**: 433-509.

Beecher, W. J. 1951. Adaptations for food-getting in the American Blackbirds. Auk **68**: 411-440.

Beer, J. R. & D. Tibbits. 1950. Nesting behavior of the Red-winged Blackbird. Flicker **22**: 61-77.

Bent, A. C. 1958. Life histories of North American Blackbirds, Orioles, Tanagers, and Allies. U. S. Nat. Mus. Bull. 211.

Brody, S. 1945. Bioenergetics and growth. New York. Reinhold.

Brown, L. H. 1958. The breeding of the Greater Flamingo *Phoenicopterus ruber* at Lake Elmenteita, Kenya Colony. Ibis **100**: 388-420.

Chaney, R. W. 1951. Prehistoric forests of the San Francisco Bay area. In: Geologic Guidebook of the San Francisco Bay Counties. Cal. Div. Mines Bull. 154.

Cole, L. C. 1954. Population consequences of life history phenomena. Quart. Rev. Biol. **29**: 103-137.

Davis, D. E. 1942. Number of eggs laid by Herring Gulls. Auk **59**: 549-554.

Dawson, W. L. 1923. The Birds of California. Vol 1. (Book-lovers' ed., San Diego, Los Angeles, San Francisco, South Moulton Co.).

Dickey, D. R. & A. J. van Rossem. 1922. Early nesting

Mailliard, J. 1900. Breeding of *Agelaius tricolor* in Madera Co., California. Condor 2: 122-124.

──────. 1914. Notes on a colony of Tri-colored Redwings. Condor 16: 204-207.

Martin, N. D. 1960. An analysis of bird populations in relation to forest succession in Algonquin Provincial Park, Ontario. Ecology 41: 126-140.

Maynard Smith, J. 1958. The theory of evolution. London: Pelican.

Mayr, E. 1939. The sex ratio in wild birds. Amer. Nat. 53: 156-179.

McIlhenny, E. A. 1940. Sex ratio in wild birds. Auk 57: 85-93.

Neff, J. 1937. Nesting distribution of the Tri-colored Redwing. Condor 39: 61-81.

Nero, R. W. 1956a. Redwing nesting in bird house. Auk 73: 284.

──────. 1956b. A behavior study of the Red-winged Blackbird. Wilson Bull. 68: 5-37, 129-150.

Orians, G. H. 1960. Autumnal breeding in the Tri-colored Blackbird. Auk 77: 379-398.

Paynter, R. A., Jr. 1949. Clutch-size and the egg and chick mortality of Kent Island Herring Gulls. Ecology 30: 146-166.

Pearson, O. P. 1950. The metabolism of hummingbirds. Condor 52: 145-152.

──────. 1954. The daily energy requirements of a wild Anna Hummingbird. Condor 56: 317-322.

Pitelka, F. A. 1959. Numbers, breeding schedule, and territoriality in Pectoral Sandpipers of northern Alaska. Condor 61: 233-264.

Rand, A. L. 1948. Glaciation, an isolating factor in speciation. Evolution 2: 314-321.

Roberts, A. 1940. The birds of South Africa. London: H. F. & G. Witherby, Ltd.

Schenk, J. 1929. Die Brutinvasion des Rosenstares in Ungarn im Jahre 1925. Verh. Vi Int. Orn. Kongr. 1926: 250-264.

──────. 1934. Die Brutinvasionen des Rosenstares in Ungarn in den Jahren 1932 und 1933. Aquila 38-41: 136-153.

Selander, R. K. 1958. Age determination and molt in the Boat-tailed Grackle. Condor 60: 355-376.

──────. 1960. Sex ratio of nestlings and clutch size in the Boat-tailed Grackle. Condor 62. 34-44.

Serebrennikov, M. K. 1931. Der Rosenstar (*Pastor roseus* L.), seine Lebensweise und ökonomische Bedeutung in Uzbekistan (Turkestan). J. f. Orn. 79: 29-56.

Sherman, A. R. 1932. Red-winged Blackbirds nesting in treetops near top of hill. Auk 49: 358.

Slobodkin, L. B. 1953. On social single species populations. Ecology 34: 430-434.

Smith, H. M. 1943. Size of breeding population in relation to egg laying and reproductive success in the eastern Redwing. Ecology 24: 183-207.

Stewart, R. E. & J. W. Aldrich. 1951. Removal and repopulation of breeding birds in a spruce-fir forest community. Auk 68: 471-482.

Todd, W. E. C. 1940. Birds of western Pennsylvania. Pittsburgh: University Press.

Tyler, J. G. 1907. A colony of Tri-colored Blackbirds. Condor 9: 177-178.

Williams, J. F. 1940. The sex ratio in nestling eastern Redwings. Wilson Bull. 52: 267-277.

Williams, L. 1952. Breeding behavior of the Brewer Blackbird. Condor 54: 3-47.

Wright, P. L. & M. H. Wright. 1944. The reproductive cycle of the male Red-winged Blackbird. Condor 46: 46-59.

VII
The Evolution of Territorial Systems

Editor's Comments on Papers 22 and 23

22 **Crook:** The Adaptive Significance of Avian Social Organizations
Symp. Zool. Soc. London, No. 14, 182–183, 191–195, 199, 201–202, 204–218 (1965)

23 **Brown:** The Evolution of Diversity in Avian Territorial Systems
The Wilson Bull., **76**, 160–169 (1964)

In this part we examine the factors that have been involved in the evolution not only of territoriality but of social systems in general. Wynne-Edwards (1962) in his monumental book *Animal Dispersion in Relation to Social Behaviour* hypothesized that social systems have evolved as a means of regulating populations and maintaining homeostasis. This theory has not been generally accepted. Crook (Paper 22), after rejecting Wynne-Edwards' theory, presents a theory of his own. He believes that social systems have evolved in response to the types of environments a given population of a species inhabits. Crook began his work with the exceedingly ambitious task of studying the social systems and ecology of more than 60 species of weaver birds (Crook, 1964). Crook summarizes his findings on pages 182–183 in Paper 22. He was able to trace the origins of weaver bird social organizations back through the seasonal changes in abundance and distribution of food to the basic rainfall patterns of an area. In addition, he showed that there exist secondary modifications resulting from predator pressure.

Crook's classic 1964 paper has prompted a wave of studies along the same lines, including studies of primates (Crook and Gartlan, 1966) and African bovids (Estes, in press). Estes studied bovid social organization in the same general area of Africa as was inhabited by Crook's weaver birds. It appears that territoriality occurs in these ungulates where food resources are year-round and predictable but only moderately abundant, as in forests. In the open plains, where rainfall is seasonal, ungulates form hierarchies rather than territories just as weaver birds and primates do in similar habitats.

John H. Crook is among the group of brilliant young ethologists, including Peter Marler and Jane Goodall-Van Lawick, to receive their training under the guidance of W. H. Thorpe and R. A. Hinde at Cambridge. Crook has had a tremendous influence on studies of the influence of ecological factors on the evolution of social systems. He is currently Reader in Ethology at the University of Bristol, England. His main interest is in the relation between ecology and social systems in birds and mammals, especially primates.

In Paper 23, Jerram Brown presents a theory for the evolution of diversity in avian territorial systems. Brown refutes, as does Hinde in Paper 5, the idea that territoriality has evolved with a single function such as food or population regulation. Instead he postulates that the basis of a social system is competition for limited resources. Only if these resources are economically defensible will it be profitable for an individual to expend his energy in attempting to defend them. In some species, food may be sufficiently concentrated so that it is feasible to defend it through establishment of a territory. For other species, the critical defensible resource may be nesting or courtship sites. It is the combination of economic defensibility and the degree of aggressiveness that has evolved in response to competition that determine the kind of territory, be it colonial nesting, leks, or single large territories.

Brown is Professor of Biology at the University of Rochester, New York. As did many ethologists, he began as a naturalist. Out of these boyhood field experiences came his eventual interest in behavior, ecology, and evolution. Brown says that he has found no substitute for such field experiences in his formal university training. Imitating Agassiz, his motto is "Study nature, not books." Brown was a close associate of Gordon Orians while they were graduate students at Berkeley; they have recently collaborated in writing a more general review of the evolution of social systems in all animals (Brown and Orians, 1970).

The evolution of social systems is intimately related to bioenergetics and the evolution of mating systems (presented in Part VI). Other papers of special importance that deal with these interrelationships are those by Eisenberg (1966), Brown and Orians (1970), and Orians (1969, 1971, 1973).

References

Brown, J. L., and G. H. Orians. 1970. Spacing patterns in mobile animals. *Ann. Rev. Ecol. Systematics,* **1**: 239–262.

Crook, J. H. 1964. *The Evolution of Social Organization and Visual Communication in Weaver Birds (Ploceinae). Behaviour* (Suppl. 10). 178 pp.

Crook, J. H., and J. S. Gartlan. 1966. Evolution of primate societies. *Nature* **210**: 1200–1203.

Eisenberg, J. F. 1966. The social organizations of mammals. *Handbuch der Zoologie.* 8 Band/39 Lieferung: 10(7), 1–92.

Estes, R. D., in press. Social organization of the African bovids. *Proc. Conf. Behavior of Ungulates and Its Relation to Management.* University of Calgary Press, Calgary, Canada.

Orians, G. H. 1969. On the evolution of mating systems in birds and mammals. *Amer. Naturalist,* **103**: 589–603.

Orians, G. H. 1971. Ecological aspects of behavior. *In* D. S. Farner and J. R. King (eds.), *Avian Biology,* Vol. I, 513–546. Academic Press, New York. 586 pp.

Orians, G. H. 1973. Ecoethological aspects of reproduction: Discussion. *In* D. S. Farner (ed.), *Breeding Biology of Birds,* 27–39. National Academy of Science, Washington. 515 pp.

Wynne-Edwards, V. C. 1962. *Animal Dispersion in Relation to Social Behaviour.* Oliver & Boyd, Edinburgh. 653 pp.

22

Reprinted from *Symp. Zool. Soc. London*, No. 14, 182–183, 191–195, 199, 201–202, 204–218
(1965)

The Adaptive Significance of Avian Social Organizations

JOHN HURRELL CROOK

Department of Psychology, University of Bristol

INTRODUCTION

The individuals comprising a species population are not normally distributed at random throughout their characteristic habitat. Rather they are dispersed in a typical manner, in flocks or territories for example, and make characteristic movements between locations such as roosts and feeding places, or between summer or winter ranges. This " dispersion " (Lack, 1954) is a function of the social organization of the species determined by the responses of individual birds to one another and to their common environment. These responses are the expression of a dynamic interaction between a number of behavioural variables which include at least the following:

(a) factors promoting congregation, e.g. common habitat preferences, social tendency, sexual tendency ;

(b) factors promoting dispersion, e.g. individual distance or territorial aggression, escape tendency and " avoidance " ; and

(c) factors determining the form of social units and their duration, i.e. individual relationships, the manner and duration of pair bonding, etc.

The multivariate nature of the proximate causation of social behaviour in birds (see Hediger, 1955 ; Conder, 1949 ; Emlen, 1952 a ; Marler, 1956 b ; Hinde, 1952 ; Crook, 1961 a) means that care must be taken not to use the term " social organization " as a blanket variable, obscuring the complex nature of the process it denotes. This is of particular significance in relation to adaptiveness, because evolutionary changes in social life are best considered in terms of shifts in strength within a balanced drive relationship. Selection pressures are not conceived as acting directly on social organization as such, but rather upon particular tendencies (nest site attachment, escape, sex, etc.) a shift in any one of which may entail a change in the system as a whole. The term " organization " is useful in that it denotes a complex of behavioural characteristics determining the mode of dispersion of a population and the inter-individual encounters within it. It allows quick comparisons between taxa which differ markedly in the type of social life they show.

In 1954 Lack showed that the numerical dispersion of a species throughout its habitat is related to the abundance of food. Birds tend to breed at higher densities where food or food for the young is more plentiful and the dispersion is determined primarily by the preferences of birds breeding for the first time. In a recent study Crook (1961 b, 1963 b, 1964) investigated the adaptive significance of contrasts in social organization between the closely related species of the subfamily Ploceinae (weaver birds). He found that the differences correlated closely with contrasts in ecology. Thus dispersed breeding, active pursuit in courtship and monogamy, correlated with insect eating and low light values in the forest environments of tropical Africa while flocking, relatively static courtship with visual advertisement of nests in colonies and polygamy, went with the high light values of the savannah. Aerial advertisement was performed only by those species that hid their dispersed nests in grassland. Marked adaptiveness in social organization was a characteristic feature of the evolutionary radiation of the group and the study strongly supports the ethological view that contrasts in behaviour are discrete adaptations to particular social and environmental conditions and express the interaction, through time, of a number of ultimate determinants, often tending to impose contradictory demands on the organism (Hinde & Tinbergen, 1958 ; Tinbergen, 1959 b). A complex web of selection pressures fashions the social life of a species, and convergence and divergence between related and unrelated forms will be as frequent here as in morphology.

In contrast with this approach Wynne-Edwards (1962) has put forward the view that many types of avian social organization, including territorial, colonial and arena behaviour, all have essentially the same survival value in relation to ensuring the homeostatic regulation of population size (see below). The present paper attempts first to evaluate this highly unitary interpretation of social behaviour and the more general hypothesis of which it forms a part and, secondly, to survey all birds to see if the sort of adaptive trends apparent in the weaver study might not be traced with broader implication in the class as a whole.

* * * * * * *

A COMPARATIVE SURVEY OF AVIAN SOCIAL ORGANIZATIONS IN RELATION TO ECOLOGICAL VARIABLES

Suggestions regarding selection pressures responsible for a given behaviour depend upon an adequate study of its function. Comparative surveys of closely related species in this respect then provide evidence of the likely course of evolution and are the basis for an analysis of phylogeny. The functions of behaviour normally concern orientations to some feature of the social or physical environment. As a first line of enquiry we should therefore attempt a survey of correlations between the social attributes of bird species and their ecology. To cover a whole class in a short publication necessarily demands both condensation of material and an unhealthy arbitrariness in the use of classificatory criteria. Nevertheless, if marked adaptive trends are present they should emerge. More precise evidence of survival value will come from close comparative study of related groups of species that differ in social organization within the same families. Adequate studies of this type remain regrettably few but they are certainly highly indicative. Studies of this kind furthermore set the stage for an experimental examination of functional significance (i.e. Tinbergen et al., 1962).

First we must provide some classification of avian social organizations and of the ecological variables likely to have played selective roles in their evolution.

Classification of avian social organizations

As the survival values of social organizations within and without the breeding season can be shown to differ very markedly the latter are listed separately. Social organizations in the non-breeding season may be classified as follows :

300

1. Organizations characterized by relatively solitary behaviour.
   (i) Overdispersion through apparent " avoidance " (p. 186). Individuals (or pairs) occur in locations scattered throughout the habitat ; territorial boundaries are not apparent.
   (ii) Contiguous territories. Individuals occur in defended locations with contiguous borders ; the territories tending to occur together in one or more parts of the available habitat forming what Fisher (1954) has described as a mosaic of territories in a " neighbourhood ". Feeding (e.g. in shrikes ; Durango, 1956) is largely restricted to the territory.

The holding of the territory may endure throughout the non-breeding season and then convert to a breeding territory in which females join males (e.g. the robin ; Lack, 1943) or it may break down in hard weather and the birds then move in flocks (e.g. the blackbird ; Snow, 1956). In winter tits tend both to maintain territories and to move in flocks of local territory holders. At night the birds normally separate to individual holes (Hinde, 1952 ; Kluijver, 1957).

2. Organizations characterized by gregarious behaviour (see Appendix).
   (i) Flocks without constant composition in which no inter-individual relationships based on personal recognition occur :
       (a) Flocks forming part of a daily " dispersal system " based upon a nightly roosting site (i.e. Crook, 1960 a ; Burns, 1957 ; etc.). There is no evidence here that the same individuals move regularly in the same parties.
       (b) Flocks performing either nomadic movements, migration or spatial routinism but not based upon a fixed roosting locality. Chaffinch flocks vary greatly in constancy and resident individuals tend to separate to roost in the old territorial area. In hard weather nomadism and the use of communal roosting often develops (Marler, 1956 a). Chaffinches do not compete for roosting holes as do tits (above) and this may account for their less marked winter territorialism.
   (ii) Flocks of constant composition in which high degree of individual recognition is achieved :
       (a) Flocks showing spatial routinism (e.g. juncos ; Sabine, 1959) within a limited range. Here amalgamation between flocks may occur and while visitors may enter established flocks they remain low in the peck order.
       (b) Flocks showing territorial defence of their foraging range (e.g. Crotophaginae ; Davis, 1942).
       (c) Flocks of variable size within which pair and family bonds are maintained (e.g. geese ; Boyd, 1953 ; Davies, 1963).

Social organizations in the breeding season may be classed as follows.

Several categorizations of behaviour have been used previously. Variations in pair bonding (duration of bonds, monogamy–polygamy–polandry, location in pairing in flock, territory or lek, division of labour between sexes in parental care, e.g. Lack, 1940) and contrasts in modes of intersexual signalling (releasers present in one or both sexes, etc. ; Lorenz, 1935 modified by Tinbergen, 1939) have been utilized as criteria while Huxley (1938 b) and Lack (1940) have listed displays of differing function (threat, i.e. antaposematic, advertisement between sexes—i.e. gamosematic, precopulatory, i.e. epigamic and post-nuptual displays maintaining pair bonds). For our present purposes in correlating behaviour with ecology it is best to classify the behaviour in terms of contrasts in dispersion within the habitat.

1. Overdispersion due to solitary breeding or territorialism (territories with contiguous boundaries either filling the habitat or occurring in " neighbourhoods " within it). The cryptic nests are well dispersed. In polygamous birds several nests may be located in the same territory.

2. Colonial sites containing numerous small territories crowded closely together. The colonies are easily located by predators but the nests within them are commonly difficult of approach due to the siting of the colony.

3. Communal nesting. A group of birds are closely associated in reproductive activities and a varying degree of " cooperation " occurs between them irrespective of pair bonding. Territorialism is commonly reduced within the group and occasionally inter-group territorialism occurs.

4. Arena behaviour. The males perform dramatic displays at individual " courts " within a small area called the arena. In some species, where the crowding is less marked, the general area in which display occurs is known as an " exploded arena " (Gilliard, 1963). Females come to the courts for mating (in some species they only mate with the dominant male), but they nest elsewhere. Males play no part in parental care. Variations in the extent of this behaviour may be found between species in the same family.

Classification of ecological variables and the manner of food exploitation by birds

Three main groups of parameters are adequate for our purpose : these are contrasts in habitat and type of food ; dispersion of food supplies ; and nest sites.

Habitat and food type

1. Carnivorous terrestrial birds :
    (a) Food mainly other vertebrates.
    (b) Food mainly other insects and invertebrates : (i) ground feeders ; (ii) aerial feeders ; and (iii) arboreal feeders.
    (c) Scavengers.
2. Sea- and open-water birds—food mainly fish :
    (a) Surface feeders.
    (b) Divers from the air.
    (c) Swim-divers.

3. Shore and shallow-water birds feeding on invertebrates, plant materials and/or small vertebrates.
4. Vegetarian–omnivorous feeders—mostly ground or waterside birds :
   (a) Open-country or forest-floor ground feeders.
   (b) Forest arboreal feeders.
   (c) Waterside birds.
5. Fruit- or leaf-eaters in forests.
6. Flower birds—feeders on nectar and insects.
7 Seed- and bud-eaters of woodland fringe, secondary woodland and " savannah ".

Food dispersion

1. Dispersed, cryptic food difficult to locate and difficult to catch.
2. Locally abundant foods found sporadically in patches in the habitat.
3. Food more or less overdispersed (Salt & Hollick, 1946) throughout the habitat.

By (1) is meant prey which must be located and captured with a high degree of individual skill. In (2), once the food is located, there are typically few problems of exploitation. Examples of (3) are evenly spread surface plankton at sea or insects in the air as caught by birds sampling the stock without having to search widely for it. Cases of (3) are probably not frequent. Thus fish shoals, for example, are probably not at all evenly dispersed at the surface and plankton likewise is doubtless concentrated by currents. These categories are broad generalizations suitable for a survey of the width attempted, but not for more precise analysis at lower taxonomic levels.

Nest sites

1. Open ground, camouflaged or not camouflaged, of easy access.
2. Open ground nests protected by features of siting from approach by predators (e.g. on islands, etc.).
3. Cliffs, ledges or clefts providing increased inaccessibility to predators.
4. Trees, bushes :
   (a) Sites offering no particular protection from approach.
   (b) Sites offering protection from approach.
   (c) Sites in which a particular mode of construction confers protection either against approach or against nest entry.
5. Tall standing grass or reeds conferring crypticity, and if islanded in water some degree of protection from approach as well.
6. Cavities, i.e. holes in trees, ground or earth banks etc. whether natural or excavated by the birds.

An important feature of these sites is the contrasting types of protection they offer against predation. These may also be categorized here :
A. Protective quality primarily inaccessibility.
   (i) Nests sited in inaccessible places where there is no shortage of sites, e.g. 2, 3 and 4b above.

(ii) Inaccessible sites irregularly or sparsely dispersed in the habitat, e.g. some cases of 3, 4b, 5 and 6.

(iii) Nests as in (i) above, whose accessibility is further reduced by the form of the nest structure itself, e.g. 4c—weavers, icterids, etc.

(iv) Nests (usually 4a or 4b) protected additionally by either the size or aggressive character of the owners. i.e. many predators and the very aggressive Tyrannid flycatchers and drongos.

B. Protective qualities primarily concerned with crypticity and dispersion.

(v) Nests whose accessibility is reduced by reason of their structure, i.e. globular nests in vegetation, canopies or covers placed over nest during parents' absence, etc.—but which are otherwise protected by being both well hidden and dispersed, e.g. 4c.

(vi) Nests without a protective structure, protected only by their relative infrequency in the habitat and by crypticity, e.g. 1, 4a.

Relations between social organizations and ecology

Correlations between social organizations and ecological variables are shown in Table 1 (pp. 196–199). The families are listed in the modified Wetmore order of Austin (1962). Only families for which adequate information is available have been included and it is admitted that much that is accepted appears to be based on relatively casual observation.

Of the 141 families in the table (90 non-passerine and 51 passerine) territorial breeding occurs in 79 (47+32), colonial (or communal) breeding in 27 (25+1+1 ?) and arena behaviour in 9 (4+5, also in some species in two other families). In twenty-six families marked adaptive radiation in social organization has occurred within the family. For ease of discussion these are ignored for the moment and only those families with a single type of social organization well described are considered. Of the seventy-nine families with territorial breeding, forty are likewise territorial in the non-breeding season, although some of these contain species (usually those with the broadest diets) that may flock in bad conditions seasonally, or which commonly join mixed-species foraging groups in woodland (see below). There are thirty-nine families territorial when breeding and regularly gregarious at other seasons and twenty-four families colonial when breeding and gregarious at other times. These single type families are shown numerically in Table 2 (see p. 200) in relation to their food habitats, the type of dispersion of their food and the type of protection provided for the nests. Communal and arena breeders will be discussed separately.

The most persistently territorial birds occur in some twenty-eight primarily insectivorous families, two carnivorous ones (Sagitariidae, and Tytonidae), four fish eaters (Pandionidae, Podicipedidae, Gaviidae and Alcedinidae), two nectar feeders (Trochilidae, Philepittidae) and certain more or less omnivorous ground birds (Scopidae, Turnicidae). Almost all these birds must hunt or search for their food and success depends on a combination of stealth (cryptic approach), speed and skill. This is especially true for the carnivores and the majority of the insectivores. It seems likely, therefore, that the solitary nature of these species is an adaptation to their mode of food exploitation. Gregarious behaviour would not in general be of value, for it would interfere with individual methods of food acquisition. This cannot, however, be the complete story. Thus the feeding behaviour of grebes and divers differs little from that of cormorants, but the latter are gregarious and usually breed colonially in protected sites. By contrast, divers and most grebes normally disperse their nests and grebes cover their eggs on leaving. Dispersion and crypticity of nests here correlate with territorial isolation. On Lough Neagh a change in lake level has forced the great crested grebe to breed in crowded conditions in the few remaining suitable sites. The lessened crypticity in unprotective localities is undoubtedly a major factor in the heavy mortality, through predation, that these birds now have to endure (Mylne, 1963, personal communication ; see Nero *et al.*, 1958). In kingfishers the nest site is a hole, usually in a river bank. There seems at first sight no reason why nesting in so protected a manner should not become colonial and indeed in one species (*Ceryle rudis*) the birds do nest in small groups. For the majority, however, a major factor must be competition for food along local stretches of water wherein the food supply must be limited. Colonial breeding would tend to overexploit the areas within easy distance of the colony. Here also, individual methods of hunting probably ensure greater success per bird than group attacks on easily alarmed fish shoals (or on insects) could allow.

Certain vultures, eagles and owls sometimes form colonies in which more than one species coexist. This can occur only when there is sufficient food in the general area to support the increased local population. Among vultures, solitary hunting is valuable when small cadavers provide most of the food but the ability to congregate at large ones is also clearly advantageous. Little is known of the Scopidae or Turnicidae.

The majority of families that are territorial when breeding, but gregarious at other times, are unspecialised feeders with a broad diet. In particular over half of the insectivores in this class are relatively tolerant feeders. Among such birds food supplies tend to be richer in certain localities in the general environment than in others, so that by congregation the majority of birds accumulate over the richest grounds more effectively than mere aggregation would allow. Congregation is thus advantageous here in increasing the probability that an individual will locate and exploit food effectively. Flocking probably also confers some protection against predators through an increased awareness of approach by the group as a whole. This may be particularly significant in the grouse and such-like birds, since the groups move through ground vegetation preventing easy scanning of local environment. The switch from flocking to territorial behaviour at the start of the breeding season seems primarily due to the nature of the preferred nesting sites, which cannot confer protection against the approach of predators. Thus in the majority of these families the nests are broadly dispersed, a social mechanism reducing the frequency with which searching predators can locate nests (Tinbergen, 1952, 1953 ; Simmons, 1956). In addition, however, many species (e.g. Fringillidae) show diet changes in spring or at least take contrasting food items for their young. This may also necessitate a shift in dispersion if the food has to be exploited in a different way.

No less than fourteen sea-bird families are colonial and gregarious. They breed on sandbanks, islands, coastal stacks or cliffs, all difficult of access to terrestrial predators. Ashmole (1963) points out that many sea birds have a tendency to congregate tightly within a colony and this may promote social activities which increase the synchronization of breeding activities. This would tend to narrow the spread in time of breeding, a procedure of value, if seasons of food abundance or accessibility were limited, and which would shorten the period during which predation was experienced and reduce its effectiveness (i.e. Darling, 1935 ; Cullen, 1960 a). Lack (1954) points out that juveniles tend to settle near the parental sites and that their familiarity with a locality must be advantageous. Furthermore, Ashmole (1962) and Lack (personal communication) both consider that, once colonies are established, it will be adaptive for juveniles to go to occupied sites since, in general, traditional places are more likely to be safe from predators than untested sites nearby. We have seen that in at least some colonial weavers (p. 188) the young males set up colonies distinct from those of adults, but that they spend a " practice " year in them before breeding. Ibises, spoonbills, flamingos, storks, etc. also nest in protective sites in vegetation over water (Morel & Morel, 1961) on islands in lakes, swamps or lagoons, or on mud drums built in shallow water over mud (Brown, 1958). Storks thrive around human habitation so long as they are protected by sentiment. These birds thus utilize protected sites much in the manner of sea birds. In all these cases the food is dispersed in manner (2) & (3) and congregation through gregariousness is probably a major factor increasing the effectiveness of individual food exploitation.

However, even in colonial birds local dispersion may correlate with contrasts in siting, the nature of predation and responsiveness to attack. Thus Tinbergen (1952) showed that herring-gull colonies, to which terrestrial predators had gained access, contained nests more widely spaced than others. Cullen (1960 a) compared the dense nesting of sandwich terns with the more dispersed nesting of common and arctic terns, all three species being " colonial ". Sandwich terns only nest under conditions of low predation (e.g. good protection) and desert easily if disturbed. They do not show behaviour increasing crypticity (removal of eggshells, defaecation away from the nest, chicks leave nest to defaecate, etc.) that are a regular part of the anti-predator system of the other terns and also the black headed gull (Tinbergen et al., 1962). In addition, the breeding period of the sandwich tern is shorter than that of the other birds and this correlates with early pairing prior to the arrival at the colony. They tend to sit tight on the eggs, when aerial predators approach, thereby protecting the eggs under them. The other terns, relying on the relative crypticity of their nests, attack and pursue predators.

While many birds are gregarious outside the breeding season and territorial within it, none appears to show the opposite condition. This suggests a special advantage in the change to territorialism for breeding. The factors may become clearer when the many insectivores are examined separately (Table 3, p. 203).

The table reveals the following points :

(i) 29/48 (60·4 per cent) of the families represented are largely territorial at all times. 16/48 (33·3 per cent) only when breeding and only 3/48 (6·3 per cent) are both gregarious and colonial.

(ii) 12/16 (75 per cent) of those families that change from flocking to territorial behaviour when breeding nest cryptically. Of the other families the Paridae utilize either holes or complex globular nests similar to those of Nectariniidae and Diceaedae and which are often pendant from inaccessible twig ends. Some Tyrannids place their nests in protected tree sites (thorn bushes, wasps' nests, etc., see below).

(iii) Of the twenty-nine all-territorial families, twelve utilize cryptic, dispersed nests, but seventeen have inaccessible sites, mostly in holes.

(iv) Gregariousness in the non-breeding season characterizes " swift " type feeders (not the dusk flying Caprimulgiformes, nor usually the bulky more ground-pouncing rollers), many nectar feeders, including those families with the broadest diet (i.e. Zosteropidae, Parulidae), and certain arboreal insectivores, including those with least restricted diets such as Paridae and Vireonidae. It is not found among specialised " flycatchers ", nor among ground feeders in woods. With the exception of the dippers it is typical of open-country ground feeders. Thus in general, the most territorial of the insectivores are the " flycatchers ", arboreal birds and nectar feeders of specialised diet and sylvan ground feeders.

Among insectivores territorial maintenance, individual search and hunting seem most appropriate in their forest environments where the density of undergrowth and low light values lessen the value of flocking. Nevertheless many tend to form mixed-species flocks that travel around relatively limited circuits (i.e. spatial routinism) in forests or open woodland (Moynihan, 1962 ; and quoted). In these groups synchronization and integration are poor, enabling individuals to hunt largely by themselves, while remaining members of a loose flock. In a few families (Meliphagidae, Tyrannidae, Campephagidae) such behaviour is only abandoned when the birds disperse to build their cryptic style nests in territories. While Moynihan (1962) stresses the effect of mutual protection against predators through group alertness and the presence of some aggressive birds (e.g. Tyrannids), Short (1961) suggests that foraging in mixed flocks prevents repeated sampling of trees and bushes already visited and thus improves the likelihood of successful search by all participants. The behaviour may be of especial value in times of food shortage which may well occur in the relatively constant climate of tropical woodland where the numbers of insectivores may often be in danger of over-exploiting supplies. Again as we have seen, flocking is most typical of species with diets including items, such as berries, etc., commonly dispersed in the habitat in manner (2) and not requiring individual skills in their collection. However, if the food supplies are sufficient to enable a bird to find enough food alone, territorial maintenance seems preferable, because the bird is already in possession at the start of the next breeding period.

Insectivores which are normally gregarious take up territories for breeding primarily when the nests are of a type that need to be dispersed. This is especially clear when the Artamidae and Hemiprocnidae, with unprotected nests, are compared with other very similar groups (e.g. Apodidae). However, the fact that birds building secure pendant nests on twig ends do not form colonies (as do granivorous weavers but not insectivorous ones) and that some Tyrannids may form inter-specific but not intra-specific colonies in protective trees (Smith, personal communication) suggests that isolation is important here for a reason additional to nest protection. Furthermore, among those families that are almost always territorial the majority utilize inaccessible sites. In at least those birds building complex nests on twig ends and for certain types of hole nest, there seems no valid reason why the siting should prevent colony formation should it be selected on other grounds—although this would not apply to holes infrequently dispersed in the habitat, nor to those in relatively insecure substrata. It is suggested that the primary reason for territorialism remains the need for individual food exploitation. Since this is now centred on the nest site and both interference in foraging and a tendency to over-exploit food within easy reach would occur at colonies, territorialism is selected. In cases where selection favours both the use of cryptic nests and dispersion in relation to food, territorialism will be especially marked.

W. J. Smith (personal communication) writes that certain Tyrannid, Cotingid and other species in Panama are primarily flycatching insectivores but may also take fruit. They are normally territorial but different species may breed together colonially, in sites, many of which, are clearly protective against predators (thorn bushes, ant nests, wasps, etc.). In addition some participants chase aerial predators in concerted attacks of considerable ferocity thus, not only increasing the security of their own nests, but those of other species present as well. The species seem to exploit the local environment for food in different ways so that competition between them may be markedly reduced or absent. It seems then that these birds are inter-specifically colonial when breeding but remain intra-specifically territorial, probably due to competition between conspecies for a limited local food supply, or the need for individual hunting within the preferred food-space or both.

Moynihan (1963) has described another complex inter-specific social organization among certain Andean Thraupidae. Several closely related forms appear to exploit almost identical niches in forest and show mutual avoidance at food sites rather than the inter-specific territorialism (Simmons, 1951) that might have been expected. While a similar relationship has been suggested between the weavers *Ploceus melanocephalus* and *P. jackson* (Crook, 1964), in both cases further work is needed to substantiate the explanations given. Inter-specific colonialism is not uncommon among birds normally gregarious (e.g. terns, herons, ibises, etc.). Among weavers competition for food appears to be absent in most cases (White, 1951) but may occur in mixed colonies of *P. cucullatus* and *P. nigerrimus* in Africa and *P. bengalensis* and *P. manyar* in India (Crook, 1963 b). *Ploceus megarhynchus* nests commonly in colonies in trees in which a drongo has also built. The aggressive behaviour of the drongo is undoubtedly protective and furthermore these weavers behave likewise to approaching " predators " including man, a behaviour they may have learnt from the drongos. The relative ages of birds in the nests suggests that the weavers seek the drongos and not vice versa (Ali & Crook, 1959).

Interim conclusions on the adaptiveness of dispersion types

(i) Relatively solitary breeding and/or territorialism occur where the nest sites are easily accessible to predators ; the nests are thus protected by being cryptic and dispersed. Overdispersion within the habitat furthermore permits individuals to feed near their nests thus obviating the need for travel and search at a distance. This may be of especial significance during the breeding season in areas poor in food wherein socially breeding birds would rapidly exhaust supplies in the colonial area.

In addition it allows individuals to exploit certain food sources without social interference detrimental to success (N.B. especially certain insectivores).

Dispersed territories provide other advantages, such as non-interference from conspecies in mating and nest building, reduction of spread of diseases, etc. (Hinde, 1956) but as these also apply to territories in most colonies they cannot be significant in determining contrasts in population dispersion. Learning of the local topography occurs just as well in flock movements around a colony as it does in individual movements in a large territory.

(ii) Group foraging develops in insectivorous birds that are primarily territorial so long as the group integration remains loose enough to permit individual food acquisition. It then probably increases both the food-finding frequency of each participant and confers an increased awareness of the approach of predators. Many species joining such flocks have relatively broad diets and the behaviour may be especially advantageous in relatively poor feeding conditions. In better conditions non-breeding territorialism is probably preferable in that the birds are then in possession of a territory at the start of the breeding season.

(iii) Where food occurs in patches of local abundance, foraging in tightly integrated flocks allows improved efficiency in food finding for individuals within them, so long as the manner of exploitation permits close congregation. Flocking is also associated with conditions imposing migration and nomadism.

(iv) Gregarious birds ((iii) above) become territorial for breeding, primarily when the preferred nest sites are relatively accessible to predators and dispersion and crypticity thus confer some measure of protection. In addition, shifts in diet or the special food requirements of the young may necessitate condition (i). Thus, while some species may forage in flocks away from their territories while breeding (e.g. *Euplectes* spp.) others will not do so (*Fringilla coelebs, Parus major*).

(v) Gregarious birds breed in colonies when nesting sites are relatively inaccessible to predators and large enough to contain numerous nests.

There must be a food supply around the colony sufficient to enable at least some recruitment to the population to occur.

(vi) Complex inter-specific social organizations may develop firstly, when a relative absence of food competition between species allows the formation of mixed-species colonies in protective sites, secondly, in inter-specific territorialism and, thirdly, possibly, when selection favours an avoidance of territorial combat, through the development of tolerance behaviour of an inhibitive kind.

(vii) The population dispersion of a species is thus considered a function of (a) the preferred food, its abundance and manner of distribution in the environment and (b) nest site selection in relation to predator approach. The overall density of a species depends primarily upon the abundance of food exploited. Seasonal fluctuations in supply necessitate shifts in patterns of population dispersion depending upon a balance of advantages in relation to several factors. Contrasts in the nature of food and its dispersion likewise correlate with differences in the dispersion and foraging behaviour of the exploiting populations. The protective qualities of nest sites differ according to whether predators are denied access by inaccessibility or by crypticity. These likewise impose differences in population dispersion. Social organization thus expresses the complex evolutionary response of a species to numerous variables of the environment.

Communal social organizations and cooperation

Communal social organizations crop up sporadically in the class in two main lines of evolutionary development ; firstly in cases where close congregation into colonies has led to cooperative nest building, but in which individual territorial defence of nest chambers remains undiminished and, secondly, in cases where several individuals participate together in reproductive activities with an absence of territorial antagonism. In some of these latter cases group defence of a territory around the communal breeding site has been recorded. In addition to these examples, cooperation has developed in certain non-reproductive situations—for example in the cooperative fishing of cormorants and pelicans (Bartholomew, 1942).

Communal " lodges " with individually defended nest chambers

In two Ploceid species vast structures composed of grass and/or spiny twigs are built in thorn trees in African savannah. *Philetairus socius* (Friedmann, 1950) is monogamous but *Bubalornis albirostris* (Crook, 1958) is polygamous, the males defending the greater part of the large lodges at which the harem assembles and of which several occur together in the same tree. In both cases the nest holes are lined and defended individually, while the bulky " lodge " is added to by the same group as a whole. This is done especially on the upper sides away from the nest entrances. In the dry season *Bubalornis* lodges look dilapidated although the birds roost in them. Material is added haphazardly to the roof, territories are no longer defined and the birds visit several different nest holes before settling for the night. The continued maintenance is presumably dependent on low level nest building carried out by the group throughout the dry weather.

Other lodge builders are a parakeet (*Myopsitta monarchus*) of Argentina (Friedmann, 1935), the palm chat (*Dulus dominicus*) of Hispaniola (Wetmore & Swales, 1931), an oven bird, Furnariidae (Austin, 1962), building a columnar lodge containing about five separate compartments placed around a branch high above ground level, and a Ploceine weaver *Malimbus rubricollis* (Crook, 1964). Little is known of any of these species but, in all cases, the basic behavioural shift seems to have been a major decrease in territory size permitting neighbours to build on parts of each other's nests. As this strengthens the fused structure as a whole, it has survival value for each participant. In addition, the increased congregation into protective colonial sites permits a larger population to utilize them, probably benefitting also from the mutual stimulation they receive there (see p. 190). Fusion of densely crowded nests, by chance building on the house next door, has been reported in certain weaver species which otherwise do not show cooperative behaviour. Furthermore *Dinemellia*, a close relative of *Bubalornis*, nests in pairs, building structures of a form resembling a single *Bubalornis* nest chamber. These occur separately or, occasionally, slightly fused within the same tree (Crook, unpublished work).

Communal reproduction with loss of pair territoriality

Communalism of this type varies greatly in the extent to which otherwise individual functions are performed by a group. Cases occur sporadically throughout the class.

In two authenticated cases communalism is associated with defence of a group breeding territory. Among the Crotophaginae, Davis (1942) described the ani (*Crotophaga ani*) living in groups of fifteen to twenty-five birds and defending an area around the communal nest in which several females lay eggs. Young birds remain with the flock and may assist in rearing the next brood, they may also remain to breed within the colony in the following season. While individuals may leave a group they can only join another after a period of fighting. In other species of false cuckoo less advanced types of communalism are known. The Australian magpie (*Gymnorhina tibicen*, Cracticidae) (Carrick, 1963 and in preparation) holds group territories throughout the year and fights in teams, cock with cock and hen with hen. Intense competition occurs between these groups (up to eleven birds) for the occupation of territories in the best feeding areas. The least successful teams fail to breed.

Loss of pair territorialism and the development of mutual care of nests and young occurs to various extents in other species : i.e. laying of eggs in the same nest by several females, ostriches, rheas, megapodes ; additional adults participating in incubation and care of young, Mexican jay (Brown, 1963), Mexican tanager (Snow & Collins, 1962), acorn woodpecker (Leach, 1925 ; Ritter, 1938 quoted in Wynne-Edwards, 1962), several Timaliid species (Marquis Yamashina, 1938 ; Friedmann, 1935 ; MacDonald, 1959), helmet shrikes (Prionopinae, i.e. Mackworth-Praed & Grant, 1955), two species of Australian Graliniidae, a broadbill (Eurylaimidae) in Thailand, and perhaps the *Hypocolius* (Bombycillidae) of Iraq ; none of which has been adequately studied. In several cases (and also in certain species without communal breeding, Skutch, 1935) the assistance of juveniles in rearing young in a succeeding brood, or in the following year and, even their remaining to breed with the adult group in subsequent years, has been reported. All these birds live in small groups, sometimes called " sisterhoods ", and both contact behaviour and allopreening are often pronounced. Flocking of this type may occur, however, without communal breeding, as for example in the Coliidae and Phoeniculinae.

Davis (1942) discusses the conditions under which communalism in the Crotophaginae may have arisen. Three situations occurring together seem to have been operative. Firstly a weakening of pair territorialism permitting the group to remain together, secondly a reduction in sexual fighting permitting looser bonding and a tendency towards promiscuity attributed to colonial living and thirdly, the fragmented nature of the preferred habitat (small copses) necessitating the clumping of individuals in small groups. He believes that communalism in these birds is an offshoot of an evolutionary trend towards cuculine parasitism.

All cases discussed (except Ratites and Megapodes) have certain features in common. The birds occur in small groups with reduced mutual antagonism and some development of cooperation in reproductive activities. In several the habitat tends to be fragmented. Finally the young are often associated with the parents during breeding. All differ greatly, however, from the communal " lodge " builders which move in much larger flocks, whose pair or harem territories are defended with vigour and among which antagonism remains much in evidence. Furthermore, apart from nest building, there is no cooperation among them. There is thus not necessarily a correlation between colony life and cooperative behaviour, and the two types of community are likely to have had entirely distinct origins.

Snow & Collins (1962) point out that mutual assistance, in combination with long breeding seasons and a spread of breeding by individual pairs, may be of great survival value in areas of relatively short food supply. Similarly assistance given by juveniles is likely to increase the chances of survival of young in the nest under such conditions. It seems possible that the marked absence of aggression in these groups is an adaptation enabling assistance to be given in rearing young under difficult circumstances of food finding. This could arise, moreover, without undue difficulty if the breeding groups were in fact " clans " in which dispersal of the young had been reduced in favour of their participation in communal activities. Marked mutual habituation, contact behaviour and the establishment of inter-individual relationships would favour social facilitation of activities and group antagonism to strangers leading to group territorialism. Clearly longitudinal studies of family relations and dispersal in species with this breeding system will be essential before the problem is clarified.

Arena behaviour

Arena behaviour is recorded primarily among ground birds of open country and woodland, many of them vegetarian—omnivores, and arboreal fruit or berry eaters of forests. The birds involved, descriptive accounts and discussion are provided by Armstrong (1947, 1964) and Gilliard (1963) who mention species in twelve families, i.e. Scolopacidae, Otidae, Tetraonidae, Icteridae, Ploceidae, all more or less open country ; Menuridae, Phasianidae, Ptilonorhynchidae, ground birds of woodland ; Pipridae, Cotingidae, Paradisaeidae and Trochilidae of forests.

Both authors include under this heading not only typical lekking birds in which the males display in company, but also species with similar habits in which the males are much more dispersed although still within hearing distance of one another. Gilliard calls the latter " exploded arenas ". The justification for this explanatory term requires further examination as the field data on the majority of these species remain far from complete.

While it seems that the " court " is an extremely small territory and the " arena " a tightly compressed " neighbourhood ", the behaviour is perhaps more likely to have developed from a form of pre-breeding group display, such as occurs on the " assembly " grounds of gulls or lapwings. Males with enhanced signal characteristics achieve more matings than others, but the trend towards promiscuity can only give rise to arena behaviour if the female can rear a brood by herself, thereby enabling the male to spend so much time on the court. Furthermore, since elaborate display is likely to attract predators, it is advantageous for females to nest away from the display grounds. So long as females can rear some young alone, any promiscuous male which mates with more females than his rivals must, in a long breeding season, produce more offspring. In bowerbirds the provision of an architectural, rather than a plumage stimulus, probably emerged from the male's performance of vestigial nest building at the court, coupled with the tendency to enhance visibility at the site by clearing it. Once sexual selection of the architectural stimulus began, selection against predation reduced the brilliance of the male's dress until it resembled that of the female. Gilliard (1963) notes that certain bowerbird males still perform head movements in display typical of related crested species, although they have long ago lost their colourful plumes. Selection against hybridization, or for display contrast, may explain the variety of display and plumage among related sympatric arena-birds (Paradisaeidae for example) but not the origin of the complexity itself. The displays of the forest birds in particular are distinguished by their bizarre nature. They are presumably derived from " territorial " posturings through intra-sexual selection at defended display sites and which have become functional in attracting females and thus further modified by inter-sexual selection. The low light values in the forest then ensure the development of much enhanced signals providing good communication.

Among the forest birds Snow (1962 a, 1962 b) has studied two *Manacus* species (Pipridae). Here arena behaviour is associated with a small clutch size, the female rearing the young in the nest alone. The birds furthermore have a low reproductive success due to high predation, probably mainly by snakes. Snow agrees with Skutch (1949) that the low clutch size of these birds is advantageous, in that the nests require visiting less frequently and that they are therefore less conspicuous to predators. If this were so and the frequency of nest visiting was dependent on the number of mouths requiring food, one would expect the greatest mortality to occur with young in the nest. However Snow's results with *Manacus manacus* reveal the heaviest mortality to occur before hatching. This suggests that frequency of nest visiting may not be vitally important and the low clutch size may thus not be a function of high predation.

The breeding peak in *Manacus manacus* follows the rains, suggesting that it coincides with a flush of insect food upon which the young are fed. It follows the peak in the variety of fruits utilized by adult birds. Thus, irrespective of whether food supplies are super-abundant or not, the timing of the occurrence of young in the nest appears to coincide with optimum food availability for young rather than parents. Even so it is improbable that in a tropical forest environment, where fluctuations in food supply are in general less than in temperate zones, the females working alone would be capable of rearing more than two young. As Snow points out, arena birds of forests are all primarily fruit eaters (manakins, cotingas, birds of paradise), and arena behaviour rarely occurs among insectivores. The relative ease with which adult males can stuff themselves with berries may be a major factor permitting long hours in the court. Furthermore, the absence of the male from the female's foraging ground might be said to conserve her food supplies, but not that of the young, since she is feeding them on insects—possibly in competition with other insectivores in the habitat. The low clutch size is thus probably linked to the development of the male's role in arena display and his consequent absence from parental care.

The above account suggests that arena displays arise through sexual selection but only under ecologically permissive circumstances enabling the female to rear a brood without male participation in either nest building, territorial defence or care (see also p. 187, discussion of polygamy).

Divergence in social organization within families

In Table 1 some twenty-seven families show marked divergence in social organization within the taxon. These families are particularly interesting since the systematic relations of the species are either well known from morphological and ethological studies, or are at least wide open to future research. Furthermore, comparisons between species allow a greater precision than a general survey.

Few families have been studied intensively in the terms of this discussion. The studies of Tinbergen (1959 a) and co-workers E. Cullen (1957), J. M. Cullen (1960 a, 1960 b) and Manley (1960) on the Laridae, Crook (1964) on Ploceinae, Immelmann (1960, 1962 a, 1962 b) on Australian grassfinches and Meliphagidae, Stonehouse (1953, 1960), Sladen (1958) and other workers on penguins, are the most comprehensive and were the inspiration for the present review. Studies of other families remain fragmentary but, even so, the picture suggests relationships similar to those discussed above with variant themes expected. The contrasts between the gregarious, insectivorous hawks (both the Accepitridae and Falconidae) and their flesh-eating relatives, has received little attention, nor has anyone attempted a survey of colonial life in Sturnidae. Studies of the Timaliidae and Laniidae should prove especially valuable in yielding further information on communal behaviour. The same applies to families in which arena behaviour occurs. Thus while red grouse and ptarmigan are monogamous and occupy territories in the normal way with the males playing a role in parental care, the black cock, sage grouse and prairie chicken males remain at the " lek " and fail to occupy territories (Scott, 1950). Jackson's whydah (Ploceinae—van Someren, 1945), Pipridae, Menuridae, Paradisiaedae, etc., all require similar attention.

In several families isolated studies require collating into systematic surveys. This would be especially promising in the Icteridae, which parallel in a remarkable way many of the social organizations shown by the Ploceidae. In spite of the many individual studies (i.e. Chapman, 1928 ; Nero, 1956 ; Selander & Giller, 1961 ; Orians, 1961 b) they remain a rich mine for future work. Likewise comparison between different species of sparrows (*Passer*) i.e. Summers-Smith (1963), Deckert (1962), Kunkel (1959) and Gabrilov (1962), suggests a similar, though more closely-knit, radiation here. Immelmann's (1962 b) Australian studies need expanding to include the rich Estrildine fauna of Africa where both solitary and flocking species occur in contrasting environments. These last species furthermore offer material for a comparative field study of " contact " behaviour. Certain well known and accessible groups still lack adequate comparative treatment in the field. Thus the following studies among others are an open invitation to further research : Johnsgard (1962, and quoted) on the Anatidae, Emlen (1952 b) on Hirundinidae, Lorenz (1938) on Corvidae, Dilger (1960) on Psittacidae, Meyerriecks (1960) on certain Ardeidae, Ficken & Ficken (1962) with Parulid warblers, Dilger (1960) on the *Agopornis* species, Snow *et al.* (1963) on Turdidae and Hinde (1955–56) on Fringillidae. These are long term projects of the utmost fascination for the future.

This paper has stressed the close relationship between studies of population dynamics and ethology in understanding the survival values of contrasting types of social behaviour. The same will be true in future work and team studies in which both aspects are treated together (as in the recent B.O.U. Centenary expedition to Ascension Island) will provide the most meaningful information in this quest. Furthermore, in order to unravel the complex ontogeny of communal life, techniques of longitudinal study will be necessary.

SUMMARY

The main findings of a survey of social organizations in relation to ecological variables are presented on p. 205. Communal behaviour involving cooperation in reproductive activities occurs sporadically in the class and appears to have evolved under two types of circumstance :

(*a*) Where colonialism has led to a marked reduction in territory size consequent upon the advantages of crowding as many birds into protected sites as possible. Under these circumstances individuals may extend nest building on to neighbouring nests so long as territorial infringement of the nest entrance does not occur. Where mutual construction on fused nests strengthens the structure as a whole, and/or renders it even less accessible to predators, selection will rapidly encourage the development of " lodges ". In these cases individual territorialism and agonistic behaviour are not lost and cooperation is limited to building.

(*b*) Where extension of parent–child bonding has led to the participation of the latter in brood care in a succeeding generation and such behaviour improves the chances of survival of the chicks, selection may favour an increase in mutual participation, especially if the breeding season is long and the breeding is staggered. Group participation in incubation and chick-care may occur at single nests or at groups. Such communal bird parties are probably " clans ". In a few cases group territorialism has developed.

Arena behaviour occurs where reproductive advantage falls to those cocks providing the most " attractive " displays over the longest period. The emergence of the typical behaviour depends on, firstly, the presence of territorial/courtship display in social groups not connected with nest site selection, secondly the performance of nest construction and incubation by the females, thirdly, loose pair bonding permitting polygamy and promiscuity, fourthly, food availability such that females working alone can rear a brood (possibly of diminished size) and which enables the male to spend little time in feeding. Lastly the remarkable display structures of the males renders them a risk at the cryptic nests thereby favouring their separation from them. Among bowerbirds a transfer of decoration from the male plumage to the bower has occurred. Arena behaviour is thus a function of inter- and intra-sexual selection under ecologically permissive conditions. This sequence of evolutionary events parallels closely those suggested for the emergence of polygamy in certain other groups (p. 188). Comparisons between species in families within which social organization has diverged, support the above arguments with a wealth of detail.

$$* \quad * \quad * \quad * \quad * \quad * \quad *$$

REFERENCES

ALI, S. (1931). The nesting habits of the baya (*Ploceus philippinus*). *J. Bombay nat. Hist. Soc.* **34** : 947–964.

ALI, S. & CROOK, J. H. (1959). Observations on Finn's baya (*Ploceus megarhynchus* Hume) re-discovered in the Kumaon terai, 1959. *J. Bombay nat. Hist. Soc.* **56** : 457–483.

ALLEE, W. C. (1931). *Animal aggregations.* Chicago.

ALLEE, W. C., EMERSON, A. E., PARK, O., PARK, T. & SCHMIDT, K. P. (1949). *Principles of animal ecology.* Philadelphia, London.

AMBEDKAR, V. C. (1961). The ecology and breeding biology of the Indian weaver birds with special reference to the baya weaver bird (*Ploceus philippinus* (Linn.)). M.Sc. thesis, Bombay University.

ARMSTRONG, E. A. (1947). *Bird display and behaviour.* London.

ARMSTRONG, E. A. (1955). *The wren.* London : Collins.

ARMSTRONG, E. A. (1964). *Lek display.* In Thomson, A. L. (ed.). *New dict. birds.* London & New York.

ASHMOLE, N. P. (1962). The black noddy *Anous tenuirostris* on Ascension Island. Part 1. General biology. *Ibis* **103b** : 235–273.

ASHMOLE, N. P. (1963). The regulation of numbers of tropical oceanic birds. *Ibis* **103b** : 458–473.

AUSTIN, O. L. (1962). *Birds of the world.* London.

BARTHOLOMEW, G. A. A. (1942). The fishing activities of double crested cormorants on San Francisco Bay. *Condor* **44** : 13–21.

BENDELL, J. F. (1959). Food as a control of a population of white-footed mice, *Peromyscus leucopus noveboracensis* (Fischer). *Canad. J. Zool.* **37** : 173–209.

BOYD, H. (1953). On encounters between wild White-fronted geese in winter flocks. *Behaviour* **5** : 85–129.

BROWN, J. L. (1963). Social organization and behaviour of the Mexican jay. *Condor* **65** : 126–153.

BROWN, J. L. (1964). The evolution of diversity in avian territorial systems. *Wilson Bull.* **76** : 160–169.

BROWN, L. H. (1958). The breeding of the greater flamingo *Phoenicopterus ruber* at Lake Elmenteita, Kenya Colony. *Ibis* **100** : 388–420.

BURNS, P. S. (1957). Rook and jackdaw roosts around Bishop's Stortford. *Bird Study* **4** : 62–71.

CARRICK, R. (1963). Ecological significance of territory in the Australian magpie, *Gymnorhina tibicen.* *Proc. 13th int. orn. Congr.* : 740–753.

CHAPMAN, F. M. (1928). The nesting habits of Wagler's oropendula (*Zarhynchus wagleri*) on Barro Colorado Island. *Bull. Amer. Mus. nat. Hist.* **58** : 123–166.

COLLIAS, N. E. & COLLIAS, E. C. (1959). Breeding behaviour of the black-headed weaverbird *Textor cucullatus graueri* (Hartert) in the Belgian Congo. *Ostrich Suppl.* no. 3 (*Proc. 1st Pan-Afr. orn. Congr.*) 1959 : 233–241.

CONDER, P. (1949). Individual distance. *Ibis* **91** : 649–655.

COOMBS, C. J. F. (1960). Observations on the rook *Corvus frugilegus* in southwest Cornwall. *Ibis* **102** : 394–419.

COOMBS, C. J. F. (1961). Rookeries and roosts of the rook and jackdaw in southwest Cornwall. *Bird Study* **8** : 32–37, 55–70.

CROOK, J. H. (1958). Études sur le comportement social de *Bubalornis a. albirostris* (Vieillot). *Alauda* **26** : 161–195.

CROOK, J. H. (1960 a). Studies on the social behaviour of *Quelea q. quelea* (Linn.) in French West Africa. *Behaviour* **16** : 1–55.

CROOK, J. H. (1960 b). Studies on the reproductive behaviour of the baya weaver (*Ploceus philippinus* (L.)). *J. Bombay nat. Hist. Soc.* **57** : 1–44.

CROOK, J. H. (1961 a). The basis of flock organization in birds. In *Current problems in animal behaviour*, edited by W. H. Thorpe & O. L. Zangwill. Cambridge.

CROOK, J. H. (1961 b). The fodies of the Seychelles Islands. *Ibis* **103a** : 517–584.

CROOK, J. H. (1963 a). Comparative studies on the reproductive behaviour of two closely related weaver bird species (*Ploceus cucullatus* and *Ploceus nigerrimus*) and their races. *Behaviour* **21** : 177–232.

CROOK, J. H. (1963 b). The Asian weaver birds : problems of co-existence and evolution with special reference to behaviour. *J. Bombay nat. Hist. Soc.* **60** : 1–48.

CROOK, J. H. (1964). The evolution of social organization and visual communication in the weaver birds (Ploceinae). *Behaviour* suppl. no. 10.

CULLEN, E. (1957). Adaptations in the kittiwake to cliff-nesting. *Ibis* **99** : 275–302.

CULLEN, J. M. (1960 a). Some adaptations in the nesting behaviour of terns. *Proc. 12th int. orn. Congr. Helsinki* 1958 : 153–157.

CULLEN, J. M. (1960 b). The aerial display of the Arctic tern and other species. *Ardea* **48** : 1–37.

CULLEN, J. M. (1963). Allo-, auto- and hetero-preening. *Ibis* **105** : 121.

DARLING, F. (1938). *Bird flocks and the breeding cycle. A contribution to the study of avian sociality.* Cambridge.

DAVIES, S. J. J. F. (1963). Aspects of the behaviour of the magpie goose, *Anseranas semipalmata. Ibis* **105** : 76–97.

DAVIS, D. E. (1942). The phylogeny of social nesting habits in the Crotophaginae. *Quart. Rev. Biol.* **17** : 115–134.

DECKERT, G. (1962). Zur Ethologie des Feldsperlings (*Passer m. montanus* L.). *J. Orn.* **103** : 428–486.

DILGER, W. C. (1960). The comparative ethology of the African parrot genus *Agapornis*. *Z. Tierpsychol.* **17** : 649–685.

DISNEY, H. J. DE S. & HAYLOCK, J. W. (1956). The distribution and breeding behaviour of the Sudan dioch (*Quelea q. aethiopica*) in Tanganyika. *E. Afr. agric. J.* **21** : 141–156.

DURANGO, S. (1956). Territory in the red-backed shrike *Lanius collurio*. *Ibis* **98** : 476–484.

EMERSON, A. E. (1960). The evolution of adaptation in population systems. In *The evolution of life*, **I**, edited by Sol Tax. Chicago : University Press.

EMLEN, J. T. (1952 a). Flocking behaviour in birds. *Auk* **69** : 160–170.

EMLEN, J. T. (1952 b). Social behaviour in nesting cliff swallows. *Condor* **54** : 177–199.

EMLEN, J. T. (1957). Display and mate selection in the whydahs and bishop birds. *Ostrich* **28** : 203–213.

EMLEN, J. T. & LORENZ, F. W. (1942). Pairing responses of free living valley quail to sex hormone pellet implants. *Auk* **59** : 369–378.

FICKEN, M. S. & FICKEN, R. W. (1962). The comparative ethology of the wood warblers : a review. In *The living bird*. Cornell : Laboratory of Ornithology, 103–122.

FISHER, J. (1954). Evolution and bird sociality. In *Evolution as a process*. London.

FRIEDMANN, H. (1935). Bird societies. In *Handbook of social psychology*, edited by Carl Murchison. Clark University.

FRIEDMANN, H. (1950). The breeding habits of the weaver birds. A study in the biology of behaviour patterns. *Ann. Rep. Smithson. Inst.* **1949** : 293–316.

GAVRILOV, E. I. (1962). Biology of the Spanish sparrow (*Passer hispaniolensis* Temm.) and their control in Kazakstan. (In Russian.) *Trud. Inst. Zashch. Rast., Alma Ata* **7** : 459–528.

GIBB, J. A. (1960). Populations of tits and goldcrests and their food supply in pine plantations. *Ibis* **102** : 163–208.

GIBB, J. A. (1962). L. Tinbergen's hypothesis of the rôle of specific search images. *Ibis* **104** : 106–111.

GILLIARD (1963). The evolution of bowerbirds. *Sci. Amer.* August : 38–46.

GLAS, P. (1960). Factors governing density in the chaffinch (*Fringilla coelebs*) in different types of wood. *Arch. néerl. Zool.* **13** : 466–472.

HEDIGER, H. (1955). *Wild animals in captivity*. London.

HINDE, R. A. (1952). The behaviour of the great tit (*Parus major*) and other related species. *Behaviour* suppl. no. 2.

HINDE, R. A. (1955–56). A comparative study of the courtship of certain finches (Fringillidae). *Ibis* **97** : 706–745 ; **98** : 1–23.

HINDE, R. A. (1956). The biological significance of the territories of birds. *Ibis* **98** : 340–369.

HINDE, R. A. & TINBERGEN, N. (1958). The comparative study of species-specific behaviour. In *Behaviour and evolution*, edited by A. Roe & G. G. Simpson. Yale.

HUNTINGTON, C. E. (1963). Population dynamics of Leach's petrel *Oceanodroma leucorhoa*. *Proc. 13th int. orn. Congr.* : 701–705.

HUXLEY, J. S. (1938 a). The present standing of the theory of sexual selection. In *Evolution—essays presented to Professor E. S. Goodrich*, edited by G. R. de Beer. Oxford.

HUXLEY, J. S. (1938 b). Threat and warning coloration in birds. *Int. orn. Congr.*, **1934** : 430–455.

IMMELMANN, K. (1960). Contributions to the biology and ethology of the red-eared firetail (*Zonaeginthus oculatus*). *W. Aust. Nat.* **7** : 142–160.

IMMELMANN, K. (1962 a). Beiträge zur Biologie und Ethologie australischer Honigfresser (Meliphagidae). *J. Orn.* **102** : 164–207.

IMMELMANN, K. (1962 b). Beiträge zu einer vergleichenden Biologie australischer Prachtfinken (Spermestidae). *Zool. Jb. (Syst.)* **90** : 1–196.

IMMELMANN, K. (1963). Tierische Jahresperiodik in ökologischer Sicht. *Zool. Jb. (Syst.)* **91** : 91–200.

JENKENS, D. (1963). Population control in the red grouse (*Lagopus l. scoticus*). *Proc. 13th int. orn. Congr.* : 690–700.

JENKENS, D., WATSON, A. & MILLER, G. R. (1963). Population studies on red grouse *Lagopus l. scoticus* (Lath.) in N.E. Scotland. *J. anim. Ecol.* **32** : 317–376.

JOHNSGARD, P. A. (1962). Evolutionary trends in the behaviour and morphology of the Anatidae. *Rep. Wildf. Tr.* **1960–61** : 130–148.

KIKKAWA, J. (1961). Social behaviour of the white-eye *Zosterops lateralis* in winter flocks. *Ibis* **103a** : 428–442.

KLUIJVER, H. N. (1957). Roosting habits, sexual dominance and survival in the great tit. *Cold Spring Harbor Symp. on Quant. Biol.* **22** : 281–285.

KLUIJVER, H. N. & TINBERGEN, L. (1953). Territory and the regulation of density in titmice. *Arch. néerl. Zool.* **10** : 266–287.

KUNKEL, P. (1959). Allgemeines and und soziales Verhalten des Braunrucken-Goldsperlings (*Passer* (*Auripasser*) *luteus* Licht.). *Z. Tierpsychol.* **18** : 471–489.

LACK, D. (1940). *Pair formation in birds. Condor* **62** : 269–286.

LACK, D. (1943). *The life of the robin.* London.

LACK, D. (1954). *The natural regulation of animal numbers.* Oxford.

LEACH, F. A. (1925). Communism in the California woodpecker. *Condor* **27** : 12–19.

LEHRMANN, D. S. (1958). Induction of broodiness by participation in courtship and nest building in the Ring dove (*Streptopelia risoria*). *J. comp. phys. Psychol.* **51** : 32–36.

LOCKIE, J. D. (1956). Winter fighting in feeding flocks of rooks, jackdaws and carrion crows. *Bird Study* **3** : 180–190.

LORENZ, K. (1935). Der Kumpan in der Umwelt des Vogels. *J. Orn.* **83** : 137–213, 289–413.

LORENZ, K. (1938). A contribution to the comparative sociology of colonial nesting birds. *Proc. 8th int. orn. Congr.* **1934** : 207–218.

MACARTHUR, R. H. (1955). Fluctuations of animal population and a measure of community stability. *Ecology* **36** : 533–536.

MACDONALD, M. (1959). Communal nesting feeding in babblers. *J. Bombay nat. Hist. Soc.* **56** : 132–134.

MACKWORTH-PRAED, C. W. & GRANT, C. H. B. (1955). *Birds of eastern and north-eastern Africa* **2**. London : Longmans, Green & Co.

MANLEY, G. H. (1960). The swoop and soar performance of the black-headed gull, *Larus ridibundus* L. *Ardea* **48** : 37–51.

MARLER, P. (1956 a). Territory and individual distance in the chaffinch *Fringilla coelebs*. *Ibis* **98** : 496–501.

MARLER, P. (1956 b). Studies of fighting in chaffinches (3). Proximity as a cause of aggression. *Brit. J. anim. Behav.* **5** : 23–30.

MARLER, P. (1956 c). Behaviour of the chaffinch (*Fringilla coelebs*). *Behaviour* suppl. no. 5.

MARSHALL, A. J. (1959). Internal and environmental control of breeding. *Ibis* **101** : 456–477.

MEYERRIECKS, A. J. (1960). Comparative breeding behavior of four species of North American herons. *Publ. Nuttall orn. Cl.* no. 2 : 1–158.

MILLER, A. H. (1947). Panmixia and population size with reference to birds. *Evolution* **1** : 186–190.

MOREL, G. & BOURLIÈRE, F. (1956). Recherches écologiques sur les *Quelea quelea quelea* (L.) de la basse vallée du Sénégal. II. La réproduction. *Alauda* **24** : 97–122.

MOREL, G. & MOREL, M. Y. (1961). Une ᵀᵀeronnière mixte sur le bas-Sénégal. *Alauda* **29** : 99–117.

MOYNIHAN, M. (1962). The organization and probable evolution of some mixed species flocks of neotropical birds. *Smithson. Misc. Coll.* **143** (7) : 1–140.

MOYNIHAN, M. (1963). Inter-specific relations between some Andean birds. *Ibis* **105** : 327–339.

MYLNE, C. K. (1963). Film on great crested grebes. *Ibis* **105** : 425.

NERO, R. W. (1956). A behavior study of the red-winged blackbird. I & II. *Wilson Bull.* **68** : 5–37, 129–150.

NERO, R. W., LAHRMAN, F. W. & BARD, F. G. (1958). Dry land nest site of Western grebe colony. *Auk* **75** : 347–349.

ORIANS, G. H. (1961 a). Social stimulation within blackbird colonies. *Condor* **63** : 330–337.

ORIANS, G. H. (1961 b). The ecology of blackbird (*Agelaius*) social systems. *Ecol. Monogr.* **31** : 285–312.

PERRINS, C. (1963). Survival in the great tit. *Proc. 13th int. orn. Congr.* : 717–728.

RICE, D. W. & KENYON, K. W. (1962). Breeding cycles and behaviour of Laysan and black-footed albatrosses. *Auk* **79** : 517–567.

RICHDALE, L. E. (1963). Biology of the sooty shearwater *Puffinus griseus*. *Proc. zool. Soc. Lond.* **141** : 1–117.

ROSENZWEIG, M. L. & MACARTHUR, R. H. (1963). Graphical representation and stability conditions of predator prey interactions. *Amer. Nat.* **97** : 209–224.

SABINE, W. S. (1959). The winter society of the Oregon junco : intolerance, dominance and the pecking order. *Condor* **61** : 110–135.

SALT, G. & HOLLICK, F. S. J. (1946). Studies of wireworm populations. II. Spatial distribution. *J. exp. Biol.* **23** : 1–46.

SCOTT, J. W. (1950). A study of phylogenetic or comparative behaviour of three species of grouse. *Ann. N.Y. Acad. Sci.* **51** : 1062–1073.

SELANDER, R. K. & GILLER, D. R. (1961). Analysis of sympatry of great-tailed and boat-tailed grackles. *Condor* **63** : 29–86.

SHORT, L. L. (1961). Interspecies flocking of birds of montane forest in Oaxaca, Mexico. *Wilson Bull.* **73** : 341–347.

SIMMONS, K. E. L. (1951). Interspecific territorialism. *Ibis* **93** : 407–413.

SIMMONS, K. E. L. (1956). Territory in the little ringed plover *Charadrius dubius*. *Ibis* **98** : 390–397.

SKEAD, C. J. (1956). A study of the red bishop-bird, *Euplectes oryx*. *Ostrich* **27** : 112–126.

SKEAD, C. J. (1959). A study of the red-shouldered widow-bird *Coliuspasser axillaris axillaris*. *Ostrich* **30** : 13–21.

SKUTCH, A. (1935). Helpers at the nest. *Auk* **52** : 257–273.

SKUTCH, A. (1949). Do tropical birds rear as many birds as they can nourish ? *Ibis* **91** : 430–455.

SLADEN, W. J. L. (1958). The Pygoscelid penguins. *Sci. Rep. Falkland Is. Depend. Surv.* no. 17 : 1–97.

SLOBODKIN, L. B. (1961). *Growth and regulation of animal populations*. New York.

SNOW, D. (1956). Territory in the blackbird *Turdus merula*. *Ibis* **98** : 438–447.

SNOW, D. W. (1962 a). A field study of the black and white manakin, *Manacus manacus*, in Trinidad. *Zoologica, N.Y.* **47** : 65–104.

SNOW, D. W. (1962 b). A field study of the golden headed manakin, *Pipra erythrocephala* in Trinidad. *Zoologica, N.Y.* **47** : 183–198.

SNOW, D. & COLLINS, C. T. (1962). Social breeding behaviour of the Mexican tanager. *Condor* **64** : 161.

SNOW, D. W. & SNOW, B. K. (1963). Breeding and the annual cycle in three Trinidad thrushes. *Wilson Bull.* **75** : 27–41.

SOMEREN, V. D. VAN (1945). The dancing display and courtship of Jacksons' whydah (*Coliuspasser jacksoni* Sharpe). *J. E. Afr. nat. Hist. Soc.* **18** : 131–141.

SOUTHERN, H. N. (1959). Mortality and population control. *Ibis* **101** : 429–436.

SPARKS, J. H. (1963). Social structure of the red avadavat (*Amandava amandava*) with particular reference to clumping and allopreening. *Anim. Behav.* **11** : 407.

STEWART, R. E. & ALDRICH, J. W. (1951). Removal and repopulation of breeding birds in a spruce-fir forest community. *Auk* **68** : 471–482.

STONEHOUSE, B. (1953). The emperor penguin *Aptenodytes foresteri* (Grey). I. Breeding behaviour and development. *Sci. Rep. Falkland Is. Depend. Surv.* no. 6.

STONEHOUSE, B. (1960). The king penguin *Aptenodytes patagonica* of South Georgia. I. Breeding behaviour and development. *Sci. Rep. Falkland Is. Depend. Surv.* no. 23 : 1–181.

SUMMERS-SMITH, D. (1963). *The house-sparrow*. London : Collins.

TINBERGEN, L. (1960). The dynamics of insect and bird populations in pine woods. *Arch. néerl. Zool.* **13** : 259–473.

TINBERGEN, N. (1939). The behaviour of the snow bunting in spring. *Trans. Linn. Soc. N.Y.* **5**.

TINBERGEN, N. (1952). On the significance of territory in the herring gull. *Ibis* **94** : 158–159.

TINBERGEN, N. (1953). *The herring gull's world*. London.

TINBERGEN, N. (1956). On the functions of territory in gulls. *Ibis* **98** : 401–411.

TINBERGEN, N. (1957). The functions of territory. *Bird Study* **4** : 14–27.

TINBERGEN, N. (1959 a). Comparative studies of the behaviour of gulls (Laridae) : a progress report. *Behaviour* **15** : 1–70.

TINBERGEN, N. (1959 b). Behaviour, systematics and natural selection. *Ibis* **100** : 318–330.

TINBERGEN, N., BROEKHUYSEN, G. J., FEEKES, F., HOUGHTON, J. C. W., KRUUK, H. & SZULC, E. (1962). Egg shell removal by the black-headed gull, *Larus ridibundus* L. ; a behaviour component of camouflage. *Behaviour* **19** : 74–117.

VERNER, J. (1964). Evolution of polygamy in the long billed marsh wren. *Evolution* **18** : 252–261.

WETMORE, A. & SWALES, B. H. (1931). The birds of Haiti and the Dominican Republic. *U.S. Nat. Mus. Bull.* **155** : 345–352.

WHITE, C. M. N. (1951). Weaver birds at Lake Mweru. *Ibis* **93** : 626–627.

WRIGHT, S. (1960). Physiological genetics, ecology of populations and natural selection. In *The evolution of life*, **I**, edited by Sol Tax. Chicago : University Press.

WYNNE-EDWARDS, V. C. (1955). Low reproductive rates in birds, especially sea-birds. 11*th int. orn. Congr., Basel* **1954** : 540–547.

WYNNE-EDWARDS, V. C. (1962). *Animal dispersion in relation to social behaviour.* Edinburgh, London : Oliver & Boyd.

YAMASHINA, MARQUIS (1938). A sociable breeding habit among Timaliine birds. 9*th int. orn. Congr., Rouen* **1938** : 453–456.

23

Reprinted from *The Wilson Bull.*, **76**, 160–169 (1964)

THE EVOLUTION OF DIVERSITY IN AVIAN
TERRITORIAL SYSTEMS

Jerram L. Brown

WHAT are the conditions which facilitate or hinder the evolution of territoriality? No generally accepted solution to this problem has yet been found—perhaps because too specific an answer has been sought for too general a question. Instead, the *diversity* of systems of territorial and other aggressive behavior has come to be well appreciated, as evidenced in recent reviews of territoriality (e.g., Kuroda, 1960; Carpenter, 1958; Hinde, 1956), and the impossibility of providing a specific answer applicable to all types of territoriality is now realized.

Arguments over which are the primary selection pressures leading to certain types of territoriality continue, however, as shown in the recent contributions bearing on the "function" of territoriality by Stenger (1958), Wynne-Edwards (1962), Kalela (1958), Kuroda (1960), Peters (1962), and others.

The present paper offers a new orientation to the problem by presenting a general theory for the evolution of territoriality with special reference to its diversity among species. Since most of the previous theories have already been shown to be untenable or severely limited (see especially Carpenter, 1958; Tinbergen, 1957; and Hinde, 1956, for criticism of them), little attention will be given to them here.

GENERAL THEORY

A theoretical framework for the consideration of some of the mechanisms promoting and limiting the evolution of territorial behavior is outlined in Fig. 1.

Aggressive behavior is generally employed by individuals in the acquisition of goals which tend to *maximize individual survival and reproduction*. Natural selection should favor aggressive behavior within a population when these goals are consistently and easily accessible to individuals through aggression but should not favor it when they are not accessible. For example, when a food supply cannot be feasibly defended, because of its mobility or transient nature, generally no territorial system is evolved to defend it; and the territory, if present, may be restricted only to the nest and the area reachable by the parents on the nest. Such cases are found in colonial sea birds, nomadic and social feeding passerine species, and aerial feeders. In these species the goal of increased or guaranteed food supply is unlikely to be attained through aggression.

On the other hand, if the individual depends for its nesting requirements,

160

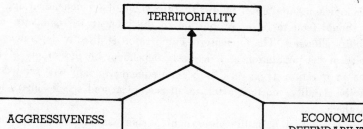

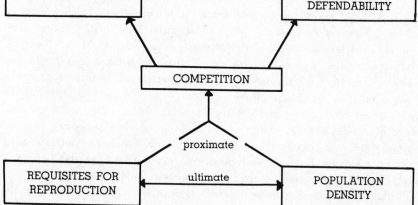

Fig. 1. A general theory of the evolution of diversity in avian intraspecific territorial systems.

food supply, and attraction of a mate on a relatively fixed and well-defined area, then this all-important area is typically defendable and becomes the classical territory. In short, *defendability* of the food supply, mate, mating place, nest, or other requisite for reproduction or survival is one of the most important determinants of the system of territorial behavior which is attained through natural selection. "Defendability" should be conceived in terms of the time and energy budgets of an individual as well as in purely physical terms.

Since intraspecific aggressiveness is primarily a behavioral response to competition for ecological requisites in short supply, *the predominant single factor tending to increase aggressiveness through natural selection should be competition*. Competition, as used in this discussion, may be said to exist when any ecological requisite exists in a quantity less than optimal for the total number of individuals which exploit it. Competition may exist for mates, food, roosting spots, breeding space, or any other necessity for reproduction in short supply. Competition is not necessarily expressed through aggression or threat but it frequently engenders such behavior.

On this logical assumption it follows that the value of site-dependent aggressiveness should tend to be in proportion to the intensity of competition—defendability allowing. The intensity of competition is directly dependent on the density of the population and inversely dependent on the supply of the requisites in question (Fig. 1). It is, consequently, complexly related to productivity, natality, mortality, and to all ecological and species characters affecting them.

Too much aggression in the absence of a short supply of the disputed requisite would eventually be detrimental. Consequently, a balance must be achieved between the positive values of acquired food, mate, nesting area, protection of family, etc., and the negative values of loss of time, energy, and opportunities, and risk of injury. Where this balance may lie in any particular species is influenced by a great variety of factors—to name a few: population density, physiological limitations and susceptibilities of the species, nest construction and site requirements, distance to food from nest, stage of development of young at birth, foraging time necessary to raise young, clutch size, time necessary to protect young, reaction of potential mate to too much or too little aggressiveness, conspicuousness to predators, migration, climate, weather, size of bird, and richness of food supply.

Within the population those *individuals* with the *optimal balance* of the genetic factors working for and against a particular form of aggressiveness would leave the most surviving and reproducing offspring; the type and degree of aggressiveness exhibited by these individuals would become, through natural selection, the norms for the population.

In short, it is argued that the type of territoriality evolved in a species depends on the types of requisites for which competition exists and upon the degree to which they are economically defendable in terms of balance between advantages and disadvantages of such defense to individuals (not the population). The problem for a particular species then becomes that of demonstrating which requisites are in short supply, which are not, and how it is economical for certain ones to be defended and not others.

APPLICATIONS OF THE THEORY

The general applicability of a theory based on competition and economic defendability to species exhibiting diverse types of territoriality may be illustrated with the following examples.

Colonial nesters.—A simple form of territoriality is exhibited by the Brandt's Cormorant (*Phalacrocorax penicillatus*), which was studied by Williams (1942). This species nests along the Pacific Coast of North America on islands and cliffs. At the start of the breeding season males begin giving an advertising display in a small area a little larger than the size of the future nest; copulation occurs at the nest. The territory consists of the nest and a barren area extending a few feet or more around it. It is used in the attraction of a mate, for copulation, and defense of the family. All food is obtained from the sea under conditions which make the defense of a feeding area completely impractical if not impossible. Consequently, no matter how intense competition for food might be, the evolution of a territory used for feeding would be blocked through lack of defendability. On the other hand, the small area used for mating and family defense is feasibly defendable, and competition for the often limited optimal nesting space probably intensifies the necessity of defense of the nesting territory in this species.

Leks.—For the special evolutionary problems offered by the lek type of social organization the Sage Grouse (*Centrocercus urophasianus*) serves as an example. The data below have been taken from the extensive study by Patterson (1952). Sage Grouse live for most of the year in loose social groups of predominantly one sex. At the start of the breeding season cocks defend small display territories in a communal display area. Within the group of displaying males are a few dominants, each surrounded by a few subordinate "guard cocks." The females come to the display ground for copulation, usually choosing a dominant male. Aggressiveness is important for a male to achieve a dominant position; fighting and birds with blood-stained plumage are commonly seen on the lek. Nesting is performed by the female alone, who generally chooses an area well away from the lek where a richer supply of food, water, and cover exists. After the last egg has hatched the chicks leave the nest and are led by the hen to areas of suitable food and cover sometimes as much as 460 yards away. In summer and fall males and hens which were unsuccessful nesters move to areas of richer food supply, either higher altitudes or crop lands (up to 1 to 5 miles away).

According to the theory outlined here, the form of territoriality evolved in a species is determined primarily by competition and defendability. It is necessary, therefore, to relate the lek system to the environmental requirements of the Sage Grouse and to determine those requisites for which competition does and does not exist and whether or not they are economically defendable. Food, in the opinion of Patterson, was not a limiting factor on his study areas. He wrote, ". . . environmental deficiencies in the form of food, cover, and water are believed to be practically non-existent as sage grouse decimating factors, once the breeding season has been inaugurated" (p. 139). Consequently, "there seems to be no competition between individuals for the essentials of daily survival such as food, cover, or water" (p. 176). Patterson estimated juvenile mortality as 95% of the total mortality for the population and considered that, "losses to natural enemies probably constitute the greatest source of juvenile mortality" (p. 139). Although a richer food supply might theoretically allow a higher population density and the occupancy of an increased area of suitable habitat, *for the individuals which are alive during the reproductive period*, food availability apparently does not limit reproduction. Consequently, competition for a food supply for the young in this species appears to be negligible, and any time or energy devoted to intra-specific defense of a food supply would be a net loss to the individuals concerned. The food supply may be considered as physically but not economically defendable under these conditions.

Furthermore, since protection against predators capitalizes on protective coloration and immobility of the precocious young, inconspicuousness of the family is necessary. Defense of an area around the nest would be detrimental by attracting predators, and the absence of the male from the nesting area is advantageous by decreasing conspicuousness of the family, and by reducing the potential prey population there (even if he were protectively colored). Furthermore, since the young do not have to be fed by the parents, the presence of the male is not necessary for that purpose.

Thus freed from the responsibilities of protection and care of nest and young, the males have full freedom of competition for the fertilization of females. To this end have evolved the elaborate and conspicuous plumage and display in the males and the lek system of mate selection. Once evolved, the lek system tends to perpetuate itself through the demonstrated preferential success of the dominant males within the lek (74% of 174 observed matings). Copulations at the periphery of the lek or outside of it are rare.

Summarizing, in the Sage Grouse although a food supply for the young might be physically defendable, it would not be economically defendable by the male during the breeding season because of the absence of competition for food at that season and the importance of predation in reducing productivity. Consequently, no large feeding and breeding territory is maintained by natural selection; competition among males for females has intensified, and, *together with other characteristics of the species and physical environment*, made possible the lek type of social organization. A similar explanation in principle for the evolution of the lek system in the Black and white Manakin (*Manacus manacus*) was given by Snow (1962).

Large territories.—The type of territory in which feeding, mating, and rearing of the young are all carried out together poses the most difficult problem for any theory of the evolution of territoriality, for the evidence is as contradictory as are the opinions of the many authors who have treated the subject. The fact that large territories occur only in species which utilize them for feeding would suggest that this type of territoriality has evolved in response to competition for food. This viewpoint is favored by Stenger

(1958) and Pitelka (pers. comm.) but opposed by Lack (1954) and Hinde (1956).

If this type of territory had evolved and were maintained in response to competition for food for the young, it would first be necessary to show that the nestling and fledgling mortality were commonly and in most populations of the species attributable ultimately to food shortage and only proximately to predation. However, the evidence presented by Lack (1954) on the causes of such mortality in thrushes favors stark predation uncomplicated by food shortage. There is but little reliable evidence bearing directly on this point in other species.

Despite the small amount of actual evidence that competition specifically for a food supply for the young commonly exists during or before the period when the young are being fed, the nature of the evolution of clutch size suggests that food may frequently be in short supply at that time. Clutch size probably tends to be increased through natural selection to the most productive number (in terms of eventual reproduction of the young produced) that the environment allows. Since the environmental limit to productivity in nests not affected by predation or parasitism is probably set primarily by the rate at which food can be brought to the young, it seems possible that competition for food for the young would frequently exist.

Another type of evidence offered in defense of food shortage as the primary cause for the evolution of large territories is the correlation between territory size and food supply. It is generally known that territorial (and nonterritorial) species have denser populations (and usually smaller feeding areas) in habitats where their food supply is better. This has been demonstrated quantitatively by Kluyver (1951) for the Great Tit (*Parus major*) and by Stenger (1958) for the Ovenbird (*Seiurus aurocapillus*). But if territory size is adjustable within limits to the breeding density in these species (as it apparently is), the correlation between territory size and food supply could be wholly a result of the normal habitat preference of the species and not directly related to the evolution of territoriality.

A more universal and easily demonstrable reason for the evolution of this type of territoriality is that it is dependent on competition for the *opportunity to breed*, as determined by ownership of a suitable area (in terms of feeding and nesting habitat). It may be debated whether the food density at the time the young are fed is adequate or not, but there is no question for many species with large territories, that possession of a territory is a prerequisite for the opportunity to mate and begin nesting. Even in a nidifugous species for which food is more than ample for the reproductive effort of all the individuals in any one area (assuming static clutch size), competition for space may result in restriction of the breeding population to those who by their aggressiveness are capable of holding a territory in an area of habitat acceptable to both sexes (e.g., certain Parulidae during high densities of spruce budworm, Stewart and Aldrich, 1951; Hensley and Cope, 1951). This would result in restriction of the maximum breeding density to the most aggressive birds. Such restriction has been indicated to occur in tits (Kluyver and L. Tinbergen, 1953; Gibb, 1956), Red-winged Blackbirds (Orians, 1961), Song Sparrows (Tompa, 1962), and strongly suggested to occur in many passerine species by studies of repopulation of artificially depopulated areas (Hensley and Cope, 1951; Stewart and Aldrich, 1951) and numerous other observations on the rapid remating of marked birds upon loss of their mate (e.g., Magpies, Minton, 1958; Shannon, 1958).

It should not be inferred that if the competition is not for food that it must be for mates, for many passerine species with large territories are monogamous with as many females as males in the breeding population.

The aggressiveness necessary to establish a large, exclusive territory may gain relatively little in terms of food, cover, and mates when they are already in adequate supply for the population as a whole; but by mere possession of an opportunity to breed, the territory owners would leave more reproducing offspring than the nonowners. As long as counter selection against aggressiveness were weak, *aggressiveness per se would be maintained in the population merely by the exclusion of less aggressive birds from breeding.*

The fact that the peak of territorial defense in some species (in terms of area and behavior) occurs before the young must be fed and often before the female arrives (e.g., Odum and Kuenzler, 1955) tends to support this idea. The males can afford to devote excess energies to territory defense during the period when they have little else to do but forage for themselves. After the mate arrives there is, of course, a selective advantage to protecting her from other males, but this could be done more efficiently by accompanying her and would not require a territory.

The correlation between large territories and their utilization for feeding might also be explainable on the basis of competition for space in which to breed. If aggressiveness were maintained in the population mainly by the exclusion of less aggressive individuals from breeding, the usage of the territorial space in foraging would be secondary to the fact that an aggressive individual was spending 100% of his time in a discrete area and defending it.

It seems likely that *both* limited food and exclusion by aggressiveness per se have been important selective agencies in the evolution of large territories. Under conditions of limited food density and medium to high population densities competition both for food and for space per se may be expected to be operative. Under the unusual conditions of high food density and low population density, neither type of competition would constitute an effective selective force and territory defense would be absent or minimal. If both food density and population density were high, exclusion by aggressiveness would be the primary factor (e.g., Bay-breasted Warbler, *Dendroica castanea*, during outbreaks of the spruce budworm). If food density and population density were low, then defense of the food supply would be the primary factor.

Regardless of whether competition in this specific type of territoriality is for opportunity to breed, food, mate insurance, or some combination of factors, the general theory proposed in this paper would apply. For the object of the competition is not necessarily specified in the general case—only that it be economically defendable.

POPULATION CONTROL

Since territoriality appears in some species to participate in the control of population density (e.g., Kluyver and L. Tinbergen, 1953; Gibb, 1956; Tinbergen, 1957; Orians, 1961; Tompa, 1962), the hypothesis has been advanced (Wynne-Edwards, 1962) that territoriality and much of the ritualized agonistic behavior which characterizes it in many species have evolved to serve as mechanisms of population control. The argument fails primarily because it does not take account of the fact that changes in gene frequency are the result of competitive advantages accruing to individual genotypes rather than to the group as a whole.

It is not sufficient to demonstrate that genetic changes in some individuals in the direction of increased territoriality and efficiency of population control benefit all members of the population *equally*, including those individuals lacking these genetic changes. For, if the benefits of territoriality were equally distributed among all members of the population, then according to the Hardy–Weinberg equation the frequencies of the genes determining the increased territoriality would remain unchanged in successive generation rather than increasing. Consequently, it is impossible to account for the evolution within a population of territoriality, "epideictic displays," and population control on such a basis, notwithstanding the massive documentation assembled by Wynne-Edwards (1962). His proposal does not give a solution to the problem of how individuals in which territoriality is more strongly developed than others in the same population are adaptively superior to them.

The proposal that territoriality in a species may have evolved through extinction of nonterritorial populations and survival of territorial ones (Wynne-Edwards, 1962) is an insufficient explanation for two reasons. In the first place, the proposal does not explain how territoriality evolved in the original territorial populations. Secondly, the magnitude of the differences in territoriality which occur between local populations of a species and between closely related species make it seem probable that such differences can evolve rapidly within a population in response to local conditions and do not usually require the processes of extinction of whole populations and invasion by others. The vast differences in territoriality exhibited by such closely related pairs of species as the Red-winged and Tricolored Blackbirds (Orians, 1961) and the Scrub and Mexican Jays (Brown, 1963) support this view.

SUMMARY

Recognition of the diversity of systems of territoriality among species has clearly indicated that an understanding of the evolution of territoriality requires a theory which accounts for the diversity according to more general ecological principles than those which have been proposed in the past.

A general theory of territoriality is proposed which depends upon the influence of two primary variables, *competition* and *economic defendability*, and on the adaptive value of aggressiveness under various conditions of these variables. Examples of application of the theory in different types of social systems (colonies, leks, and large territories) are given.

It is suggested that in species with large territories used for both feeding and nesting, territoriality might, under certain conditions, be maintained or selected for in a population merely through the exclusion of less aggressive individuals from the opportunity to breed in a suitable habitat. Such exclusion would, however, be limited by counter-selection pressures when aggressiveness became too detrimental to reproduction.

ACKNOWLEDGMENTS

I wish to thank Drs. W. J. Hamilton III, G. H. Orians, and F. A. Pitelka for arousing and sustaining my interest in territoriality by their ever-stimulating and illuminating discussions of the subject during the period of our common residency at the Museum of Vertebrate Zoology, University of California, Berkeley (1956–60).

LITERATURE CITED

BROWN, J. L.
　　1963　Social organization and behavior in the Mexican Jay. *Condor*, 65:126–153.
CARPENTER, C. R.
　　1958　Territoriality: a review of concepts and problems. Pp. 224–250. *In* Roe, A.
　　　　　and G. G. Simpson, Behavior and evolution. New Haven, Yale University
　　　　　Press.
GIBB, J.
　　1956　Territory in the genus *Parus*. *Ibis*, 98:420–429.
HENSLEY, M. M., AND J. B. COPE
　　1951　Further data on removal and repopulation of the breeding birds in a spruce-
　　　　　fir forest community. *Auk*, 68:483–493.
HINDE, R. A.
　　1956　The biological significance of the territories of birds. *Ibis*, 98:340–369.
KALELA, O.
　　1958　Über ausserbrutzeitliches Territorialverhalten bei Vögeln. *Ann. Acad. Sci.
　　　　　Fenn.* (A), 4(42):1–42.

KLUYVER, H. N.
   1951   The population ecology of the Great Tit, *Parus m. major* L. *Ardea*, 39:1–135.
KLUYVER, H. N., AND L. TINBERGEN
   1953   Territory and the regulation of density in titmice. *Arch. neerl. Zool.*, 10:265–289.
KURODA, N.
   1960   An essay on bird territoriality. *Misc. Reports Yamashina's Inst. for Ornith. and Zool.*, 2:133–137.
LACK, D.
   1954   The natural regulation of animal numbers. London, Oxford University Press. 343 pp.
MINTON, C. D. T.
   1958   Magpie's rapid replacement of a dead mate. *Brit. Birds*, 51:309.
ODUM, E., AND E. J. KUENZLER
   1955   Measurement of territory and home range size in birds. *Auk*, 72:128–137.
ORIANS, G. H.
   1961   The ecology of blackbird (*Agelaius*) social systems. *Ecol. Monogr.*, 31:285–312.
PATTERSON, R. L.
   1952   The Sage Grouse in Wyoming. Wyoming Game and Fish Commission. 341 pp.
PETERS, D. S.
   1962   Gedenken zum Revierproblem. *Ornith. Mitteil.*, 14:161–171.
SHANNON, G. R.
   1958   Magpie's rapid replacement of a dead mate. *Brit. Birds*, 51:401–402.
SNOW, D. W.
   1962   A field study of the Black and White Manakin, *Manacus manacus*, in Trinidad. *Zoologica*, 47:65–104.
STENGER, J.
   1958   Food habits and available food of Ovenbirds in relation to territory size. *Auk*, 75:335–346.
STEWART, R. E., AND J. W. ALDRICH
   1951   Removal and repopulation of breeding birds in a spruce-fir community. *Auk*, 68:471–482.
TINBERGEN, N.
   1957   The functions of territory. *Bird Study*, 4:14–27.
TOMPA, F. S.
   1962   Territorial behavior: the main controlling factor of a local Song Sparrow population. *Auk*, 79:687–698.
WILLIAMS, L.
   1942   Display and sexual behavior of the Brandt Cormorant. *Condor*, 44:85–104.
WYNNE-EDWARDS, V. C.
   1962   Animal dispersion in relation to social behavior. New York, Hafner Publ. Co. 653 pp.

DEPARTMENT OF BIOLOGY AND CENTER FOR BRAIN RESEARCH, UNIVERSITY OF ROCHESTER, ROCHESTER, NEW YORK, 27 AUGUST 1963

VIII

The Experimental Approach to Territory

Editor's Comments on Papers 24, 25, and 26

24 **Van Den Assem:** Territory in the Three-Spined Stickleback *Gasterosteus aculeatus* L.
Behaviour (Suppl. 16), 2–4, 44–52, 81–83, 90–93, 144–153, 158–159 (1967)

25 **Alexander:** Aggressiveness, Territoriality, and Sexual Behavior in Field Crickets (Orthoptera: Gryllidae)
Behaviour, **17**, 131–132, 172–180, 183, 211–214 (1961)

26 **Watson and Moss:** A Current Model of Population Dynamics in Red Grouse
Proc. 15th Intern. Ornithol. Congr., 134–149 (1972)

Field studies of the functions of territory have been severely handicapped by the multitude of environmental variables that cloud relationships and the single, short breeding season of most species. For these reasons, fish and invertebrates are proving especially valuable as study animals; the territorial and breeding behavior they show in captivity is almost the same as that in the field. They can be studied over many months and under conditions where environmental factors can be tested separately. Van Den Assem's classic study (Paper 24) has analyzed how each step in the reproductive and parental cycles of the three-spined stickleback is affected by territory. He finds that not only is there a limit to the number of territories a fixed area will support, but that each step in reproduction is affected by density. Of special interest is his finding that a given habitat will support more territories when those territories are established simultaneously than when they are taken up successively over a period of time.

Johannes Van Den Assem is a graduate of the University of Leiden, where he is currently Senior Scientist in the Ethology Department. Van Den Assem has been involved, as have so many European ethologists, in many aspects of behavior, including reproductive behavior of the Sandwich tern, orientation behavior of a digger wasp, and host-finding behavior of parasitic insects. Following a period of study in the tropics, he is now studying the behavior of parasitic wasps.

Invertebrate zoologists have been slow in turning to the study of territoriality, which is somewhat surprising in view of the ease with which these animals can be kept in confinement. R. D. Alexander's monograph on cricket behavior (Paper 25) is important for its theoretical implications. First, he shows that male crickets begin with a hierarchical social system that turns into a territorial one only after males have been able to create a crevice to defend. Males sing more after winning an encounter, and singing releases song in neighboring males. Since females are attracted by song, this would favor aggregations of territorial males and also give an advantage to individual males singing the most. This then could lead to behavior similar to leks in birds. Alexander suggests that the development of site attachment and territorial defense is one process by which social behavior evolves in insects.

Alexander is Professor of Zoology and Curator of Insects at the University of Michigan, where he teaches behavior and evolutionary ecology. Since receiving his Ph.D. from Ohio State University in 1956, he has traveled throughout the United States, Australia, Mexico, Hawaii, and Fiji, studying the systematics and behavior of crickets, katydids, and cicadas. In 1961, he received the AAAS–Newcomb Cleveland

Prize for his paper, "The role of behavioral study in cricket classification," and in 1971, the Daniel Giraud Elliot Medal from the National Academy of Sciences for his work on the systematics, evolution, and behavior of crickets and other insects.

Perhaps the most comprehensive work on territory is that pursued for almost 20 years by members of the Nature Conservancy of Scotland on red grouse. These were begun by David Jenkins and continued by Adam Watson and others. In Paper 26, Watson and Moss summarize these studies. These studies have employed a number of experimental approaches. The nutritive and physical properties of heather, the red grouse's staple food, has been manipulated by burning and fertilizer. Grouse have been removed at specific times from their territories to see under what circumstances the vacant territory will be reoccupied. Males have been injected with testosterone to observe the effects of territorial behavior. The combination of this experimental approach and the naturally occurring variables over such a long period has produced perhaps the best understanding of territoriality in a free-living population. Watson and Moss report that territory size is not determined primarily by density of the grouse population; hence they do not subscribe to the elastic disc theory. Instead they find that territory size is a function of soil fertility, the nutritive level of heather, ground-level visibility, and differences in the aggressiveness of grouse from year to year. The study is unusual for its synthesis of the effects of ecological, behavioral, and genetic factors upon territory size and population density.

Adam Watson was born and raised in Aberdeenshire, Scotland. After a brief period in Canada, he returned to the University of Aberdeen to study behavior of ptarmigan and, later, red grouse. He is currently in charge of the Nature Conservancy Mountain and Moorland Research Station near Aberdeen. He is a Fellow of the Royal Society of Edinburgh and Senior Principal Scientific Officer for Special Merit in Research. I first met Watson in Aberdeen, where his shaggy red locks and beard and colorful Scottish kilt immediately caught my attention. I was soon to find that his critically keen mind was an even better reason to be attracted to him. Robert Moss, a plant ecologist at the University of Aberdeen, has brought his deep understanding of plant–soil relationships to bear on the total picture of red grouse ecology.

Reprinted from *Behaviour* (Suppl. 16) 2–4, 44–52, 81–83, 90–93, 144–153, 158–159 (1967)

TERRITORY IN THE THREE-SPINED STICKLEBACK
GASTEROSTEUS ACULEATUS L.

AN EXPERIMENTAL STUDY IN INTRA-SPECIFIC COMPETITION

BY

J. VAN DEN ASSEM

A i m   o f   t h e   s t u d y.

The aim of this study has been to present some objective evidence about the function or functions of territorial behaviour. When writing on the significance of territory in blackbirds, Snow (1956, p. 446) commented that only "after very many years' observation it might be possible to analyse, for example, nest-success and the maintenance of the pair-bond in relation to such variables as size of territory and distance of the nest from other nests, and so to test objectively some of the suggested functions of territory, but available data so far are quite insufficient for this".

In contrast to blackbirds (and most other bird species) the Three-spined Stickleback seems to be a more favourable subject for quantitative work on territory problems: the animal is easily available, the males hold a well-defined territory, its reproductive behaviour is well-known, the animals can be studied in the laboratory under more or less natural conditions where various experiments can be carried out without severely interfering with the animal's activities. Lastly, the reproductive cycle is short enough to permit the collection of a reasonable amount of data in a reasonable time. Therefore, the present study was undertaken, partly based on previous work on this topic by van Iersel (1958). Though no claim is made that all functions of stickleback territory are understood, some objective evidence has been found for at least some functions, as the reader may judge from the following chapters.

The topic under consideration has both ethological and ecological aspects. A point is made here that it is important to consider certain ecological concepts — competition, regulation of numbers, population structure, *etc.* — from an ethological point of view as well. Since any population is composed of living individuals, their behaviour must play an important role in producing the effects in which ecologists are most interested, and the study of relevant behaviour patterns may help in a better understanding.

In reviewing the literature I realized that my opinion is not an original one. TINBERGEN (1949) stressed a similar point. More recently, LACK (1966, p. 280) made a similar plea in stating that in the future experimental techniques, based on a sound knowledge of the 'natural ecology' of a species could be rewarding. I suspect that LACK implies that an understanding of a species' behaviour under natural conditions is a worthwhile part of ecology. PARK (1961), quoted by LACK (*l.c.*) phrases the same idea as "natural history is one of the prime sources of insight and knowledge for the modern ecologist". In fact, CHITTY (1960) and CHRISTIAN (1959) have long advocated this approach.

Definition of territory.

Because different kinds of territory have been found, different definitions have been proposed. A descriptive definition should contain the two elements indicated in the first paragraph: a) spatial restriction of (some) behaviour, and b) intolerance, leading to defence and resulting in isolation after the local removal or repulsion of conspecifics.

Among many formulations given, NOBLE's simple description (1939) — *a territory is any defended area* — seems satisfactory for the present paper since under appropriate conditions there is hardly ever any doubt as to whether or not a stickleback will defend an area.

In practice only the effects of defence may be manifest, *i.e.* the exclusion of conspecifics from an area by a (temporary) resident. The way in which this defence is accomplished may be difficult to assess; sometimes the mere presence of an owner inside his holding may suffice.

Most kinds of territories can be grouped as breeding territories. This applies also to the Three-spined Stickleback where territories are exclusively connected with breeding activities. It should be clear that a territory in itself is non-existent, it always is inseparably connected with appropriate behaviour of its owner. Territorial behaviour in the strict sense is such behaviour as serves to establish and maintain a territory.

The term territory as used in this paper, essentially following NOBLE, is described as follows: the territory of a Three-spined Stickleback is a topographical area of limited extent where a male-owner spends almost all of his time, and where conspecifics are excluded. Conspecifics are chased as soon as they trespass; the only exceptions are ripe females and, what I may call, sneaking males — both under specified conditions to be described later (see p. 6, 81, 100).

Outline of the paper.

This paper will be concerned first with an account of the behaviour of solitary males; a correlation is established with territory characteristics such as dimensions and structure of the environment; aspects studied are the way in which a male claims an area as his territory, nest-building related to available space, choice of a nest site, and development of his pattern of moving about in the post-nestbuilding period, in relation to the environment.

Next the state of affairs in a solitary situation is compared with the more complicated interrelations in a rival situation. Rival males appear to influence each other in settling success, in choice of a nest site, in building proper, and in activity-pattern characteristics of the post-building phases. Aggression shown by males present in a rival situation is correlated with the respective sizes of their holdings, and something like a rank order is established.

In a later section courtship-success is analysed, in solitary situations first, then compared with the scores in situations where rival males have to compete for a female. Territory size again appears to be of primary importance for having success within a group. In the last section of the paper development of eggs present in nests of solitary males is compared with development in rival nests. In the latter situation egg yield is obviously reduced, most often because of egg-stealing. Moreover, the development of eggs in rival situations is noticeably slower, and correlated with certain behaviour elements of the corresponding nest-owners. Territory size is also of importance in this last phase of the reproductive cycle.

* * * * * * *

V. SETTLING AND NEST-BUILDING IN RIVAL SITUATIONS

All results described up to this page were obtained with solitary males. However, the presence of conspecifics (rivals) across a boundary seems to be an important characteristic of any stickleback territory; the remaining part of this paper will deal with the effects of interactions of rivals during different phases of the breeding cycle as related to territory phenomena.

Two types of rival situations will be distinguished: *simple* (only two males present) and *complicated* (more than two males present).

NEST-BUILDING SUCCESS IN SIMPLE RIVAL SITUATIONS

The effects of a continuous presence of another male stickleback on nest-building were investigated in the following set-up.

Tanks of 60×30 cm^2 were divided into two 30×30 cm^2 compartments by a glass partition fitting tightly across the middle. Each tank was surrounded by a cardboard mantle to standardize the situation. In a number of tanks a row of plants (*Elodea* sp.) was present along the partition at both sides (tanks with row), in others vegetation was absent (bare tanks). All compartments were provided with nesting material, uniformly spread in small tufts. One male was introduced into each compartment; in any one tank the neighbours got in at exactly the same moment. In control tanks one compartment was left empty.

In these experiments no male had to fight to claim an area, the glass partition was kept in place all the time and prevented an invasion by a neighbour. (This is a major difference from experiments to be described in a later section, where males could only be successful in claiming an area by their own activities).

Following introduction aggressive interactions were often observed between neighbours, in spite of the glass partition; they were most frequent in bare tanks, and less abundant but still frequent enough in tanks with a row. Initially biting against the glass occurred most, but later on threatening became predominant. At that time the situation had stabilized by the formation of a provisional 'boundary'. A really stable situation is not obtained here, because not all males are able to maintain their holding after removal of the partition. (This is not relevant in the present experiments, however).

Two groups of fish were used in these experiments: A. males which still had a low level of reproductive activity (at the start of the season, see p. 25), and B. pre-treated males (16 hours of light daily at 18° C for at least one week in a 120 cm store tank).

Only complete nests were scored. A nest was judged as complete from its external appearance (*i.e.* provided with a good entrance, and large enough to hold a female). An owner of such a nest will court upon confrontation with a female. Various aspects of the results of nest-building of these males are represented in Tables 29, 30, 31.

Table 29 represents the accumulated overall nest-building success in different situations over periods up to 72 hours after introduction. In the A group the nest yield in rival situations was significantly higher than among solitary males (see also p. 25); within rival situations the score in tanks with a row was better than in bare tanks. From these results I conclude that early in the season the presence of a conspecific has a stimulating effect on nest-building. This effect has also been observed by WUNDER (1930), who was of the opinion, however, that only females (Animierweibchen) could produce it.

In the B group results were opposite to those in A. Within the first 12 hours after introduction no difference was apparent between situations; after that period, however, the group of solitary males took a significant

TABLE 29

Nest-building success in simple rival situations, glass partition present

A. males with low initial level of reproductive activities

situation	numbers of nests completed within a period of hours after introduction (represented as total numbers and as a percent.)			
	12	24	48	72
♂ | × 1) *	0/30 0%	0/30 0%	2/30 7%	5/30 17%
	n.s.	0.01	<0.05	0.05
♂ | ♂ 2)	2/38 5%	7/38 18%	11/38 29%	15/38 39%
	n.s.	0.05	<0.05	n.s.
♂ × ♂ 3)	5/20 25%	9/20 45%	12/20 60%	12/20 60%
	0.01	<0.001	<0.001	<0.01
♂ | × *	0/30 0%	0/30 0%	2/30 7%	5/30 17%

B. males with advanced level of reproductive activities

situation	12	24	48	72
♂ | × ♂ × × *	9/31 29%	22/31 71%	24/31 77%	28/31 90%
	n.s.	<0.01	0.05	n.s.
♂ | ♂	12/60 20%	22/60 37%	33/60 55%	43/60 72%
	n.s.	n.s.	n.s.	n.s.
♂ × ♂	6/28 21%	12/28 43%	18/28 64%	23/28 82%
	n.s.	0.05	n.s.	n.s.
♂ | × ♂ × × *	9/31 29%	22/31 71%	24/31 77%	28/31 90%

1) Tanks with one compartment left empty (control);
2) tanks with one male in each compartment, separated by a glass partition;
3) situations as in 2), but with a row of plants in addition.
*) Controls have been represented twice in the table; p-values were calculated with Fisher's Exact Probability test, or with a χ^2 test.

lead and arrived at the highest score recorded after 72 hours. In rival situations the scores after 24 and 48 hours (bare tanks only) were significantly lower. After 72 hours the scores were still lower than for the solitary group, but differences were no longer significant.

These results indicate an inhibition, be it temporary, on building in rival situations owing to the presence of a conspecific.

* * * * * * *

In Table 31 a survey is represented of building activities in individual tanks. In any one tank one of three combinations may be found: both neighbours build a nest (+ +), one out of the couple builds (+ —), or neither builds (— —). The building-results of solitary males, scored cumulatively after 24, 48 and 72 hours, were used for calculating an expected frequency of building in rival situations 1). Differences between expected

1) The expected frequency of the three combinations (++), (+—), and (——) has been calculated as: $(++) = a^2$. N; $(+—) = 2\,a\,b$. N; $(——) = b^2$. N; a = the proportion of builders in corresponding control experiments, b = the proportion of non-builders, and N = the number of rival couples tested.

and observed frequencies were tested with the Kolmogorov-Smirnov one-sample test (SIEGEL, 1956).

In the A group the observed frequencies of nest-combinations deviated significantly from the expected frequencies.

In bare tanks a great surplus of (+ —) combinations was obtained. No significant difference could be found between bare rival situations and controls for the total nest yield in periods later than 24 hours following introduction (Table 30). However, the frequency of nest combinations expected to be found over that period (calculated from the nest yield in solitary situations over the same period) differed significantly from the observed frequency (bare tanks with rivals: observed, (+ +) 0, (+ —) 7, (— —) 5; expected, (+ +) 0.05, (+ —) 1.5, (— —) 10.5; D = 0.45, p = 0.01). It follows that rival males affect one another (as was concluded earlier) but only one male was stimulated, the other was inhibited. In this way an inhibiting effect is demonstrated in the A group as well (apart from the stimulating effect, cf. Tables 29, 30). It becomes more clear in the course

TABLE 31

Nest combinations in individual tanks

A. group

		hours after introduction					
		24		48		72	
situation	combination	obs.	exp.	obs.	exp.	obs.	exp.
	++	1	0	1	0	1	0.5
♂│♂	+—	6	0	10	2.4	13	5.4
	——	12	19	8	16.6	5	13.1
	p level	0.01		<0.01		<0.01	
	++	3	0	4	0	4	0.3
♂×♂	+—	3	0	4	1.3	4	2.8
	——	4	10	2	8.7	2	6.9
	p level	<0.01		<0.01		<0.01	

B. group

	++	3	15.1	9	18.0	14	24.5
♂│♂	+—	15	12.4	16	10.5	15	5.2
	——	12	2.5	5	1.5	1	0.3
	p level	<0.01		0.01		<0.01	
	++	4	7.0	6	8.4	9	11.4
♂×♂	+—	4	5.8	6	4.9	5	2.4
	——	6	1.2	2	0.7	0	0.2
	p level	0.05		n.s.		n.s.	

of time. (In this respect it has to be noted that the state of affairs in the A group 72 hours after introduction resembled that of B at the start of the experiment).

I conclude that a kind of dominant-subordinate relationship was established between neighbours — in spite of the glass partition —, probably already within the first 24 hours following introduction. A dominant status gives a male a better chance of success in building than can be achieved by a solitary male at that time, a success obtained at the expense of the neighbour, however.

The interrelations between neighbours were less clear in tanks with a row. Stimulation appeared to occur (Table 29), but compared with bare tanks there was no distinct surplus of (+ —) situations. On the contrary, the (+ +) situations were relatively more frequent, which may indicate that stimulation was mutual. (The relative surplus of (+ +) combinations in tanks with a row compared with bare tanks is significant, Fisher's Exact Probability test, p = 0.03, tested for 72 hours after introduction). Apparently this two-way stimulating effect was due to the presence of the row of plants. This row may act directly in this respect, adding an extra stimulating effect. On the other hand, it could act in a more indirect way as well: as a good boundary between two territories it might prevent the establishment of a marked dominant-subordinate relationship between the opponents.

In the B group the differences found for the bare tanks and the tanks with a row, 24 hours after introduction of the fish, are significant. The low success in building was characterized by a deficit for (+ +) and an equal surplus of (— —) situations. This changed in time (48-72 hours) into a surplus of (+ —) and a deficit of (+ +) situations. Before all, the low success was brought about by the deficit of (+ +) situations, most probably caused by a dominant-subordinate relationship as found in the A group.

In tanks with a row the significant difference after 24 hours disappeared later in time, though a small deficit of nest yield (Table 29) remained apparent, caused by too few (+ +) and too many (+ —) scores.

The conclusion to be drawn from the results of nest-building in simple rival situations is that three different mechanisms were found to be involved. a. One having a stimulating effect, apparent in the A group in the first 24 hours of the experiment. b. Another one having an inhibiting effect, apparent in the later stages of the A group, and leading to a surplus of (+ —) situations. In the B group this factor was apparent in the first 24 hours only. A stimulating effect could not be found in the latter group. The ultimate surplus of (+ +) situations in B (compared with the results of nest-building of solitary controls) originated in another way than the surplus of $(+ +)$ situations in A. In the former the inhibiting mechanism was no longer effective later in time, but stimulation did not occur; in the latter a stimulating effect was still apparent. c. An additional effect of the row of plants was found which provided an extra stimulating effect (in A), or, besides, a decrease of inhibition earlier in time (in B).

* * * * * * *

SETTLING AND BUILDING IN COMPLICATED SITUATIONS

The condition of males which arrive at their breeding habitats in early spring will be more or less comparable to that of males of the A group, described in the preceding section of this chapter. A point is made here that stimulating effects of conspecifics and vegetation (as were found in the experiments described above) might cause the establishment of breeding concentrations: males would not spread randomly over a suitable area but rather would settle in groups. A simultaneous settling and nest-building is probably the general rule under natural conditions in sticklebacks; they move into freshwater to their breeding habitats in large groups. SEVENSTER (*in litt.*) witnessed settling in a lake in southern Finland in 1966 and recorded a mass-nesting to take place spread over a few days only. Concentrations of breeding animals have been recorded for many species, especially birds, though the factors which cause these concentrations were never, to my knowledge, fully investigated. Experiments dealing with a possible attraction by conspecifics in sticklebacks will be dealt with elsewhere (see also Discussion, p. 147).

Success in settling and nest-building of rival males in open competition was estimated in the ditch (600×100 cm^2), or in parts thereof; vegetation was absent in all situations. I used variable numbers of males in areas of similar dimensions, and similar numbers of males in areas of variable sizes to investigate mutual influences by rivals.

Males were introduced into the experimental situation in either of two ways: simultaneously or (partly) successively. In simultaneous introductions all males were released at the same time and their nesting-success was recorded at 24 hour intervals. In a successive introduction, one male (to be called further male a) was allowed to claim the entire area as his territory. Twenty four hours after completion of his nest one or more males (to be called b males) were introduced and male a had to defend his holding against these intruders; at different times again success in building of b males was recorded.

Simultaneous introductions.

'Winter-condition' males.

Males collected randomly from the basement store tanks were used for experiments in 150 × 100 cm² compartments of the ditch. These males were of the same sample as those used in the A group in simple rival situations (p. 44, Table 29); the same controls have been used for comparison. Data are represented in Table 32.

TABLE 32

Nesting-success of rival males starting from 'winter condition', 150 × 100 cm²

number of rival males	number of duplicates	number of complete nests built within a period of hours after introduction, and corresponding nesting-success							
		24		48		72		120	
2	8	1	6%	7	44%	8	50%	10	63%
5	8	3	8%	9	23%	13	33%	19	48%
10	8	4	5%	11	14%	21	26%	29	36%
controls (Tab. 29)	30	0	0%	2	7%	5	17%		

2 males *vs* controls, 24 h., n.s.; 48 h., Fisher's Exact Prob. t. $p<0.01$; 72 h, $\chi^2 = 5.7$, $p<0.01$; 10 males *vs* controls, 24 h., n.s.; 48 h., Fisher's Exact Prob. t. $p = 0.05$; 72 h., $\chi^2 = 3.9$, $p<0.05$; 2 males *vs* 10 males, 48 h., $\chi^2 = 6.1$, $p<0.02$.

The same conclusion as was drawn for rival males separated by a glass partition is valid for rival males in open competition: among rivals a higher rate of nest-building was scored than for solitary males, again an indication of the stimulating effect of conspecifics on nest-building (cf. Table 29 and p. 45).

It has to be noted that the area available to the controls was only 30 × 30 cm², and thus differed from the area used in the present rival experiments (150 × 100 cm²). It was found earlier, however, that tank dimensions — within a range used here — did not influence success in building of solitary males (cf. Table 15). The results of the 2-male situations provide an additional argument that size of the available area did not play a role here. The success in building of males set free (present situation) did not differ from that of males separated by a glass partition (preceding section of this chapter).

Pre-treated males.

The results of nest-building of pre-treated males are represented in Table 33. In the two-male experiments in a large area (600 × 100 cm²) both fish nested in most trials. In smaller areas (150 × 100, 120 × 40 cm²), however, success in nesting was 50% because (+—) situations only were obtained. (The same applies to tanks of 60 × 30 cm² when two males are used in free competition. Within 24 hours almost always one male becomes dominant and prevents the other one from claiming a suitable area to build as well).

Fewer males were able to complete a nest within 24 hours when a larger number of competitors was present at the beginning (experiments in

TABLE 33

Nesting-success of rival males in areas of different sizes at different densities

size of exp. area in cm²	number of males	number of duplicates	number of complete nests built within a period of hours after introduction, and corresponding nesting-success			
			24		72	
600 × 100	2	14	25	89.3%	—	—
600 × 100	120	1	24	20.0%	27	22.5%
150 × 100	2	6	6	50.0%	7	58.0%
150 × 100	10	18	77	42.8%	115	63 %
150 × 100	20	6	37	30.8%	42	35 %
150 × 100	30	5	43	28.7%	33 (4)	27.5%
120 × 40	2	19	19	50.0%		
60 × 30	2	x	½x	50.0%		

2 males, 600 *vs* 150, 24 h., Fisher's Exact Prob. test, p level 0.05; 150 × 100, 10 *vs* 20, 24 h., $\chi^2 = 3.87$, p = 0.05; 72 h., $\chi^2 = 38.1$, p$<$0.001; 150 × 100, 10 *vs* 30, 24 h., $\chi^2 = 7.0$, p$<$0.01.

150 × 100 cm²). (The differences found between 2 and 20 or 30 males are not significant owing to the small sample of 2-males available). Stated otherwise, with an increasing pressure among males, the proportion that is able to settle and build a nest will decrease.

After 72 hours hardly any change was observed in the 2-male situations: the dominant — subordinate relationship did not change in time. Among larger numbers of males a distinct increase of success in building was scored for the 10-male situations, hardly an increase for the 20-male ones, while success of 30-male groups had remained constant. Apparently, in the latter situations a minimum territory size had been reached which set a limit to a further increase of nesting. For similar densities in different areas (compare 120 males in 600 × 100 and 30 males in 150 × 100 cm) similar results were obtained, which indicates that these results depended on the initial density of males.

* * * * * * *

VII. TERRITORY AND COURTSHIP-SUCCESS

The owner of a complete nest is willing to court a female that happens to enter his territory; a female that appears before the nest is ready is treated aggressively. During the parental cycle a male with a strong parental tendency will no longer court a female; she is then also treated aggressively.

In general, a nest-owning male is not able to lead a female to his nest at once, nor will he tolerate her in its direct vicinity immediately upon her arrival. Between appearance of the female and leading her to the nest a sequence of more or less obligate behaviour elements has to be shown, called courtship. In the Three-spined Stickleback it consists basically of the following elements: zigzag approach, nest-visit, creeping through, zigzag approach, bumping, nest-visit, fanning, glueing, zigzag approach, leading female to nest, showing nest-entrance. Creeping through, bumping, fanning and glueing may be absent. For a short description of courtship see p. 6; more detailed descriptions in TER PELKWIJK & TINBERGEN, 1937a, b; VAN IERSEL, 1953.

Females used in experiments were always carefully selected for willingness, i.e. for adopting a persistent courtship attitude and a tendency to follow upon leading, to be sure that the condition of females was as homogeneous as possible. Differences found between situations could be attributed to differences between males tested, and conclusions could be drawn for smaller samples than when females had been chosen randomly.

A courtship sequence started when the male approached the female — almost always immediately upon her introduction into the experimental area —, it usually ended with the female entering the nest. Such a courtship will be called positive.

Not in all cases, however, a female entered the nest once she had arrived in front of it (upon following the leading male); sometimes she nosed into the entrance and subsequently fled. The male then started chasing her, but if she re-adopted a courtship attitude the male usually started to court her again. At the end of this next bout of courtship she could still enter the nest. Such a courtship, consisting of more than one, of what I may call courtship cycles, will also be called positive, or successful.

If a female did not enter a nest within a period of 30 minutes, though she did demonstrate her willingness to court all the time, a negative courtship was scored.

Trials in which the female started to flee wildly after some time were stopped thereupon and again a negative courtship was scored. When tested immediately afterwards in another situation such a female could again court optimally and enter a nest.

A courtship sequence may have an untimely end; e.g. if the female enters a nest 'under her own power'. Very occasionally this happened when she was very willing and probably on the verge of spawning spontaneously. In such a case she searched for a nest herself, and appeared to be able to recognize the 'entrance', a mechanism not due to previous experience with nest characteristics (SEVENSTER, pers. comm., observed it in laboratory-reared, completely inexperienced females). Another relevant observation in this respect is that on some occasions females tried 'to enter' the circular dark upper surface of a markation peg, level with the sand bottom, even when the male was showing the real entrance nearby.

Most cases of nest-entering without the male showing the entrance were observed when the male crept through his nest during courtship; at the moment he wriggled through the tunnel the female dashed for the entrance and crept after him (see also sneaking males, p. 101). In these cases the male started to quiver (p. 6) as soon as he had left the nest; the female then spawned within a few minutes.

For calculating the average duration of a courtship sequence all these aberrant cases were left aside (in all experimental categories they formed only a minor part).

In very small territories a courtship sequence was sometimes abnormal in so far as leading dropped out: a male performing on his nest, *e.g.* fanning, could suddenly start showing the entrance, whereupon the female approached and then entered the nest.

In this chapter the term *courtship-success* will be used frequently. Courtship-success of males in specified categories was assessed by two different measures:

a. as the number of positive courtships, expressed as a percent of all courtships in that category of experiments, and

b. as the average duration between introduction of a female into the experimental situation, and the moment she started to enter the nest.

In all trials individuals were used that had not built before during the breeding season; males were used once in an experiment.

COURTSHIP-SUCCESS OF SOLITARY MALES IN TERRITORIES OF DIFFERENT SIZE

Twenty-four hours after a solitary male had completed his nest a selected female was introduced into his territory. The dimensions of territories used were: 10 × 10 and 20 × 20 cm² (wire cages in the ditch), 30 × 30 and 60 × 30 cm² (glass-walled tanks), and 150 × 100 and 300 × 100 cm² (compartments of the ditch). Results are represented in Table 54.

Two groups of territories can be distinguished, as it follows from the experimental results.

In the first group — with territory dimensions ranging from 300 × 100 to 30 × 30 cm² — the measures for success have rather constant values:

TABLE 54

Courtship-success of solitary males in territories of different size

territory size	number of courtship sequences			average duration pos. court- ship in sec.	average duration ♀ stays in nest in sec.	courtships in which ♂ CT
	n	n positive	n pos. as %			
10 × 10	41	25	61%	308	49	54%
20 × 20	41	25	61%	353	49	59%
30 × 30	40	34	85%	84	66	30%
60 × 30	46	40	87%	91	50	50%
150 × 100	15	14	93%	91	47	1×
300 × 100	21	19	90%	99	47	33%

CT — creeps through the nest.

Percent positive courtships, 20 × 20 *vs* 30 × 30, $\chi^2 = 4.8$, $p < 0.05$; 30 × 30 *vs* 300 × 100, n.s.

Duration of courtships, 10 × 10 *vs* 30 × 30, Mann-Whitney U test, $p < 0.001$; 30 × 30 *vs* 300 × 100, Mann-Whitney U test, $p > 0.5$.

85-90% of all courtships ended with the female entering the nest, and the average duration of a positive courtship sequence was between 84 and 99 seconds.

In the second group, comprising only the very small territories, the percentage of positive courtships is very significantly lower (61%), and the duration of positive courtships amounts to about a four-fold (308-353 seconds).

I conclude that in a solitary situation courtship-success is maximal within a wide range of territory sizes, starting from a minimum size of 30 × 30 cm². Courtship-success decreases sharply in territories of smaller dimensions.

* * * * * * *

Minimum size of a territory, success in courtship compared to success in building.

It has been noted above (p. 83) that the same critical dimensions for a territory as found for success in courtship hold as well for success in building. If the available surface was less than 30 × 30 cm² building appeared to be inhibited to some extent.

There are good arguments to state that a relatively high level of aggression is the main factor to explain the results of courtships in small territories. This relatively high level of aggression could very well be responsible for the results of experiments on nest-building as well. (The inhibiting effect of aggression on nest-building found in rival situations, p. 48, provides an additional argument in this respect).

It has been pointed out above that the nest site can be understood as the aggression centre of the territory from where an aggression gradient spreads out to the boundaries (p. 63). In case the available space is sub-minimal it can be reasoned that the fish is constantly in the close vicinity of its (potential) nest site, and therefore uninterruptedly at the aggressive peak level. This is in contrast to males which can repeatedly move out further away, as is possible in tanks of a larger size. In the former situation the 'build-up' of the tendency to build a nest may be more difficult than it is in the latter situation, which may result in differences between situations as were found. An experiment to test this hypothesis would be to confine males to a sub-minimal area (*e.g.* a 10 × 10 cm² wire cage in a larger tank) but allow them periodically to move out for some period. Success in nesting of these males should be higher than of males kept confined continuously. Such experiments were tried but technical difficulties were not overcome satisfactorily, leaving the results ambiguous and the hypothesis still unproven.

COURTSHIP-SUCCESS IN RIVAL SITUATIONS

The introduction of a female into a rival situation immediately gives rise to a renewed and very intense competition among males; usually all males present start courting. Territory boundaries provide some restriction in this respect — often some hesitating can be observed before trespassing — but they definitely do not prevent males from zigzagging towards a female outside their holding. Such a frequent trespassing of boundaries leads to aggressive interactions between males.

Two alternating patterns (or, in case more than two males are present, two interwoven patterns) can therefore be observed: courtship, sometimes mixed with overt aggression, to the female; bouts of chasing, biting, spine-fighting, or attacks with threatening and sand-jerking between males. The proceedings in a rival situation under laboratory conditions agree with those observed in the field.

After introduction, the female will usually react to several males in succession — though some females seem to ignore rival males completely once they stay near a particular male: in spite of intense zigzagging around her, her attention remains fixed on that one male, even if he is not continuously in her immediate neighbourhood. This fixation mechanism, if it may be called so, was not further investigated and it is therefore unknown why it seems to operate in certain cases, and not in others.

As already stated above (p. 81) a male is not immediately ready to lead a female to his nest upon her entering his territory. He has to perform a sequence of behaviour elements first, both near the female and on his nest. This state of affairs is the setting of an intricate pattern of interactions between males. The rule is that a female, reacting to a certain male, is lured away by a rival as soon as the latter comes in a position to do so. This happens most often when a male pays a visit to his nest and has left the female waiting at some distance away. At that moment a rival may draw her attention by zigzagging near her, and make her come over to his own territory. Arriving at his boundary, however, the rival now finds himself in a relatively disadvantageous position, because, being himself not yet capable of leading, he has to pay a visit to his own nest first, and has to leave the females at some distance. At this point the proceedings may be repeated a number of times.

Even if a rival does not succeed in attracting the female, his intrusion into another territory has an adverse influence: his appearance calls forth aggressive behaviour in the territory-owner that interrupts his courtship with an attack on his opponent. The probability of leading a female immediately thereupon is set back by this aggressive behaviour for either male; as can be deduced from observations, leading is clearly postponed so that the overall duration of positive courtships (p. 81) is distinctly prolonged. A good female will immediately follow once a male starts leading.

Courtship experiments to be reported here were done in two different situations:

A. *Simple rival situations, i.e.* only two territorial males present. The majority of these experiments were done in 60 × 30 cm² tanks. An opaque partition was used to have both males build a nest (a situation hard to realize in open competition, see p. 52). After the males had completed their nest the partition was removed and the situation was left to stabilize. Other simple situations were realized in the ditch without the help of a partition.

B. *Complicated rival situations, i.e.* more than two territorial males present. All these experiments were done in compartments of the ditch.

The results of courtship experiments in rival situations are represented in Table 62. Trials with only one courting male have been grouped separately since no obvious competition was observed in those cases. Trials in which the female approached a nest and entered it on her own have been left aside. The following conclusions are drawn from these results:

a. In all cases the durations of one-male courtships are significantly shorter than those with two or more males involved. Competition for a female appears to prolong courtship duration considerably — as was to be expected.

TABLE 62

Courtship in rival situations

simple situations

exp. set-up	1 courting male number *)	average duration pos. courtships	2 courting males percent positive	average duration pos. courtships
60 × 30 bare	7	134	56% (19/34)	801
60 × 30 row	11	118	77% (31/40)	340
300 × 100	10	63	75% (15/20)	233
600 × 100	6	62	(1 2)	130

complicated situations

	1 courting male		2 or more courting males	
150 × 100	8	73	75% (45/60)	290**)
300 × 100	7	89	70% (26/37)	379
600 × 100	5	82	55% (5/9)	296

*) In cases with one courting male only, courtship success was always 100%; if '♀ a.e.-1' was followed by turning away, courtship by rivals would follow (those cases were scored under '2 or more courting males'.

**) This figure is not an average of a random sample; relatively too many introductions of the female into territory (1) are included, see p. 97.

Durations of positive courtships in 60 × 30 cm², one-male *vs* solitary males, Mann-Whitney U test, $p < 0.01$, two-tailed.

Probability of positive courtship, solitary sit. *vs* simple rival sit., $\chi^2 = 11.8$, $p < 0.001$; solitary sit. *vs* compl. rival sit., $\chi^2 = 13.6$, $p < 0.001$.

Effect of a row in 60 × 30 cm², percent positive courtships, bare *vs* row, $\chi^2 = 3.9$, $p < 0.05$; duration of courtships, bare *vs* row, Mann-Whitney U test, $p < 0.001$.

Effect of territory size, 60 × 30 *vs* 300 × 100 simple, percent positive courtships, $\chi^2 = 1.97$, $p < 0.2$; duration of positive courtships, Mann-Whitney U test, $p < 0.001$.

b. The durations of one-male courtships in 150, 300 and 600 × 100 cm² are not significantly different from one another, and moreover not different from the duration of courtships of solitary males (cf. Table 54). Obviously it is actual competition that causes longer durations of courtship; rival males are not permanently in a condition which affects their tendency to lead adversely. Courtships in 60 × 30 cm² tanks were different in this respect, however; duration of one-male courtship is significantly longer than duration of courtship of solitary males. Apparently, a rival male present at a short distance does have an effect apart from actual interruption.

c. The probability of a positive courtship is greater in a solitary situation than it is among rivals. Courtship-success in rival situations thus differs from that in solitary situations (in territories of 30 × 30 cm² or larger) in its two measures: a lower score for positive courtship, and a longer duration before the female enters the nest.

d. A row of vegetation has a positive effect on courtship-success: compare simple situations 60 × 30 cm² 'bare' and 'row' (see also p. 94). The presence of a row does not affect the probability of competition (in 7 out of 41, and in 11 out of 51 courtships one male only courted in 'bare', resp. 'row'), but it apparently does affect the way in which this competition will occur. Both percentage positive courtships and courtship duration are more favourable in situations with vegetation present.

e. Territory size seems to have an effect on courtship-success: compare simple situations 60 × 30 bare, and 300 × 100 cm² (see also p. 96). Percentage positive courtship and courtship duration are different between these categories though only the latter one significantly so. A larger territory size is correlated with a more favourable outcome.

In larger territories percentage positive courtship and duration of positive courtships do not differ between simple and complicated situations, or, once males compete for a female the number of males involved does not seem to be very important.

f. The degree of spacing-out appears to be correlated with the probability of a one-male courtship. One-male courtships in simple rival situations increase in numbers in tanks of larger sizes (60 × 30: 7 out 41, 17%; 300 × 100, 10 out 30, 33%; 600 × 100, 6 out 8, 75%). A similar trend is apparent in complicated situations.

IX. DISCUSSION

No territory, be it topographically fixed or not, comes into being without relevant behaviour of an owner. Relevant behaviour has to be understood as any behaviour which results in an animal claiming and maintaining a holding. This actual claiming and maintaining of a holding will be called territorial behaviour. For a limited period of time a territory may continue to exist even in the absence of the owner, because his neighbours have learned about it. A continuous reinforcement of the processes involved remains necessary, however, since the removal of an owner eventually results in a change in the *status quo*.

Strict territorial behaviour comprises agonistic behaviour — frequently intra-specific only — such as attack, or attack intentions, or threat. In fact, we may include in the term 'territorial behaviour' anything that can be shown to reduce a competitor's aggression. Territorial behaviour in this sense is restricted, however, to a social situation. A solitary male will not perform such behaviour. (He may perform aggressively motivated appetitive behaviour

— patrolling his area in search of other males — but this type of behaviour cannot be distinguished unambiguously from behaviour that is differently motivated). Nevertheless, it certainly makes sense to consider the area held by a solitary male a territory — at least in the Three-spined Stickleback — because a solitary male will show overt aggression upon experimental confrontation with a conspecific.

NOBLE's (1939) definition of a territory — *any defended area* — provides for most of these considerations, but should perhaps be modified to include solitary situations. It could be rephrased as *any area that will be defended* — if an appropriate stimulus is offered.

The term 'defence' should be used in a functional sense. As it was already pointed out, it will often be difficult to prove defence. The results of an owner's defence, however, can often be determined without much difficulty. Therefore, it would perhaps be preferable to redefine a territory as *any area where the mere presence or the behaviour of an owner (or owners) excludes or would exclude the simultaneous presence of conspecifics.*

On theoretical grounds, HUXLEY (1934) supposed that there has to be a minimum size for any territory. If a territory boundary is pushed back in the direction of the territory centre, defence will become so intense at a certain point that it resists further compression. A territory seems to be compressible only up to a certain point; HUXLEY used the well-known analogy of rubber discs as an illustration of this model (p. 64).

In experiments with the Three-spined Stickleback the existence of a minimum size has been proven. The density of territories did not rise above a certain value and no additional males could be accommodated (p. 54). In solitary situations a functionally minimum size was found which affected success in reproduction (*i.e.* 30 × 30 cm²). A solitary male may defend an area of smaller dimensions than this critical minimum — and therefore such a sub-minimal holding should still be called a territory — but it is no longer functioning properly. In free rival situations territory dimensions hardly ever dropped below this critical value; territory dimensions as such usually did not set a limit to reproduction in these situations. In solitary males success in nest-building and courtship appeared, within a considerable range of dimensions, to be independent of territory size. If the dimensions dropped below the critical value, nest-building was largely inhibited (p. 26). In the same situation success in courtship is significantly reduced for two reasons. First, a much longer time is needed before the female can be led to the nest (p. 84), or, in other words, duration of pair formation — which is in sticklebacks a very temporary affair — is considerably increased. Secondly, the chances of the female entering the nest, once she has arrived at the entrance, is significantly decreased (p. 86). Thus, the chances of obtaining a clutch, and hence, of fathering offspring, is reduced.

This reduction of success is probably due to the male being made aggressive by the female's sudden and continual presence near the nest, which is the aggression centre of the territory (p. 89). In such a situation a prolonged time seems to be needed to shift the owner's aggression-sex balance to a relatively sexual state. Therefore, it seems to be not just the presence of a female, but *a female at a certain distance from the nest* that represents an important factor in pair formation, at least at the beginning of courtship.

The percentage of females entering the nest was much smaller in sub-minimal territories than in those of larger size (p. 86). Presumably, the aggression-sex balance required for a successful courtship is more unstable in sub-minimal territories, because it may switch back immediately to a relatively aggressive state if the female, following her arrival at the entrance to the nest, does not enter at once. An alternative explanation, however, may be that in sub-minimal territories the male's greater aggressiveness during courtship may have induced in the female a correspondingly greater tendency to flee. This in turn may have affected her willingness to enter the nest, even when she had been courted up to this point for the same length of time as in otherwise similar situations that had led to successful entering — see Table 59.

Once a female had entered a nest, the males were exactly similar in their ability to make the female spawn (p. 76).

Theoretically there is also a maximum size of a territory. An unlimited space is by definition no territory; an area from which other animals are excluded, implies that it has to be of some limited extent. The maximum size of a stickleback's territory has not been investigated, but it is worth considering the issues involved.

A territory-holding male frequently visits his nest, even though it is empty; in the largest available area (600×100 cm^2) he visited it on the average once every two minutes. To all appearances the time between leaving the nest and starting to return has a relatively stable value, independent of actual territory size. This time is available for the outward trip, and thus sets a limit to the radius of the territory. Size seems to be important only in so far as it has a direct bearing upon the duration of the return trip: the larger the territory the longer it takes to get back (p. 41). Although the behaviour of the male obviously determines to a large extent the size of a territory in general, it would be quite wrong to suppose that this is the only factor. There are many factors, among which we might mention clearness of the water, and heterogeneity of the habitat, which will also have a marked effect upon territory size.

Territory-owners do not act indifferently towards each other; on the contrary, they interact continuously. Such interactions between candidates for the same territory start during the settling phase, and will lead to a rather regular spacing-out of the participating males (p. 63). The effects of spacing-out were found to be still maximal in the largest area used (600 × 100 cm²), since if we start our experiment with one male at the end of the tank, and introduce the second one 24 hours later, the latter always built at the greatest possible distance away. Additional males, introduced later still, settled and built according to the same pattern, constructing their nests at a site almost equidistant and maximally away from the two nearest neighbours. A description of the pattern of nesting, and a possible explanation of how it is brought about, is given on p. 59, 65.

An area of limited dimensions cannot accommodate an unlimited number of males. Once a certain density has been reached the numbers remain constant. Such an area is said to be fully occupied (p. 54). The density of a fully occupied area does not have a strict numerical value; like all parameters which depend on biological values it is variable, and in stickleback, it seems to depend on the way in which the area is settled. When an area is settled simultaneously, its territorial density is considerably higher than it is when there is a succession in settling. This result follows from it being easier for a male to obtain a territory if he has to compete only with males that are also looking for territories than with those that are already settled. In the former case such a male has to divide his aggression among several opponents. Moreover, aggression is not yet maximal at this stage; it is after the nest has been built that aggression reaches its peak level (VAN IERSEL, pers. comm.).

The highest densities observed in a fully occupied area — in HUXLEY's model the point where a further compression of territories is no longer possible — agrees rather well with the minimum size for a territory as established for success in building and courtship (p. 55).

* * * * * * *

Breeding in a territory-cluster may lead to greater reproductive success, in spite of the proven disadvantages (see below). In the Three-spined Stickleback nest-building may be stimulated by conspecifics under certain conditions

(p. 51). In other experiments, not here reported, there was a difference in the sex-aggression ratios between the social and the solitary situation (a shift towards a relatively more sexual state in the former situation). Besides, females may find a group of males more attractive than they find a solitary male.

Stickleback males settled in a rival situation possess territories of variable size (p. 95). Such males act as competitors; the individual success in reproduction can be affected significantly by their interactions.

When a ripe female was introduced into an experimental rival situation, the duration of a successful courtship (measured as the time lapse between introduction and entering) was considerably prolonged by aggressive interactions between males. At the same time the chances of nest-entering had decreased, compared with those in solitary controls (p. 99). Clearly, the aggressive interactions between males interfered with the necessary change in the balance between the sexual and aggressive tendencies, resulting in a state of affairs very much like that in sub-minimal territories in solitary situations.

Males in rival situations did not have equal chances of leading a female to a nest and fertilizing the clutch. Owners of the largest territories had a distinct advantage (p. 97). A correlation was found between the total surface of territories and the numbers of females that entered the nest (p. 98). The notion of a distribution of females according to a mere probability seems obvious here: the probability of staying in a certain area will be correlated with its size. However, owing to continuous interactions between males, and between males and females, courtship procedures may be far from simple in complicated situations. Females seldom stuck to one courting male but on the contrary, changed partners frequently. The duration of her stay in a certain area did not seem correlated with the area's surface in a simple way, but unfortunately sufficient detailed records are not available.

In the experiments reported on, one female was introduced at a time. The degree of interactions, and the effects of introducing more females at a time are therefore unknown. In nature such a situation may be quite usual.

In the case of those males that were unable to lead a female to their own nest, the ability to sneak is another factor adding to the chances of fertilizing eggs (p. 100). The tendency to sneak could not be correlated with the rank number of the performing males; sneaking was about equally probable for males of any rank number. Males owning small territories in complicated rival situations are more likely to have their eggs fertilized by a sneaker than owners of large territories are (p. 107).

Having fertilized a clutch of eggs, a male enters a refractory period and is unable to lead a female to his nest again. This period lasts about half an hour and depends on the presence of eggs inside the nest (SEVENSTER-BOL, 1962). During this period those neighbours that did not fertilize eggs will have a better chance of success in courtship. Moreover, one male does not collect an unlimited number of clutches; 5 seems to be about the maximum possible. The conclusion that the owner of the largest territory is most successful in reproduction is too simple, additional factors greatly complicate the state of affairs.

Mutual interactions between males also affect individual reproductive success during the last phase of the cycle (parental care). Three factors were identified that reduce success in complicated rival situations: stealing of eggs, interruption of fanning bouts, and insufficient development of the fanning tendency proper. Interruption of fanning — which results in inefficient ventilation of the clutch — and insufficient fanning cause an increased mortality and retard the development of the embryos. In addition to this, slow development may even lead to higher mortality in later life (see p. 128).

Losses due to stealing of eggs, egg mortality, and retardation were least pronounced in large territories; in small territories many eggs may be lost and development may be severely retarded (p. 121, 129).

Within a group of territorial rivals, synchronization of their reproductive behaviour appears to be of great importance for ultimate success in reproduction. In experimental situations where there is any male present without a clutch of eggs in his nest (1-0 situations, p. 118), many more eggs were lost and development was much more retarded than was the case when all males had received eggs (1-1 situations). Possibly a 1-0 situation is only an experimental artifact, and 1-1 situations are the rule under natural conditions, either because there are enough ripe females for each male to receive at least one clutch, or, because those that do not get females manage to steal enough eggs (p. 131). Males whose behaviour was synchronized (1-1 situations) did interact but the effects were far less severe (p. 119, 127).

Synchronization in a population of sticklebacks probably originates right at the start of the reproductive cycle — mutual stimulation of nest-building occurs in males which are in contact with conspecifics (p. 44). If there are enough females, synchronization will be maintained up to the conclusion of the parental cycle.

A state of affairs under natural conditions and directly comparable to a 1-0 situation in sticklebacks, was described by TOMPA (1964), who studied a population of song sparrows (*Melospiza melodia*) on Mandarte Island, B.C.

Male song sparrows trying to settle among established territory-owners interfered with their breeding. The same was true for territorial males that
remained single. These unmated individuals remained aggressive throughout
the breeding season and caused continual disturbance. A year in which total
density was average, an unusually large number of unmated, territory-
holding, males resulted in a low reproductive success of breeding pairs.
According to TOMPA this was caused by breaking of pair bonds and temporary or permanent inhibition of breeding in many pairs. In another year,
when the total density was high, a relatively low percentage of eggs hatched,
most probably caused by lack of proper incubation due to interference by
non-territorial males.

In a number of cases the adverse effects of aggressive interactions on the
ultimate breeding success may be far more complicated, and less direct.
JENKINS' (1961) findings in partridge, where the amount of interactions
between pairs during late winter was thought to determine the survival of
chicks later in the season, may be an example. Possibly some delayed mortality among young may be found in sticklebacks as well. Mortality after
hatching may be correlated with the amount of interactions the males experienced during their parental phase, but no data are available to check
this point.

Competition for a territory occurred in all experiments in which more
than one male was used. Some only of the males present were able to settle,
although the rest were perfectly capable of settling as shown by their success
once conditions became more favourable (p. 52).

In nature it is very likely that there are more males present than there is
space for them to occupy territories. The densities observed by SEVENSTER
(see p. 55), and certainly by BERTIN (p. 55) suggest that, at least locally, the
available area was fully occupied.

Competition for an area, and the existence of a minimum size for an
appropriate territory make territorial behaviour a potential agent for limiting
the number of breeding individuals, as already suggested by MOFFAT (1903),
and later authors such as KLUYVER & TINBERGEN, 1953; TOMPA, 1964;
SNOW, 1958; DELIUS, 1965; JENKINS et al., 1963. The experimental results
with sticklebacks support this hypothesis.

LACK (1966) on the other hand doubts that numbers are limited in this
way. In a population of Great Tits studied over an extensive period near
Oxford, the numbers of pairs breeding in nest boxes fluctuated between
restricted limits during 19 years. In one year, however, a sudden and very

considerable increase was observed (amounting to a 69% increase over the preceding year). LACK reasons that no such high numbers would be possible if territorial behaviour had indeed limited numbers in previous years. He derived a similar argument from TOMPA's work, where also numbers were very much higher in one year than they had been in the others.

Although settling in song birds will proceed in a more complicated way than in sticklebacks — especially because of the mixing of generations and faithfulness to a territory held in a previous year — I suggest that fluctuations in numbers as found could result from the way settling of candidates took place, with simultaneous and successive settling as extremes of a scale of possibilities. TOMPA's work may give some support to this view: a peak density was reached after a catastrophic year when many settled, older males succumbed in an unusually severe and late snowstorm. Subsequently the area was resettled by younger males that probably all settled just at about the same time, and so arrived at an exceptional density.

The occurrence of apparently very diverse densities of breeding pairs in different habitats is another argument by LACK against the limiting effect of territorial behaviour. Blackbirds in the Oxford Botanical Gardens breed at a density about 10 times higher than that in Wytham wood; the breeding density of song sparrows on Mandarte Island was likewise about 10 times higher than that in certain areas in Ohio (NICE, 1937). LACK's argument does not seem to hold. Characteristics of the habitat will certainly affect the actual size of territories. In sticklebacks it is possible to accommodate many more males in a tank with screens and a well-spaced vegetation than in an entirely bare tank (e.g. cf. WUNDER, 1930, Figs. 11 and 13).

Another of LACK's arguments (1966, p. 279) is that there is no need to account for a density limiting effect of territorial behaviour where territory already has a significance in pair formation. This reasoning does not hold either; the second possibility does not logically exclude the first, on the contrary, the two may very well be connected as in fact it seems to be the case in sticklebacks. Success in courtship was found to be greater in the larger territories, and the larger territories also had an effect on density.

KALELA (1952, 1954) and WYNNE-EDWARDS (1962) think of the territory system as an important device for regulating population density. Regulation of breeding density through territoriality can only be thought of if territorial behaviour not only prevents over-crowding by limiting numbers breeding, but also ensures in one way or another that density does not drop below a certain minimum. Such a regulatory effect, which is connected with the notion of an optimal density, will be brought about, according to KALELA and WYNNE-EDWARDS, by inter-group selection. WYNNE-EDWARDS even goes so far to suggest inter-group selection as being far more important than selection at the individual level (natural selection), stating that where the interests of a group are opposed to those of the individual, it is the latter that will be sacrificed. In my opinion, however, WYNNE-EDWARDS does not present objective evidence in favour of his view; I agree with the criticism of LACK (1966), and prefer natural selection as the simpler hypothesis.

Size of an individual territory is an important factor in determining the ultimate reproductive success of the owner. The question can then be raised whether it is some individual predisposition that makes a male fitter than his rivals to claim a larger area in the competition for a territory. Or, to go one step further, are genetical factors involved and is territory size a character selected for?

Some experiments were done to test whether any consistency in territory size is apparent among settling males. An affirmative result would permit one to conclude that a predisposition, due either to ontogenetical or to genetically controlled factors was present indeed. The data, however, do not lead to an unambiguous conclusion (p. 76). The owner of the largest territory was sometimes once again among the top-ranking males if the experiment was appropriately repeated, but sometimes he was not. A second, and much more important step, will be to investigate the chances that offspring of a male with the largest territory will claim a large territory when competing with the offspring of a male that has claimed only a small area. This point is an issue for further research.

If it were found that the ability to claim a large holding is indeed a factor that is (partly) controlled genetically, the existence of a factor which limits territory size is a necessary consequence of this. A high level of the aggressive tendency, necessary to hold a very large territory, might well be disadvantageous in a courtship situation and reduce his chances of obtaining offspring (see also TINBERGEN, 1957). Interference by rivals in a complicated situation appears to be of such a kind that a certain territory owner, that has to exhibit agonistic behaviour in defending his property against trespassers, is at the same time refraining from other behaviour patterns that would positively affect his reproductive success. For example: interactions with rivals during a courtship sequence postpone leading of the female to the nest and decrease the chances of her entering it. Agonistic behaviour towards sneaking males sometimes result in a failure to fertilize eggs. Interactions during the parental phase may result in an increased mortality or a severe retardation of the eggs. Probably some compromise is realized in such a way that a male's aggression is sufficient for defending a large enough area, yet, is low enough to allow the female at the nest within a minimum period of courtship. Possibly experiments with the offspring of males selected for different levels of aggression will give evidence on this problem.

No positive evidence is available that territory size in the stickleback is affected by the availability of food. Food for the adult might just be another of these environmental factors which affect the male directly, but it does not seem to be very probable. The territory is not likely to act as a food reserve for the young because the male chases them away soon after they have become free-swimming.

Spacing-out by the formation of territories may reduce predation, but such a factor is not very probable either in the present case — though evidence is lacking. Adult sticklebacks may have only few aquatic predators, as their spiny armament has been proven quite effective (HOOGLAND. MORRIS & TINBERGEN. 1956). SCHENK (pers. comm.) observed that sticklebacks were brought in as food by terns (*Sterna hirundo*). The density observed in nesting sticklebacks suggests that a reduction of predation by spacing-out in not very likely in this case.

It is likely that in the Three-spined Stickleback the significance of its territory, given the organization of reproductive behaviour as it has been selected for in evolution, has to be found in a reduction of interference by conspecifics during reproductive behaviour.

In my opinion this seems to be the main function for territorial behaviour. As an incidental consequence of this, territorial behaviour appears to function as a regulatory mechanism. Regulation of numbers is a by-product of selection for something else of advantage at the individual level — contrary to WYNNE-EDWARDS who states that this regulatory mechanism is selected *because* it regulates numbers.

REFERENCES

ALTUM, B. (1868). Der Vogel und sein Leben. — Wilhelm Riemann Verlag, Münster, 1st ed., p. 1-196.

BAERENDS, G. P. & BAERENDS-VAN ROON, J. (1950). An introduction to the study of the ethology of Cichlid fishes. — Behaviour Suppl. 1. p. 1-243.

BERTIN, L. (1925). Recherches bionomiques, biométriques et systématiques sur les épinoches (Gasterosteidés). — Ann. Inst. Ocean. Tome II, Fasc. I, p. 1-204.

CHITTY, D. (1960). Population processes in the vole and their relevance to general theory. — Can. J. Zool. 38, p. 99-113.

CHRISTIAN, J. J. (1959). The roles of endocrine and behavioral factors in the growth of mammalian populations. — In: Comparative Endocrinology (proceedings of a symposium held at Cold Spring Harbor, N.Y., May 1958). John Wiley & Sons inc., p. 71-97.

DELIUS, J. D. (1965). A population study of skylarks, *Alauda arvensis*. — Ibis 107, p. 446-492.

GUITON, Ph. (1960). On the control of behaviour during the reproductive cycle of *Gasterosteus aculeatus*. — Behaviour 15, p. 163-184.

HINDE, R. A. (1956). The biological significance of territories in birds. — Ibis 98, p. 340-369.

HOOGLAND, R., MORRIS, D. & TINBERGEN, N. (1956). The spines of sticklebacks (*Gasterosteus* and *Pygosteus*) as means of defence against predators (*Perca* and *Esox*). — Behaviour 10, p. 205-236.

HOWARD, H. E. (1920). Territory in bird life. — John Murray, London, p. 1-308.

HUXLEY, J. S. (1934). A natural experiment on the territorial instinct. — Brit. Birds 27, p. 270-277.

IERSEL, J. J. A. VAN (1953). An analysis of the parental behaviour of the male Three-spined Stickleback (*Gasterosteus aculeatus* L.). — Behaviour Suppl. 3, p. 1-159.

—— (1958). Some aspects of territorial behaviour of the male Three-spined Stickleback. — Arch. néerl. zool. 13, Suppl. 1, p. 384-400.

JENKINS, D. (1961). Social behaviour in the partridge, *Perdix perdix* — Ibis 103a, p. 155-188.

——, WATSON, A. & MILLER, G. R. (1963). Population studies of the Red Grouse, *Lagopus lagopus scoticus* (Lath.) in North-east Scotland. — J. anim. Ecol. 32, p. 317-376.

KALELA, O. (1952). Eläinpopulaatioiden optimitiheydestä ja ryhmien välisestä valinnasta (On the optimum density of animal populations and intergroup selection). — Arch. Soc. Vanamo 6, p. 130-136.

—— (1954). Über den Revierbesitz bei Vögeln und Säugetieren als populationsökologischer Faktor. — Ann. Zool. Soc. Vanamo 16, p. 1-48.

KLUYVER, N. H. & TINBERGEN, L. (1953). Territory and regulation of density in titmice. — Arch. néerl. zool. 10, p. 265—289.

LACK, D. (1966). Population studies of birds. — Clarendon Press, Oxford, p. 1-341.

LEINER, M. (1930). Fortsetzung der ökologischen Studien an *Gasterosteus aculeatus*. — Z. Morph. Ökol. Tiere 16, p. 499-540.

—— (1931). Der Laich- und Brutpflegeinstinkt des Zwergstichlings, *Gasterosteus (Pygosteus) pungitius* L. — Z. Morph. Ökol. Tiere 21, p. 765-788.

LORENZ, K. (1963). Das sogenannte Böse (Zur Naturgeschichte der Aggression). — Borotha Schoeler Verlag, Wien, p. 1-415.

MARSHALL, A. F. (1954). Bower Birds. — Clarendon Press, Oxford, p. 1-200.

MOFFAT, C. B. (1903). The spring rivalry of birds. Some views on the limit to multiplication. — The Irish Naturalist 12, p. 152-166.

MORRIS, D. (1952). Homosexuality in the Ten-spined Stickleback (*Pygosteus pungitius* L.). — Behaviour 4, p. 233-261.

—— (1958). The reproductive behaviour of the Ten-spined Stickleback. — Behaviour Suppl. 6, p. 1-154.

MULLEM, P. J. VAN & VLUGT, J. C. VAN DER (1964). On the age, growth and migration of the anadromous stickleback, *Gasterosteus aculeatus* L., investigated in mixed populations. — Arch. néerl. zool. 16, p. 111-139.

MÜNZING, J. (1962). Die Populationen der marinen Wanderform von *Gasterosteus aculeatus* L. an den holländischen und deutschen Nordseeküsten. — Neth. J. Sea Res. 1, p. 508-525.

NELSON, K. (1965). After-effects of courtship in the male Three-spined Stickleback. — Z. f. Physiol. 50, p. 569-597.

NICE, M. M. (1937). Studies in the life history of the Song sparrow. I. A population study of the Song sparrow. — Trans. Linn. Soc. New York 4, p. 1-247.

—— (1941). The role of territory in bird life. — Amer. Midl. Nat. 26, p. 441-487.

Noble, G. K. (1939). The role of dominance in the social life of birds. — Auk 56, p. 263-273.

Pelkwijk, J. J. ter & Tinbergen, N. (1937a). Roodkaakjes. — De Lev. Nat. 44, p. 129-137.

—— & —— (1937b). Eine reizbiologische Analyse einiger Verhaltensweisen von *Gasterosteus aculeatus* L. — Z. Tierpsych. 1, p. 193-200.

Sevenster, P. (1961). A causal analysis of a displacement activity (fanning in *Gasterosteus aculeatus*). — Behaviour Suppl. 9, p. 1-170.

Sevenster-Bol, A. C. A. (1962). On the causation of drive reduction after a consummatory act (in *Gasterosteus aculeatus*). — Arch. néerl. zool. 15, p. 175-236.

Siegel, S. (1956). Nonparametric statistics for the behavioral sciences. — McGraw — Hill — Kōgakusha Comp. Ltd, New York — Tokyo, Int. Stud. Ed. p. 1-312.

Snow, D. (1956). Territory in the Blackbird, *Turdus merula*. — Ibis 98, p. 438-447.

—— (1958). A study of Blackbirds. — London.

Symons, Ph. E. K. (1965). Analysis of spine-raising in the male Three-spined Stickleback. — Behaviour 26, p. 1-74.

Tinbergen, L. (1949). Over de dynamiek van dierlijke bevolkingen. — Openbare les, Wolters, Groningen, p. 1-15.

Tinbergen, N. (1957). The function of territory. — Bird Study 4, p. 14-27.

—— & Iersel, J. J. A. van (1947). Displacement reactions in the Three-spined Stickleback. — Behaviour 1, p. 56-63.

Tompa, F. S. (1964). Factors determining the number of Song Sparrows, *Melospiza melodia* (Wilson), on Mandarte Island, B. C. Canada. — Acta zool. fenn. 109, p. 1-68.

Watson, A. (1964). Aggression and population regulation in Red Grouse. — Nature Lond. 202, p. 506-507.

Wunder, W. (1930). Experimentelle Untersuchungen am dreistachlichen Stichling (*Gasterosteus aculeatus*) während der Laichzeit. — Z. Morph. Ökol. Tiere 16, p. 453-498.

Wynne-Edwards, V. C. (1962). Animal dispersion in relation to social behaviour. — Oliver & Boyd, Edinburgh and London, p. 1-653.

25

Reprinted from *Behaviour*, **17**, 131–132, 172–180, 183, 211–214 (1961)

Aggressiveness, Territoriality, and Sexual Behavior in Field Crickets (Orthoptera: Gryllidae)

R. D. ALEXANDER

INTRODUCTION

The males in many species of crickets exhibit specialized aggressive behavior. Violent physical combat is not unusual, and LAUFER (1927) notes that the matching of fighting crickets has actually been a popular sport in the Orient for nearly a thousand years. There has been no extensive analytical or comparative study of this phenomenon, although KATO and HAYASAKA (1958) recorded a dominance order in crickets in the laboratory, and HUBER (1955) and HÖRMANN-HECK (1957) have investigated the physiological and genetic bases for aggressive and sexual behavior in the European field crickets, *Gryllus campestris* L. and *G. bimaculatus* De Geer.

Most terrestrial crickets also excavate more or less extensive burrows, and in several species, diapausing juveniles regularly pass the winter in burrows of their own construction. The burrows of adult males are usually several feet apart, and once a male has constructed a burrow he rarely becomes separated from it. Sexually responsive females are attracted to the stationary males by their loud, rhythmical stridulations, which are inevitably distinctive among naturally sympatric and synchronic species.

This study was undertaken to determine the relationship among the three activities, aggression, territoriality, and sexual behavior, and to compare their expression in different species, both closely related and in different genera and subfamilies. Special emphasis has been placed on the role of sound as a communicative mechanism. The results have shown that territorial behavior is a surprisingly complex phenomenon in field crickets. In association with the unique sexual behavior of crickets, and with complex patterns of reinforcement in dominance-subordinance relations, it has produced in some instances the rudiments of social behavior as elaborated in the termites, and in other instances, the rudiments of phase differences as elaborated in the migratory locusts.

* * * * * * *

TERRITORIAL BEHAVIOR

INTRODUCTION

The nature of territoriality in invertebrate animals is a subject of some dispute. CARPENTER (1958) completely excluded the invertebrates from a discussion of territoriality, apparently because of the paucity of investigations which would allow adequate comparisons with the phenomenon as it is known in vertebrate animals.

Four phenomena are associated to produce most of the various kinds or degrees of territoriality in different animals. The first two of these are:

1. displays of aggressive behavior in encounters between individuals.

2. a tendency on the part of individuals to remain in unusually restricted localities (the home range) or to consistently return to specific spots (crevice, burrow, nest, marker), or both of these tendencies.

These two characteristics exist in a wide variety of arthropods, among insects in both social and non-social species. When they occur together, they result in a simple sort of territoriality in which the location of individual animals can be predicted at particular times, and in which the location of aggressive encounters between particular individuals can also be predicted, but in which the outcomes of particular encounters do not necessarily bear a relationship to their location. [1])

In many vertebrate animals, one or two additional phenomena have been found to occur.

3. the domination by an individual of encounters within its home range which it would have lost anywhere else and/or the exhibition of aggression

---

1) Although widely used, NOBLE's (1939) definition of territory as "any defended area" is difficult to apply, especially to insects; because of problems in determining what constitutes "defense," it eventually comes to mean that territoriality exists in any animal which fights or displays aggression, and that an animal's "territory" is therefore any place in which it fights. It seems clearer to specifically reserve the term "territoriality" for the evolutionary stage characterized by association with a particular area, and the first indication of this is when the individuals of a species are not only aggressive but also restrict their movements (or their aggressiveness) in some special fashion, thereby affording a degree of predictability as to where they will have most or all of their fights.

by an individual within its home range in situations in which it would not have exhibited aggression elsewhere.

4. a gradual decrease in ability to dominate particular encounters and/or a gradual decrease in the number of situations in which aggression is exhibited with an increase in the distance from the center or focal point (crevice, nest, burrow, marker) of the home range. Extremes of elaboration in this direction could result in (a) reduction and eventual elimination of physical contact, with aggressive behavior becoming simply a "threat" expressed only by individuals approached at their nests or within their "home ranges" and rarely or never reciprocated, and (b) total avoidance so that each individual always remains within his own range and never trespasses upon those of his neighbors, even when behavioral interactions resulting from population density are the prime factors limiting the size of the home range.

Apparently, neither of the last two phenomena listed above (3 and 4) has previously been clearly demonstrated in a non-social invertebrate animal, though JACOBS (1955) and KORMONDY (1959) have presented evidence that male dragonflies incorporate a degree of geographic significance in their aggressive behavior. [2]).

It seems logical that in any species in which the individuals occupy crevices or burrows for brief or extended periods of time, the mere entering of such a niche would provide a degree of invulnerability, allowing animals to resist aggression by individuals which would always dominate them in encounters in the open. However, most aggressive encounters in crayfish and crickets terminate without the sort of combat whereby the subordinate is routed through sheer force of physical strength, and this might have a bearing on how often such presumed invulnerability would actually affect the outcome of specific encounters. Even if the mere entering of a crevice increases an individual's ability to dominate or to resist aggression by others, there is still the question of whether the change is due to (1) an effect upon the inhabitant of the crevice, (2) an effect upon the individual who encounters another ensconced in a shelter, or (3) effects upon both individuals.

* * * * * * *

TERRITORIALITY IN CAGED MALE FIELD CRICKETS

When the five crickets used in the principal hierarchical studies were initially introduced into the terrarium there were no niches or crevices avail-

[2] MOORE (1952) states that PUKOWSKI (1933) has demonstrated "undoubted territorial behavior" in carrion beetles (*Necrophorus* sp.); PUKOWSKI's paper was not available during preparation of this manuscript.

able. For seven days there was no evidence that any cricket was restricting his movements around the terrarium more than any other. Fig. 16 shows the location of 1096 encounters recorded during 9½ hours of observation on these days. All of the crickets spent more time and had more encounters at the EQZ end of the terrarium (Fig. 14) than at the AMV end. Two factors, the location of the water vial and the lower light intensity caused by the two wooden walls, A-E and E-Z, were probably involved.

On the ninth day of the test (Feb. 10), the dominant male (#2) was noticed to be resting alongside the water vial at P more than usual. Then it was seen that a small depression had been hollowed out there, and #2 later dug and kicked at the sand in the bottom of the depression, gradually

Before Male No. 2's Territoriality	During Male No. 2's Territoriality
(590 min.)	(90 min.)

Encounters Involving Male No. 2

4.7	6.6	8.3	11.4	12.4
2.2	1.1	2.7	10.2	11.1
1.8	4.4	8.3	12.0	9.3

totals   585

0	0	0	24.3	11.8
0	0	1.8	30.6	10.8
0	0	0	10.8	10.0

111

Encounters Not Involving Male No. 2

2.1	4.1	7.9	11.9	9.5
1.5	3.0	3.9	10.3	5.2
1.2	3.7	12.3	13.7	9.5

totals   511

1.0	3.0	6.1	20.4	7.1
0	1.0	0	7.1	1.0
2.0	4.1	11.2	24.5	11.2

98

All Encounters

3.5	5.0	7.9	11.7	11.0
1.9	2.0	3.4	10.2	5.0
1.6	4.0	10.1	12.7	10.0

totals   1096

0.5	1.5	2.9	22.6	9.6
0	0.5	0.9	19.2	6.3
0.9	2.0	5.3	17.3	10.5

209

Fig. 16. The percentages of encounters occurring in different parts of the terrarium before and after male #2 had begun to behave territorially with regard to his burrow under the water vial (arrow).

enlarging it until a day later it was big enough that he could squeeze completely under the water vial and come out on the other side. On the morning of the tenth day, #2 remained motionless or nearly so for long periods of time (up to 10-15 minutes) in the depression under the water vial, returned to it after encounters with males nearby, and was not observed to go beyond the middle of the terrarium all day long (four separate hours of continual observation and intermittent observations all through the day). Repeatedly he walked down the wall of the terrarium to near the midpoint, then turned at about the same place each time and went back to his burrow. Subsequently, the same behavior was exhibited by all the other males as each occupied inverted rectangular pillboxes open at both ends which were introduced into the terrarium.

Before Feb. 10, encounters involving #2 occurred at different locations in the terrarium with about the same frequency as encounters which did not involve him, while after Feb. 10, nearly all of his encounters were within the four compartments at the EQZ end of the terrarium (Fig. 16). In P and Q, the compartments located precisely at the entrances to #2's crevice, #2 had 21.3 % of his encounters before Feb. 10 and 41.4 % after Feb. 10, while of all other encounters, 15.5 % occurred in these compartments before Feb. 10 and 8.1 % afterward. When #2 repelled a male in compartments P or Q, the dominated individual frequently stopped retreating at D or Y, this accounting for the increase in encounters not involving #2 at these locations. D seemed more an integral part of #2's territory than Y, for he spent much more time in that area and had more encounters there.

The nature and effect of #2's attachment to his burrow at P can be summarized as follows:

1. He spent long periods of time nestled immobile in the burrow, repeatedly examined, explored, and moved about, in, and out of it, and made unusually direct returns to it after encounters.

2. He made distinctive "sallies" or "patrols" out of the burrow, these consisting of him suddenly walking out after a long period of immobility — without external stimuli apparent — walking directly away from the burrow to about the midpoint of the terrarium, then turning about and walking directly back to the burrow (not meandering along the wall of the terrarium as was usual) and beginning another period of immobility nestled inside it. His behavior during these sallies was recognized as distinctive — and aggressive — even by persons who had never seen it before. He stalked in a quick, positive fashion, reared high on his forelegs, with the maxillary palpi drawn up and back in the same position as during fighting(cf. right male, Fig. 2, for a similar position).

3. He was never absent from the burrow for very long, and he failed to travel as far away from it at any time as he had done frequently prior to his occupation of it.

4. There was a definite shift in the location of encounters, with #2 monopolizing the area around his burrow.

These characteristics appeared in what seemed to be identical form whenever crevice occupation began. It was easy to recognize territorial behavior, and no male acted this way except when he was occupying a crevice or burrow. If a male was going to begin territorial behavior, it could usually be perceived when he entered a crevice, by the characteristic way that he examined it and then settled himself inside. Several times subordinate males began to behave in this fashion upon entering a crevice temporarily vacated, but then were routed so quickly that further elements of territoriality did not appear. CRANE (1958) found that fiddler crabs occupy burrows intermittently, apparently in association with cycles of physiological change. The temporary absence of a hungry or thirsty male cricket from a burrow might produce a similar effect, and through such temporary absence a male might lose the hyper-aggressiveness associated with burrow occupation, though retaining sufficient association with the burrow to successfully return to it (*cf.* figs 18, 19). Lone males caged with a single crevice available are rarely seen outside except at night, and in the present tests the frequent failure of subordinant males to occupy temporarily vacated burrows may have been due to their continual domination by the other males with which they were forced to come into contact rather frequently.

* * * * * * *

The above sequence of events, and the fact that a male which remains immobile inside a burrow or crevice for extended periods of time is effectively isolated from other males in the immediate vicinity, indicates that occupation of a niche and its peculiar associated behavior changes a male's ability to win specific encounters. Because a male occupying a niche rarely leaves its vicinity, the effect is that males displaying territorial behavior

Winner	Niche-Occupying Male		
	Neither	No. 1	No. 4
No. 1	0	13	0
No. 4	4	1	15

Fig. 17. The outcomes of 33 encounters between two closely matched males (#1 and #4) during times when neither or only one was occupying a crevice.

dominate a much higher percentage of their encounters than they would otherwise. Although the mechanism is quite different, the effect is very similar to that produced in vertebrate animals when a complex association with a particular area enables an individual to defend it against intruders which could defeat the defender in combat in any other locality.

Fig. 17 confirms that the mere occupancy of a niche goes along with a slight adjustment in dominance, or an ability to dominate males which without the complication of niches and territories are slightly dominant to the niche occupant. Male #1's single loss while "occupying" a niche in this series occurred just after he had entered a niche, and before he had time to show any special reaction toward it.

* * * * * * *

The occupation of a crevice or burrow by a male cricket functions in still another, more indirect fashion, in increasing the likelihood of production of the calling song. A male producing aggressive chirps after repelling another male from his crevice often does not stop chirping, but continues, gradually changing to the calling song rhythm and then sometimes keeping this up for long periods of time. Through the advertisement of calling, a male is more likely to attract and copulate with a female, this again raising his dominance temporarily. A territorial male's isolation is undoubtedly enhanced by his calling, due to the similarity of this sound to the aggressive sounds, and its functioning as a mildly aggressive stimulus. Finally, the calling of one territorial male must frequently stimulate calling in neighboring individuals, as it does in other Tettigonioidea (ALEXANDER, 1960).

* * * * * * *

TERRITORIALITY AND THE EVOLUTION OF SOCIAL BEHAVIOR IN INSECTS

The several steps in the evolution of territoriality which are evident in comparisons of different species and genera of Gryllinae bear an interesting relationship to the evolution of social behavior. Social behavior in insects has apparently evolved from two kinds of beginnings: (1) the aggregation of individuals of the same generation and of about the same age or stage of development, such as overwintering or nesting queens in Hymenoptera (compare MICHENER, 1958, and PARDI, 1948a), and (2) the attenuation of parent-offspring relations. Each of these beginnings is eventually followed by the other if social interactions continue to be elaborated, and the primary difference between them is in the order of appearance of specific behavioral interactions rather than in the kind that appear.

In the first case above, PARDI has shown how dominance-subordinance relations can play an important role. The dominant queen in a group nesting together is the one with the largest ovary; the act of subordination is to release to the dominant a drop of fluid which has nutritional value in connection with further enlargement of the ovary. The result is a re-inforcement of dominance which is directly associated with ability to lay eggs. This effects a division of labor and results in the dominant becoming the functional queen and the subordinates becoming functional workers (see also, FREE, 1955).

In the second case above, there is usually an initial isolation of the breeding pairs, and this is the kind of potential "beginning" illustrated by the Gryllinae. The steps in this process may be enumerated as follows:

1. unusual restriction of movement (cyclic or permanent)
2. attachment to a particular spot
   a. special reactions to the spot, such as unusually long periods of time spent in it, repeated examination and exploration, consistent returns.
   b. modification of the spot, such as odor deposition, hollowing out, construction of webs, modification of the entrance.
3. transportation of materials out of and into the nest
   a. removal of material during excavation.
   b. removal of fecal pellets and other waste products (or their deposition only in special locations or only when outside).
   c. introduction of materials later eaten.
   d. introduction of materials later incorporated in modification (construction).
   e. translocation within the nest of materials later eaten.
   f. translocation of eggs within the nest.
4. attenuation of parent-offspring relations.
5. increase in complexity of sibling relations.
6. division of labor.
7. development of sterile castes.

Whether this process is ontogenetic or phylogenetic, it begins with a pair of individuals and culminates in the development of a colony. Among the orthopteroid insects, the highest order of complexity in this sequence is reached in the Isoptera, and the sequence is traced out ontogenetically — doubtless with some variations from the above order — each time a new colony of termites is founded by a pair of reproductives. Elsewhere among the orthopteroids, the first four steps are probably completed in the Embioptera and the Dermaptera (MELANDER, 1902; WHEELER, 1923). Among the Gryllinae, *Anurogryllus muticus* completes the first three steps, at least to 3e; other species exhibit none of the behavioral characteristics listed, and still others—as demonstrated by the various experiments reported here—exhibit different stages of intermediacy between the extremes. There is an interesting difference in this regard between the behavior of the Isoptera, Embioptera, and Dermaptera, and that of the Gryllinae. In the former groups it is the female which initiates the sequence, while in the last one it is the male which initially becomes sedentary. In *Anurogryllus* the female becomes attached to the burrow originally begun by the male after being attracted there by his song. Probably the vegetable materials accumulated by the adults in chambers of the burrow are later eaten not only by the female but also by the young nymphs which spend one or two instars with the female in the burrow. Special behavioral interactions may occur between the adult female and the young nymphs. HAYSLIP (1943) found that *Scapteriscus acletus* Rehn and Hebard and *S. vicinus* Scudder seal their eggs in cells, but the female of *Gryllotalpa hexadactyla* Perty places her eggs in the end of her burrow and "guards" both the eggs and the young nymphs. GHOSH (1912) reports that the females of *Brachytrupes achatinus* inject their eggs shallowly into the soil at the ends of their burrows, and the young nymphs remain in the adults' burrows only two or three days after hatching.

Apparently, no one has noted any special activity of crickets inhabiting burrows with respect to the disposition of fecal pellets. Only a single instance of defecation was noted in a territorial cricket during this study. An *Acheta* male, after a long period of immobility inside an inverted pillbox, backed slowly out in a straight line, stopped a few inches away, dropped a fecal pellet, and walked immediately straight back into the pillbox and became motionless as before. No other cricket was ever seen to locomote backward in this distinctive fashion.

The brief survey given above of variations in the behavior of closely related species of crickets, especially with regard to aggressiveness, territoriality, burrowing, and various aspects of the life history, shows how the different steps in the development of social behavior such as that exhibited by termites and web-spinners can occur through related stages in the step-by-step superimposition of simple changes in behavior patterns upon the existing mode of life in solitary species. Several phenomena become elaborated along with the complexity of interactions among individuals. Thus, the development of more and more extensive burrows, the elaboration of territoriality, and the transportation into the burrow of materials which can later be utilized as food begin the process of independence from inclement seasons and place the species well upon the way toward continuous breeding. Increased attachment to the burrow and long periods of time spent in close proximity by groups of individuals inside the burrow increase the chance that selection will act to favor genetically-based changes in behavioral interactions leading toward social existence. In the proper situation the process becomes an accelerating and re-inforcing one, each modification which is successfully super-imposed upon another tending to increase the likelihood of further modification in that direction. In assessing the chances that the burrowing crickets will continue to elaborate social behavior, two facts seem significant: (1) the increased length of adult life as a result of isolation and the *Anurogryllus* female's isolation in the burrow with her eggs, and (2) the appearance of stridulatory apparatus in female mole crickets after an interchangeable sound signal had become a part of the species' repertoire.

References

Alexander, R. D. (1960). Sound Communication in Orthoptera and Cicadidae.—Animal Sounds and Communication (Lanyon and Tavolka, ed.). Symposium of The American Institute of Biological Sciences 7, p. 38–92.

Carpenter, C. R. (1958). Territoriality: a review of concepts and problems. Behavior and Evolution (Roe and Simpson, ed.).—Yale Univ. Press, p. 224–250.

Crane, J. (1958). Aspects of social behavior in fiddler crabs, with special reference to *Uca maracoani* (Latreille).—Zoologica 43, p. 113–120.

Free, J. B. (1955). The behavior of egg-laying workers of bumblebee colonies.—Brit. Jour. Animal Behaviour 3, p. 147–153.

Ghosh, C. C. (1912). The big brown cricket (*Brachytrypes achatinus,* Stoll).—Mem. Dept. Agric. India 4, p. 161–182.

Hayslip, N. C. (1943). Notes on biological studies of mole crickets at Plant City, Florida.—Fla. Ent. 36, p. 33–46.

Hörmann-Heck, S. von (1957). Untersuchungen über den Erbgang einiger Verhaltensweisen bei Grillenbastarden. (*Gryllus campestris* L. X *Gryllus bimaculatus* De Geer).—Zeitschr. für Tierpsychologie 14, p. 137–183.

Huber, F. (1955). Sitz und Bedeutung nervöser Zentren für Instinkthandlungen beim Männchen von *Gryllus campestris* L.—Zeitschr. für Tierpsychologie 12, p. 12–48.

Jacobs, M. E. (1955). Studies on territorialism and sexual selection in dragonflies.—Ecology 36, p. 566–586.

Kato, M., and K. Hayasaka (1958). Notes on the dominance order in experimental populations of crickets.—Ecol. Rev. 14, 311–315.

Kormondy, E. J. (1959). The systematics of *Tetragoneuria*, based on ecological, life history, and morphological evidence (Odonata, Corduliidae).—Univ. Mich. Mus. Zool. Misc. Publ. 107, p. 1–79.

Laufer, B. (1927). Insect-musicians and cricket champions of China.—Field Mus. Nat. Hist., Anthropology Leaflet 22, p. 1–27.

Melander, A. L. (1902). Two new Embiidae.—Biol. Bull. 3, p. 16–26.

Michener, C. D. (1958). The evolution of social behavior in bees.—Proc. Tenth Intern. Congress Ent. (1956) 2, p. 441–447.

Moore, N. W. (1952). On the so-called "territories" of dragonflies (Odonata-Anisoptera).—Behavior 4, p. 85–100.

Nobel, G. K. (1939). The experimental animal from the naturalists's point of view.—Amer. Nat. 73, p. 113–126.

Pardi, L. (1948a). Dominance order in *Polistes* wasps.—Physiol. Zool. 21, p. 1–13.

Pukowski, E. (1933). Ökologische Untersuchungen an *Necrophorus* F.—Zeitschr. Morph. Ökol. Tiere 27, p. 518–586.

Wheeler, W. M. (1923). Social Life Among the Insects.—New York: Harcourt, Brace, and Company, p. 1–375.

26

Reprinted from *Proc. 15th Intern. Ornothol. Congr.*, 134–149 (1972)

A current model of population dynamics in Red Grouse

Adam Watson & Robert Moss

The Nature Conservancy, Blackhall, Banchory, Kincardineshire, Scotland

This paper describes our present thinking about population dynamics of Red Grouse *Lagopus lagopus scoticus* and the relevance of these findings to animals in general. It includes some statements about current work for which the evidence has not yet been published.

A review written in 1967 (JENKINS & WATSON 1970) describes study areas, habitats, food and methods, and summarises results on behaviour, nutrition and populations. An essay written in 1968 (WATSON & MOSS 1971) considers problems of spacing behaviour and aggression in relation to populations and nutrition, and WATSON (1970) gives a more specialised progress report of research on the hormonal background to spacing behaviour and aggression. Readers who wish to read the earlier definitive papers or fairly detailed summaries of the evidence on which much of the present model rests, should refer to these reviews. The present paper gives only a brief statement of our main findings and summarises evidence where new work since the other reviews has led us to reject earlier interpretations. Its emphasis is on population dynamics and the roles of spacing behaviour and nutrition in population change. Our current model of population processes is presented in Figures 1–3 and is the basis for the following discussion.

THE MAIN QUESTIONS ASKED

The Red Grouse year may be conveniently divided into two: the period of reproductive gain to August when there are most full grown birds, and the period of overwinter loss up to the time of low numbers in the following spring when the birds are about to breed again. We are interested in what determines breeding success, which we define as the mean number of full grown young reared per old bird in the population, and also in what determines the size or density of the spring breeding stock. Breeding populations of animals are determined partly by adult mortality over the winter and partly by the level of recruitment, which is the number of young that are successful in becoming part of the breeding population (cock and hen Red Grouse breed in their first year). Recruitment is sometimes confused with breeding success, so in terrestrial vertebrates it is useful to stick to the definitions given above.

We can next consider breeding success and recruitment not just in one year but over several years. In particular, the question here is why do annual fluctuations occur in breeding success and in spring breeding stocks, so that on the same area there are more birds in some years than in others (i.e. differences within areas).

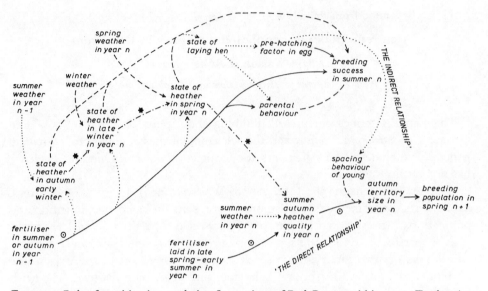

FIGURE 1. Role of nutrition in population fluctuations of Red Grouse within areas. *Explanation*: ⊙ Via fertiliser experiment. * Possible connection here not yet studied. ... Evidence based on inference, or anecdotal evidence. – – – Evidence based on associations or correlations. —Evidence based on experiments.

(1) In addition to the variation amongst year-classes shown here, there are variations amongst individuals within each year-class, some of which are inherited. Some individual variation in breeding success within year-classes can be attributed to this individual variation amongst the parents in captivity, where the food provided to all individuals is the same (KOLB 1971).

(2) Hormone experiments where nutrition was not changed show a causal relationship between behaviour and territory size.

In this figure, but not in the other ones, the various components and factors affecting them are arranged in time-sequence, with earlier times successively to the left.

Another part of our research has been to find why grouse are usually much more abundant in some places than in others (i.e. differences amongst areas).

We could have organised this paper differently by tackling a series of general topics in turn, such as the importance or lack of importance of predation, weather, food, disease etc. We did not do so because we think these topics are far too general to be of much use in framing falsifiable hypotheses that can be properly tested.

BACKGROUND INFORMATION ON NUTRITION

The shoots and flowers of heather *Calluna vulgaris* form the main food of the Red Grouse populations that we have studied, at all times of year. This simplicity of diet is a major advantage in studying nutrition, especially as heather is the predominant plant on most of the study areas and on some of them forms a virtually pure stand. However, the heather's nutrient content varies locally within

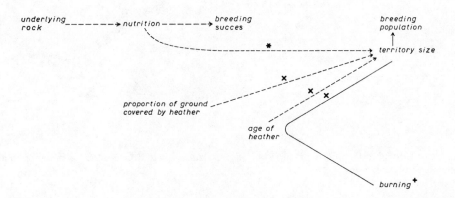

FIGURE 2. The role of nutrition in differences in mean population levels of Red Grouse amongst areas. *Explanation*: – – Evidence based on associations or correlations. — Evidence based on experiment. * See fluctuations within areas. + Probably via age of heather. × Relationship probably operating in part via nutrition.

the same area, and also varies with season, year, amount of previous grazing and the plant's age. It is therefore difficult to decide how much of the potentially edible material present would normally be in fact edible to grouse, and thus difficult even to define food, because the bird's requirements vary seasonally, and because preferences vary with season, with the heather's age and chemical composition (Moss 1967), and with the bird's sex and age. Nevertheless, nearly all the material available very locally on individual plants may be eaten at times, for instance in heavy snow when most of the heather in the area is buried and the few tips projecting through the snow become heavily grazed.

POPULATION CHANGES

Breeding success—In any year on a given area some hens rear large broods and others rear small broods or no young at all. So many variables are involved, such as predation, weather in relation to the individual hen's timing of hatch, and variable food supply for hens on different territories, that this problem is difficult to tackle in the wild. However, KOLB (1971) finds that captive hens which are all kept on the same diet and in other standard conditions lay clutches from which different proportions of the hatched chicks survive. This shows that, within a population in a given environment in any one year, variations in individual parents contribute to variations in breeding success. This leaves untouched the larger question of differences in the mean response of a whole population within an area in different years (see Fluctuations in breeding success within areas).

Breeding stock—Experiments where Red Grouse are shot (WATSON 1967a; WATSON & JENKINS 1968) show that the size of the breeding stock is determined by

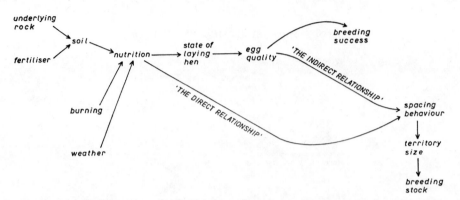

FIGURE 3. Model of the effects of nutrition on populations of Red Grouse. This is a simplified version of Figures 1 and 2.

the number of territories chosen by cock grouse in the previous autumn and by the number of hens pairing up with them (henceforth called 'territorial' hens). Mortality over the winter falls very largely both on young birds which fail to get a territory or pair up, and on old birds which fail to retain their old territory or their old pairing attachments (WATSON 1967b). Non-territorial birds will readily take territories, survive well and breed, if vacancies are made by shooting territorial birds.

On average, juvenile mortality in the first year is similar to subsequent adult mortality. As most mortality is socially induced, this means that, on average, juveniles in autumn have as good a chance of getting a territory as older birds have of keeping their territories for another year. The successful territorial cocks and hens survive very well until the next annual reorganisation of territories in the following autumn. Also, the young birds survive very well until the time in autumn when changes in strife and social organisation divide the population into those with and those without territories. The mortality overwinter is therefore secondary to prior events' that determine both the size of the spring population and the winter mortality.

Winter loss or disappearance is mostly mortality in Red Grouse, not migration. Large numbers are found dead during our studies, and recoveries of about 1300 grouse ringed as chicks and of many hundreds back-tabbed when full grown show that nearly all grouse take territories or die locally within 1 km, and that migration to settle and breed on distant moors over 5 km away is rare (JENKINS et al. 1967). This means that population dynamics depend largely on local changes in chick survival, juvenile recruitment and mortality of adults and juveniles and that large-scale immigration or emigration of breeding populations does not occur.

Fluctuations in breeding success within areas—Breeding success varies from 0.1 to 3.3 full grown young reared per old bird in different years within the same area (JENKINS et al. 1963, 1967). Most of this variation is due to mortality of chicks less than two weeks old. Mean clutch size varies only from 6 to 8 eggs. Losses of eggs up to hatching also vary amongst years, and are also small compared with variations in mortality of chicks. Variations in chick mortality, clutch size, egg losses and hatching losses all usually occur in the same direction.

Red Grouse chicks feed themselves entirely but they need brooding by the parent hen until about six weeks old. It might be thought that bad weather in the first two weeks after hatching—when most chick mortality occurs—might kill grouse chicks or prevent them getting enough food. However JENKINS *et al.* (1963, 1967) found no correlation between breeding success and certain parameters of temperature, duration of sunshine and rainfall after hatching. Instead, they and MILLER *et al.* (1966) found that breeding success was correlated with various parameters before hatching, such as the mean dimensions of eggs, the survival of adult grouse in late winter before egg laying, the proportion of green heather foliage in late winter, the timing of fresh growth of the heather in spring, the mean length of shoots grown in the previous summer, and on one area the mean temperature and duration of sunshine in the previous summer (presumably affecting plant growth).

The preliminary result about weather after hatching was tested by taking eggs laid in the wild into captivity, where the effects of bad weather or lack of food could be largely ruled out by always providing warmth and plenty of food. Survival of the chicks in captivity was correlated with that in the wild stock from which the eggs were taken. This showed that some pre-hatching factor of quality in the egg was important for breeding success (JENKINS *et al.* 1965; WATSON *et al.* 1969). As there were correlations between breeding success and certain measurements of heather before egg laying, the hypothesis was proposed that breeding success was determined by the nutritive value of the heather before egg laying. This was tested by fertilising an area. The heather's nutritive value increased and the grouse that settled on the fertilised area reared larger broods than on the untreated control area (MILLER *et al.* 1970).

Fluctuations in breeding stocks within areas—Breeding stocks fluctuate on all our study areas with amplitudes varying from twofold on some areas up to 15-fold on others. Approximately the same ground is all occupied every year by territorial birds, so changes in breeding stocks are inversely related to changes in mean territory size (WATSON & MILLER 1971).

Fluctuations in Red Grouse can be regarded as cycles, as they show (WATSON 1971) 'delayed density dependence' (VARLEY & GRADWELL 1970) which is a typical feature of population cycles. The periodicity of cycles seems irregular on most of our study areas, but on one area studied since 1957 they resemble a 'ten-year cycle'. However, any statements about periodicity are of doubtful value without a much longer run of data.

Differences amongst areas—Red Grouse are usually more abundant on some moors than on others.

(a) Differences between areas of widely different climate

(1) Breeding stocks

Breeding stocks of red grouse are much lower in western Scotland (BROWN & WATSON 1964) and western Ireland (WATSON & O'HARE 1970) than on our study areas in eastern Scotland. The western areas are very much wetter, heather is much scarcer and often sparser in growth, and the vegetation is dominated by grasses and sedges. This association between mean grouse density and dominance of heather is the basis for experiments now being done in Ireland to see if grouse stocks will increase if the environment is manipulated to improve conditions for heather growth.

(2) Breeding success

Not enough work has been done yet to know if breeding success is significantly any worse in the far west of Ireland than in eastern Scotland, but there are certainly not such profound differences as in the case of breeding densities. The main difference in autumn densities between the two regions is due to differences in breeding stocks, not breeding success.

(b) Differences amongst areas of similar climate

Most of our work is done in eastern Scotland, where heather flourishes well in the dry climate and is a dominant plant on all the study areas.

(1) Breeding stocks

On these areas, differences in mean breeding stocks or breeding densities were found by MILLER *et al.* (1966) to be correlated with the cover of heather (proportion of ground covered), thus confirming what is said in (a) above about differences between regions. MILLER *et al.* also found an inverse correlation between mean breeding density and mean age of heather (more grouse where a higher proportion of the area was covered by young heather). Heather age varies mainly according to human management by burning, and an experiment (MILLER *et al.* 1970) showed that burning of the heather in small fires up to 2 ha increased the breeding stock of Red Grouse. Burning changes the age-structure, nutrient content, accessibility and quantity of heather, as well as the physical structure of the habitat. Mean territory size increased only after three growing seasons following the first fires. Therefore the birds' response was not to high nutrient content alone, as this is highest in heather which grows in the first year after burning. The delay in response of the population might be explained by current experiments in captivity, which show that grouse find one-year-old heather unpalatable, preferring three-year-old heather most. The wild grouse might have been responding to an increase in the amount of preferred food.

The correlation in MILLER *et al.* (1966) between breeding density and the cover and age of the heather did not apply on two moors, where breeding densities were much higher than could be accounted for by the cover or age of the heather there. These areas lay over base-rich rocks and the nutritive value of the heather there was higher (MILLER et al. 1966; Moss 1969) than in heather of the same age from base-poor moors. Another feature is that the amplitude of fluctuations is much smaller on the base-rich than on the poor moors. We explain relations between populations and the complex of age, cover and nutrient content by suggesting a) that on moors over bedrock of different fertility, differences in mean breeding stocks are related to this fertility, and b) that differences in mean stocks on moors over similar rocks are related to age and cover of heather there.

The average size of territories on the base-rich moors is small. Occasionally, breeding stocks on base-poor areas rise as high as on some rich moors. In this case, territory size on the two kinds of area is the same. Our current explanation of this is that nutrition for grouse on the poor moor is temporarily as good as on a base-rich area. We assume here that the problem of mean differences amongst areas contains some elements in common with the problem of annual differences within areas, but realise that the two problems may not necessarily be the same (CHITTY 1967).

(2) Breeding success

There is no association between the cover or age of the heather (MILLER et al. 1966) on different areas and the breeding success of Red Grouse there. The only difference in breeding success between different areas is that it is usually better and more consistent on moors over base-rich rocks than over base-poor rocks.

WHAT DETERMINES TERRITORY SIZE ?

We have shown that the spring breeding stock of Red Grouse, at all densities which we have studied, is limited by their own territorial behaviour in the previous autumn. In other words, variations in density are caused by variations in territory size, and we must now ask: what determines territory size?

Within areas in any one year—Individual variation in territory size within a study area is related to physical cover or visibility at the eye level of a grouse or a man, territories being smaller where visibility from the centre of the territory is poorer, as for instance among hillocks (WATSON 1964; WATSON & MILLER 1971). However, individual variation in territory size in any one year is primarily correlated with the relative aggressiveness of the cocks towards their neighbours (WATSON 1964). Experiments with implants of sex hormones (WATSON 1970) show that androgen increases aggressiveness and territory size, whereas oestrogen makes a cock less aggressive and causes him to give up his territory. This shows that there is a causal relationship between aggressiveness and territory size within areas in any one year.

Furthermore, LANCE (1971) has found an inverse correlation within areas, between the territory size of individuals and the nitrogen content (taken to be an index of nutritive value) of the heather in each territory. The amount of variability in territory size that is accounted for ($r^2 = 25\text{-}50\%$) by nitrogen content is similar to that accounted for by variations in aggressiveness. The obvious interpretation is that individuals which are on an area of poorer nutritive value compensate by taking bigger territories.

Variations in territory size within areas amongst years—All year classes do not change their mean territory size in successive years to the same extent. On one, intensively-studied area, the fluctuation in mean territory size of old cocks amongst years was much less than that of young cocks (WATSON 1970), suggesting that the immediate or proximate cause of changes in population density is mainly a change in the mean territory size imposed by each year's crop of young cocks. Also, during the years of decline, when breeding success was poor, the young cocks that succeeded in getting territories took significantly larger territories than territorial cocks from older year classes at the same time (WATSON & MILLER 1971). In a subsequent year of increase, the territorial young cocks reared during a summer of good breeding took smaller territories and were less aggressive than territorial cocks from year classes reared during the preceding decline. This suggests that in years when breeding success is good the young cock grouse are less aggressive than usual, and vice versa. We explain the observed correlation (JENKINS *et al.* 1967) between breeding success and subsequent changes in spring breeding density by suggesting that both vary as a result of variations in a common antecedent factor. Another way of putting it is to suggest that what determines breeding success also continues to have a later influence on the determination of territory size. In our model, the initial or antecedent factor is parental nutrition.

It is not simply a matter of young differing from old birds as such, but of whole year classes varying in parameters of quality such as aggression and longevity throughout their lives. Territorial birds from one year class reared during a decline survived much better both as young birds and later as old ones, than other year classes, and took significantly larger territories both as young and later as old.

We may now ask: what determines this varying quality of year class that is associated with differences in mean territory size between years? We first consider the effects of numbers, in terms of various possible mathematical models which simply utilise data from counts and which make the implicit assumption that all individuals are equivalent. The various possibilities, which are by no means exclusive, are posed as questions.

Is territorial behaviour a density-dependent mechanism which regulates spring densities by ensuring that a higher proportion of the autumn population fails to get territories when densities are high than when low? Although there is a suggestion of a density-dependent effect in some of the data there is better evidence that the proportion failing shows so-called 'delayed density-dependence' (VARLEY & GRADWELL 1970) where the proportion failing is higher during declines than at similar densities during increases (WATSON 1971). This does not necessarily mean, as this term appears to imply, that a change in numbers is actually caused by some density-dependent effect operating the year before the change actually occurs. This mathematical relationship is obtained if declines and increases in population continue for more than one year and if the mortality factor considered, k in the terminology of VARLEY & GRADWELL (1970), is greater in the years of decline than in the years of increase. In other words, typical cycles or fluctuations give data of the sort called 'delayed density-dependent', even though there is no evidence that the density in the previous year is causally related to k. In the absence of data other than counts, the factors causing the fluctuations remain unknown.

Next, we ask does mean territory size depend on numbers of grouse available and does it change in proportion to them? This would mean that the proportion failing to get a territory would be the same in every year. This is not the case, as pointed out in the previous paragraph.

A third question involving a numerical effect is: is the aggression of the chicks conditioned by brood size, so that they are less aggressive and take smaller territories when reared in large broods and vice versa? This would result in the observed correlation (JENKINS et al. 1967) between breeding success and changes in breeding stocks. We have no direct evidence on this point. If this mechanism does occur, it probably cannot explain all the changes observed because, as will become apparent below, there is evidence for the existence of other, nutritional, effects which involve more than numbers of birds, and because the correlation coefficient in JENKINS et al. (1967) did not apply on base-rich moors, and in any case accounted for only a small proportion of the variability in breeding density.

A density-dependent numerical effect would be observed if population fluctuations were caused by a density-dependent breakdown of the stress-endocrine system as found experimentally in captive mammals by CHRISTIAN et al. (1965) at artificially high densities. This breakdown would be followed by high mortality, poor breeding, and low viability of offspring. We have no direct evidence for or against this hypothesis, but consider it unlikely to be of major importance firstly because in our data there is no large effect from density-dependent mortality and secondly because there is very little convincing evidence for the hypothesis in wild animals. This, however, does not rule out a stress effect with delayed density-dependent results. The suggestion (HÖHN 1967) that this stress-endocrine effect causes declines in populations of birds, including *Lagopus* species, is speculative. The specific anecdote mentioned by HÖHN, that many *Lagopus lagopus* at high density in northern Canada were apparently non-breeding, is a casual observation which can be explained equally well by the suggestion—which HÖHN believed highly improbable—that the hens he saw without young did breed earlier but lost all their young soon after hatching. However, this alternative simpler explanation is consistent with well documented cases of the same thing in Red Grouse and Scottish Rock Ptarmigan *Lagopus mutus*.

We next consider the role of inherited effects in population fluctuations. Taking genetic inheritance first, is there a balanced genetic polymorphism favouring more aggressive individuals when populations are more crowded (CHITTY 1967)? We have no evidence for or against this hypothesis.

Do other inherited effects occur? KOLB (1971) has evidence that maternal, non-genetic, inheritance affects certain parameters in the chicks, but not including aggression which is the main factor determining territory size. This work was done in captive conditions with parents and young on constant planes of nutrition.

Are variations in aggression inherited either by genetic or non-genetic mechanisms? We know that breeding success is pre-determined by variations in the eggs, and also that variations in breeding success are correlated with changes in numbers. The proximate cause of changes in numbers is changes in aggression and territory size, and we therefore infer that variations in breeding success are associated with variations in behaviour. Variations in breeding success are inherited and again we infer that the associated variations in aggression should also be inherited. This hypothesis is currently being tested directly by comparing the aggressive characteristics of birds reared under constant conditions in captivity and hatched from eggs taken in different years.

The question remains: what causes the variations in the eggs which result in the observed variations in breeding success and the postulated variations in behaviour? The most obvious suggestion is the plane of nutrition.

What effects do variations in nutrition have on the birds' aggression? Specifically, are the variations in egg quality noted above caused by variations in nutrition? This possibility is supported by circumstantial evidence (MOSS 1967) and by one experiment with fertiliser (MILLER *et al.* 1970). In this experiment all the grouse resident on experimental and control areas were shot in summer so that none was left by the end of July. Colonists, which had all been reared elsewhere on unfertilised heather, took the same size of territories in autumn on the fertilised heather as on the control area, although the heather's nutritive value had increased on the fertilised area. This suggested that grouse do not adjust their territory in autumn to the quality of the food at that time. Next year, the birds on the fertilised area bred better than those on the control area, autumn territory size was half that on the control area, and subsequent breeding stocks approximately double. We presume that the effect on breeding success was caused by an effect on the quality of the eggs and it is possible that changes in the eggs also contributed to the change in aggression which caused the reduction in territory size. However, an alternative explanation for the reduction in territory size was suggested by a further (unpublished) experiment.

In this case, the birds were not shot. Fertiliser was laid too late to affect egg quality via the hen's nutrition before egg laying (egg laying occurred in late

April—early May, and so, as expected, breeding success was no better on the fertilised than on the control area). During the summer the fertilised heather grew faster and was higher in nutritive value than on the control area in autumn when territories are chosen. In that first autumn, the grouse took smaller territories and so spring numbers increased on the fertilised area, but not on the nearby control area. This showed that either (a) they did adjust their territory size to the heather's nutritive value in autumn, or (b) they were predisposed to take smaller territories by a direct effect caused by a better plane of nutrition over the summer. Possibility (a) was ruled out by the first experiment, which leaves (b).

Whether the foregoing experimental results were attained by an indirect, maternally inherited effect on the aggression of the young or whether by a direct effect of nutrition on aggression, they illustrate the importance of changes in plane of nutrition for population limitation. The differences in mean grouse stocks and breeding success between 'rich' areas overlying relatively fertile rocks and 'poor' areas overlying infertile rocks (MILLER et al. 1966; Moss 1969) may be regarded as a natural analogue of the fertiliser experiments where both effects presumably operate.

This section has concentrated on the effects of plane of nutrition, and no mention has been made of the 'amount of food', a topic which has received much theoretical attention (LACK 1954; WYNNE-EDWARDS 1962). Red Grouse, like all herbivorous animals, are very selective feeders and are more selective for nitrogen and/or phosphorus when the vegetation available is poor in quality than when it is high (Moss 1967). Thus, when the plane of nutrition is high a greater proportion of the vegetation is regarded by the birds as edible than when it is low. At least three variables define the 'amount of food present': the gross quantity of plant material which falls into the same anthropocentric category as the material eaten by the birds, the nutritive value of this material, and its availability. Availability itself is a complex concept, because not all the potentially edible heather is equally available due to variations in the sward's height and density and the birds' reactions to them.

It is therefore not surprising that MILLER et al. (1966) found that differences in mean territory size in autumn were not related to differences in the mean length of current year's shoots of heather grown in the previous summer. Moss (1969) found no relation between the current year's production of heather measured by the total weight of potentially edible current year's shoots and flowers on four moors and mean spring breeding stocks of grouse; for example the area with highest production had low stocks. The hypothesis that mean territory size is adjusted to gross production of vegetation is refuted, but this does not necessarily mean that stocks are not related to the amount of food of adequate quality.

The best approximation we can make to measuring the 'amount of food of adequate quality' is to suggest that it is some function of (1) production of current year's growth and (2) chemical composition of the growth. Our present (unpublished) data indicate no relation between the chemical composition of the heather and mean territory size in any one autumn. As there is no relation to current year's production either, this indicates that grouse stocks are not directly related to the amount of food of adequate quality present in autumn.

[This is true despite LANCE's (1971) finding that the area of individual territories was inversely correlated with the nitrogen content of the heather there. It seems that some adjustment of territory size occurs in relation to food quality within any one autumn, but that variations in mean territory size are determined by earlier factors, presumably including variations in the mean level of aggression.]

DISCUSSION

The 'indirect relationship' in the model is a maternally inherited effect of plane of nutrition on aggression of the young. The crucial link between egg quality and the territorial behaviour of the young later on is still based on inference and therefore needs testing. Evidence of a 'direct relationship' between territorial behaviour and habitat features such as visibility, age of heather and proportion of the ground covered by heather, was suggested by correlations and by the fact that territory size decreased during an experiment where the heather was burned, without a prior stage of better breeding success. Evidence of a direct relationship between mean territory size and nutritive value of the heather now comes from a recent experiment with fertiliser.

We take the opportunity here of correcting an erroneous interpretation. In a recent review, CROOK (1970) referred to Red Grouse as an example illustrating population limitation by territorial behaviour, but only at high densities on rich moors. On poorer moors where breeding densities and breeding success fluctuate more, he said this suggested that the populations were determined by changes in food and not territorial behaviour. This interpretation is understandable from the Discussion in JENKINS et al. (1967), which we reject here. This indicated that breeding success was the dominant process affecting breeding stocks on poor moors, and territorial behaviour on rich moors. It also said this reconciled the views of those who believe that territorial behaviour does limit numbers and those who believe it is of doubtful importance. In fact, all the evidence we have indicates that territorial behaviour is the immediate or proximate factor determining breeding densities on both rich and poor moors in all years. We believe that associations between breeding success, food, other parameters and subsequent changes in numbers are all due to effects of these parameters on territorial behaviour, which in turn determines breeding densities.

POINTS OF GENERAL INTEREST FROM THE STUDY OF RED GROUSE

We now consider some points that we feel may be of general interest for the study of populations, behaviour and food in other tetraonids and other animals.

The Red Grouse study indicates that population dynamics and population limitation involve a complex system inter-relating behaviour, populations, nutrition, inheritance, the spatial or physical structure of the environment, and possibly competition with other species. In studying population changes and limitation, one may have to consider a complex chain of causal events, such that 'the' cause of any event can be described only with full reference to time. For example, the fact that mortality occurs at a given season does not necessarily mean that it has been caused in that season, nor does it necessarily mean that the proximate agent of mortality is of importance in the population changes. Such an agent may be unnecessary to a noted decline, it may be necessary but not sufficient or it may be both necessary and sufficient. As examples: winter predation in Red Grouse is not necessary for overwinter mortality to occur; predation is probably a necessary cause of death in Roe Deer *Capreolus capreolus* in Europe (discussed in WATSON & MOSS 1970: 201), but is not sufficient in the absence of socially induced emigration from woodland habitats which provide greater security from predation (including shooting) than surrounding farmlands; and territorial behaviour in Red Grouse is both a necessary and sufficient cause of overwinter mortality.

Losses of animals that seem important numerically may not be important if they are cancelled by subsequent compensatory changes, so that a model which assumes that successive losses are all additive (e.g. from a key factor analysis) may be mathematically correct but not be a biologically correct model. The best way to differentiate between additive losses and compensatory changes is to do experiments (WATSON 1971).

Food is a very complex concept in red grouse, as features of chemical composition, food selection, and varying nutritive requirements at different seasons or for different sexes and age classes make it difficult to define food (p. 133). This means that simplistic assumptions that food is not important because there appears to be an excess of edible green vegetation are unconvincing. There may be no absolute shortage of the total quantity of potentially edible material, but this is of much less importance than the amount of food of adequate quality.

Much current research is being done on populations in terms of energy flow. However, variations in the secondary production of many species of birds, insects and mammals may not be related to the primary production of energy at all, but may instead be associated with nutrient content or other parameters of quality. Our research on Red Grouse gives an example of this on different moors and it is likely to be true in all selective feeders, that is in all herbivores and probably many carnivores also.

Space is a resource which is not simply measurable in terms of area. In Red Grouse and in several species of animals, the physical structure of the environment affects some feature of physical cover or visibility which in turn is related to spacing behaviour (WATSON & MOSS 1970).

Individual Red Grouse within a population vary greatly in their social and aggressive behaviour, and this has important effects on their division of the habitat and on subsequent population processes. It may therefore be unwise to make simplistic assumptions that one can think of n animals in an area as all being the same, for the sake of convenience in studying populations or food. Even when grouse were all given the same diet and environment in captivity, individuals varied greatly in various aspects including reproductive characteristics (KOLB 1971).

There is good evidence of maternal non-genetic inheritance of several parameters in captive Red Grouse (KOLB 1971). This is clearly recognised to be important in domestic animals (e.g. review by COWLEY 1970), and is probably more important in wild animals than is often thought. Yet inheritance found in ecological studies tends to be explained in terms of genetic inheritance, even in cases where there is no evidence one way or the other.

Finally, the elucidation of correlations and associations between different sets of events is very useful as an aid to framing testable hypotheses. In the absence of such tests, however, hypotheses remain hypothetical and can be elevated to the status of theory only by experimental verification.

ACKNOWLEDGEMENTS

We thank Dr. D. JENKINS for helpful comments on the manuscript.

SUMMARY

Our current model of population dynamics in Red Grouse is presented.

Breeding success is pre-determined by variations in the eggs, which are expressed largely as variations in early mortality of chicks. There is circumstantial and experimental evidence that these variations are at least in part caused by differences in the nutritive value of the parents' food.

The density of the breeding stock is determined by the mean size of the territories, which are taken by the cocks in the autumn. Fluctuations in breeding stock are due to variations in territory size. This applies both to differences amongst areas and to differences amongst years within areas.

Two mechanisms, both operating via plane of nutrition, are thought to cause variations in territory size.

A change in mean territory size occurs when the heather is fertilised or burned and is thought to result from an increased plane of nutrition affecting the birds' physiology and aggressive behaviour directly. We call this the 'direct mechanism'.

Changes in territory size are inversely related to breeding success, which in turn is pre-determined in the eggs (above). Thus changes in territory size and associated levels of aggression should also be pre-determined. This 'indirect mechanism' is presently being tested.

The possibility that variations in territory size are caused by other maternally inherited or genetic inherited effects is considered. The density-dependent stress-endocrine effect postulated by HÖHN (1967) after CHRISTIAN et al. (1965) is dismissed on the grounds that density-dependent effects are not important in the population dynamics of Red Grouse, but 'delayed density-dependent' effects of stress cannot be ruled out.

REFERENCES

BROWN L. H. & WATSON A. (1964) The Golden Eagle in relation to its food supply. Ibis 106, 78–100.

CHITTY D. (1967) The natural selection of self-regulatory behaviour in animal populations. Proc. ecol. Soc. Aust. 2, 51–78.

CHRISTIAN J. J., LLOYD J. A. & DAVIS D. E. (1965) The role of endocrines in the self-regulation of mammalian populations. Recent Prog. Horm. Res. 21, 501–577.

COWLEY J. J. (1970) Food intake and the modification of behaviour. *In* Animal Populations in Relation to Their Food Resources. Ed. A. WATSON. Blackwell, Oxford.

CROOK J. H. (1970) Social organization and the environment: aspects of contemporary social ethology. Anim. Behav. 18, 197–209.

HÖHN E. O. (1967) The relevance of J. Christian's theory of a density-dependent endocrine population regulating mechanism to the problem of population regulation in birds. Ibis 109, 445–446.

JENKINS D. & WATSON A. (1970) Population control in Red Grouse and Ptarmigan in Scotland. Trans. Congr. int. Un. Game Biol. 7, 121–141.

JENKINS D., WATSON A. & MILLER G. R. (1963) Population studies on Red Grouse, *Lagopus lagopus scoticus* (Lath.) in north-east Scotland. J. Anim. Ecol. 32, 317–376.

JENKINS D., WATSON A. & MILLER G. R. (1967) Population fluctuations in the Red Grouse *Lagopus lagopus scoticus*. J. Anim. Ecol. 36, 97–122.

JENKINS D., WATSON A. & PICOZZI N. (1965) Red Grouse chick survival in captivity and in the wild. Trans. Congr. int. Un. Game Biol. 6, 63–70.

KOLB H. (1971) Development and aggressive behaviour in Red Grouse in captivity. Ph.D. thesis, Univ. of Aberdeen.

LACK D. (1954) The Natural Regulation of Animal Numbers. Clarendon Press, Oxford.

LANCE A. (1971) Telemetry studies of Red Grouse. Triennial Report, Nature Conservancy, Edinburgh.

MILLER G. R., JENKINS D. & WATSON A. (1966) Heather performance and Red Grouse populations. I. Visual estimates of heather performance. J. appl. Ecol. 3, 313–326.

MILLER G. R., WATSON A. & JENKINS D. (1970) Responses of Red Grouse populations to experimental improvement of their food. *In* Animal Populations in Relation to Their Food Resources. Ed. A. WATSON. Blackwell, Oxford.

MOSS R. (1967) Probable limiting nutrients in the main food of Red Grouse (*Lagopus lagopus scoticus*). *In* Secondary Productivity of Terrestrial Ecosystems. Ed. K. PETRUSEWICZ. Panstwowe Wydawnictwo Naukowe, Warszawa.

MOSS R. (1969) A comparison of Red Grouse (*Lagopus l. scoticus*) stocks with the production and nutritive value of heather (*Calluna vulgaris*). J. Anim. Ecol. 38, 103–112.

VARLEY G. C. & GRADWELL G. R. (1970) Recent advances in insect population dynamics. A. Rev. Ent. 15, 1–24.

WATSON A. (1964) Aggression and population regulation in Red Grouse. Nature, Lond. 202, 506–507.

WATSON A. (1967a) Population control by territorial behaviour in Red Grouse. Nature, Lond. 215, 1274–1275.

WATSON A. (1967b) Social status and population regulation in the Red Grouse (*Lagopus lagopus scoticus*). Proc. R. Soc. Popul. Study Grp, 2, 22–30.

WATSON A. (1970) Territorial and reproductive behaviour of Red Grouse. J. Reprod. Fert., Suppl. 11, 3–14.

WATSON A. (1971) Key factor analysis, density dependence and population limitation in Red Grouse. *In* Dynamics of Populations. Ed. P. J. DEN BOER & G. R. GRADWELL. Pudoc, Wageningen.

WATSON A. & MILLER G. R. (1971) Territory size and aggression in a fluctuating Red Grouse population. J. Anim. Ecol. 40, 367–383.

WATSON A. & MOSS R. (1970) Dominance, spacing behaviour and aggression in relation to population limitation in vertebrates. *In* Animal Populations in Relation to Their Food Resources. Ed. A. WATSON. Blackwell, Oxford.

WATSON A. & JENKINS D. (1968) Experiments on population control by territorial behaviour in Red Grouse. J. Anim. Ecol. 37, 595–614.

WATSON A. & MOSS R. (1971) Spacing as affected by territorial behavior, habitat and nutrition in Red Grouse (*Lagopus lagopus scoticus*). *In* Behavior and Environment. Ed. A. H. ESSER. Plenum Press, New York.

WATSON A., MOSS R., PARR R., HEWSON R. & GLENNIE W. (1969) Viability and behaviour of young Red Grouse and Ptarmigan. Grouse Research in Scotland, 13th Prog. Rep., duplicated. Ed A. WATSON, The Nature Conservancy, 12 Hope Terrace, Edinburgh.

WATSON A. & O'HARE P. J. (1970) Bringing back the grouse. Irish Times, 3 Sept., p. 10.

WYNNE-EDWARDS V. C. (1962) Animal Dispersion in Relation to Social Behaviour. Oliver & Boyd, Edinburgh.

Author Citation Index

Subject Index

Aggression in territorial behavior, 15, 18–22, 35–36, 39–42, 207, 264–265, 290–291, 352–353
  effect of nutrition on, 383–385
  and females, 269
  reasons for, 322–324
  relationship to objects, 87–90
  and song, 23–25, 35–36, 48–49, 89
  in song sparrow, 62–63
  in Uganda kob, 220

Dominance–subordinance in territorial behavior, 342–343, 366–370 (*see also* Social hierarchy and territoriality)

Fitness (*see also* Mating success and social status)
  and habitat selection, 158–178
  in hummingbird, 257–271
  in marsh wren, 272–281
Food–energy relationships in territory, 254–292 (*see also* Population regulation, by food shortage; Territory, functions of, food provision)
  in blackbird, 287–292
  in hummingbird, 257–271
  in marsh wren, 272–281

Group territory, 218–219, 310–312 (*see also* Lek behavior)
  in Australian magpie
    and ecology, 210–214
    types of groups, 207–209
  evolution of, 324–325
  in fur seal, 225–234
  in Uganda kob, 220

Habitat selection
  in blackbird, 283
  and population density, 134–141

and territorial behavior, 158–177, 207
  in marsh wren, 276–278

Lek behavior (*see also* Group territory)
  evolution of, 313–314, 325–326
  in prairie chicken, 221–224
  in Uganda kob, 220, 243–245

Mating success and social status (*see also* Fitness; Social hierarchy and territoriality)
  in Australian magpie, 202–214
  in fur seal, 225–233
  in hummingbird, 257–271
  in marsh wren, 272–281

Population regulation (*see also* Territory, functions of)
  by food shortage, 154–156 (*see also* Food-energy relationships in territory)
  nonbreeding, evidence for, in, 37, 43–48, 94, 179–190, 195, 202–214, 273–274, 284–288, 352–361
  removal of breeding birds, as test of, 43–44, 179–190, 285, 352–361
  by territoriality, 37–48, 93–97, 114–215, 284–288, 329–330, 352–361, 377–389
  theories of
    buffer mechanism theory, 95, 133–141, 168–170, 198–201
    elastic disc theory, 94–95, 115–116, 119–126, 156, 170–173, 191–201, 353, 355, 358
    group selection theory (Wynne-Edwards), 300, 359–360
    spacing theory, 95, 143–147, 173–174, 358–359

Social hierarchy and territoriality (*see also* Dominance–subordinance in terri-

397